IMPERIAL VISIONS

Nationalist Imagination and Geographical Expansion in the Russian Far East, 1840–1865

In the middle of the nineteenth century, the Russian empire made a dramatic advance on the Pacific by annexing the vast regions of the Amur and Ussuri rivers. Although this remote realm was a virtual *terra incognita* for the Russian educated public, the acquisition of an "Asian Mississippi" attracted great attention nonetheless and, indeed, even stirred the dreams of Russia's most outstanding visionaries – among them Alexander Herzen, who confidently proclaimed the annexation on Siberia's Manchurian frontier to be "civilization's most important step forward." Within a decade of its acquisition, however, the dreams were gone and the Amur region largely abandoned and forgotten. In an innovative examination of Russia's perceptions of the new territories in the Far East, Mark Bassin sets the Amur enigma squarely in the context of the *Zeitgeist* in Russia at the time. His argument is that the grand vision of Russia on the shores of the Pacific was intimately related to a number of major preoccupations of the day, including social reform, the search of *samopoznanie* or national self-understanding, Russia's relationship to the West, and the belief in a mission of universal salvation.

Written with an equally firm footing in the disciplines of historical geography and intellectual history, *Imperial Visions* demonstrates the fundamental importance of geographical imagination for the *mentalité* of imperial Russia. The work offers a truly novel perspective on the complex and ambivalent ideological relationship between Russian nationalism, geographical identity, and imperial expansion.

MARK BASSIN is Reader in Geography at University College London

T0279851

Cambridge Studies in Historical Geography 29

Cambridge Studies in Historical Geography encourages exploration of the philosophies, methodologies and techniques of historical geography and publishes the results of new research within all branches of the subject. It endeavours to secure the marriage of traditional scholarship with innovative approaches to problems and to sources, aiming in this way to provide a focus for the discipline and to contribute towards its development. The series is an international forum for publication in historical geography which also promotes contact with workers in cognate disciplines.

For a full list of titles in the series, please see end of book.

IMPERIAL VISIONS

Nationalist Imagination and Geographical Expansion in the Russian Far East, 1840–1865

MARK BASSIN

CAMBRIDGE UNIVERSITY PRESS
Cambridge, New York, Melbourne, Madrid, Cape Town, Singapore, São Paulo

Cambridge University Press
The Edinburgh Building, Cambridge CB2 2RU, UK

Published in the United States of America by Cambridge University Press, New York

www.cambridge.org
Information on this title: www.cambridge.org/9780521391740

First published 1999
This digitally printed first paperback version 2006

A catalogue record for this publication is available from the British Library

Library of Congress Cataloguing in Publication data

Bassin, Mark.
Visions of empire: nationalist imagination and geographical
expansion in the Russian Far East, 1840–1865 / Mark Bassin.
p. cm. – (Cambridge studies in historical geography; 29)
Includes bibliographical references and index.
ISBN 0 521 39174 1
1. Amur River Region (China and Russia) – History – 19th century.
I. Title. II. Series.
DK771.A3B37 1999
957′. 7–dc21 98-30355 CIP

ISBN-13 978-0-521-39174-0 hardback
ISBN-10 0-521-39174-1 hardback

ISBN-13 978-0-521-02674-1 paperback
ISBN-10 0-521-02674-1 paperback

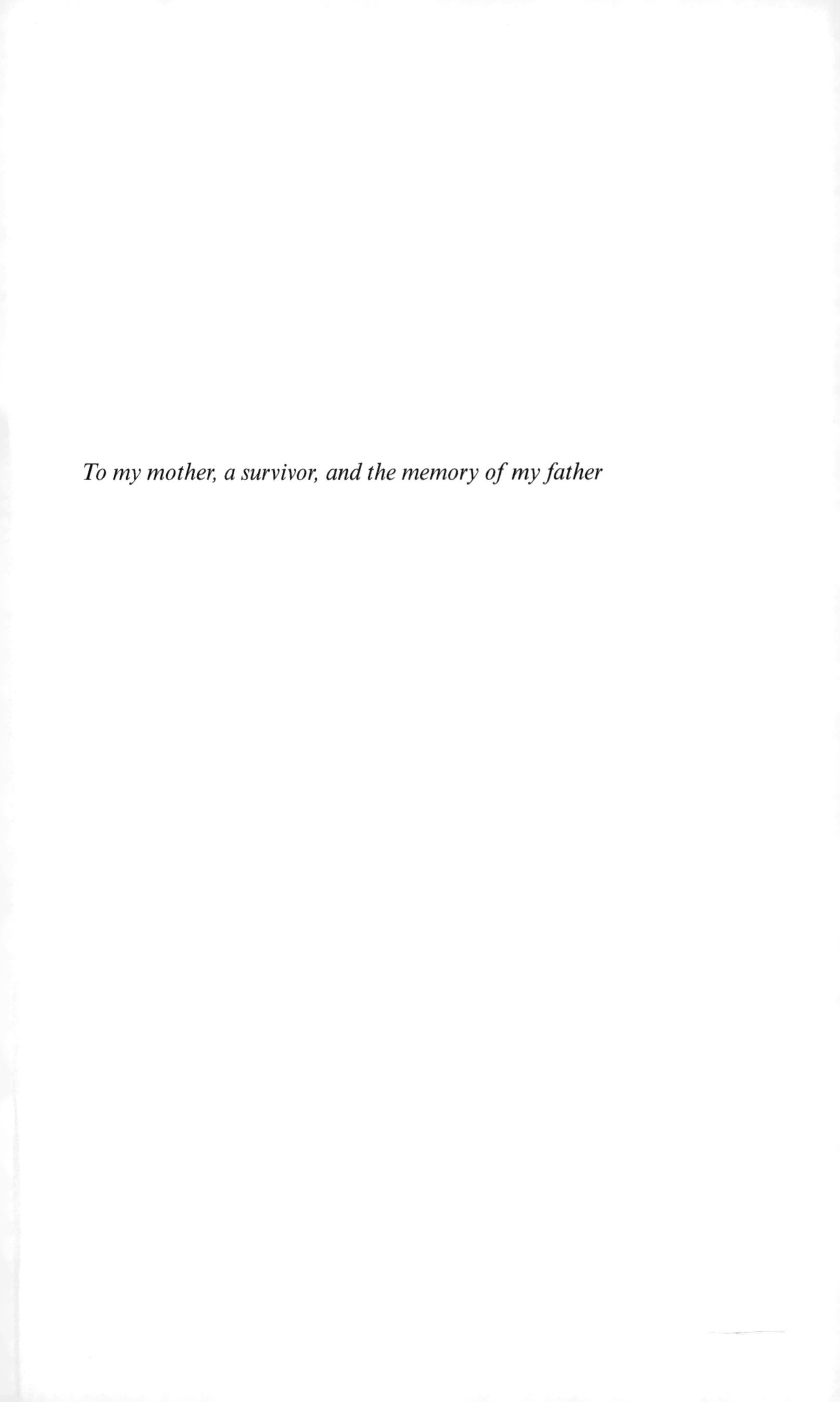

To my mother, a survivor, and the memory of my father

Contents

Foreword

I came to know Dr. Mark Bassin well during his graduate studies at the University of California in Berkeley. I remember how shortly after we met he asked me to speak Russian with him as much as possible. To be sure, he already read fluently and could use a variety of Russian sources. But he also wanted to speak the language correctly and to be as close to the Russian tongue and Russian culture as possible. So we spoke Russian, and still do when we meet. Bassin's request made it easier for me to follow over the years his progress in Russian to a very high degree of proficiency. I think that the translations from Russian in the present book, including poetry, are excellent. Apparently Bassin learned German in the same fundamental manner. The larger point is that Bassin as a scholar is the opposite of parochial. A young American who has already lived, studied, taught, conducted research, or engaged in some combination of these activities in England, Canada, Russia, and Germany, he is naturally part of the entire Western intellectual world, without fear or favor. In reference to the present work and to his treatment of Russia in general, Bassin is entirely free of the sense of unfathomable difference, mystery, or strangeness which continues to spoil so much Western scholarship on Russia.

Mark Bassin is both a geographer and an intellectual historian, and he is very well aware of his special position and allegiance to both disciplines. Without presuming to judge Bassin as a geographer, except to state that I have no criticisms to offer in that connection, I do think that he, and the present volume in particular, have much to contribute to intellectual history. One asset is the richness of detail and an apposite discussion of many individuals very little known in scholarly literature, figures who are smoothly integrated with Peter the Great or Dostoevskii. It can be argued that the book is most valuable for its fragments. Yet these fragments also form a connected and clear narrative with a beginning, a middle, and an end. Especially praiseworthy is the author's focus on the nature and structure of ideas, which keeps the surging flow of disparate details together and constitutes the skeleton, so to speak, of the book. (That focus, characterisitic of effective intellectual historians, is present in all of Bassin's writings – for example in his treatment of Eurasianism, which in other hands has recently become a hopelessly vague and even self-contradictory term.)

Imperial Visions: Nationalist Imagination and Geographical Expansion in the Russian Far East, 1840–1865 can be read, and read correctly, as a story of a romantic vision destroyed by a better acquaintance with reality. The Amur river and area, which for ages had been outside the bounds of Russian history, rather suddenly began to attract Russian attention in the nineteenth century – in the 1830s, 1840s, and culminating in the 1850s. The "Siberian Mississippi" seemed to offer enormous possibilities of development, and even to promise a new epoch for Russia. The Russian version of such romantic visions was shared, interestingly enough, by certain non-Russians, including some overwhelming American enthusiasm. Yet before long the vision lost its luster and in fact disappeared. Even the Trans-Siberian railroad, when it was finally built before the end of the century, cut across Chinese territory rather than follow the Amur river: dream could not be turned into reality. Besides, I might add, a more pragmatic and positivistic intellectual orientation largely replaced romanticism.

Still, this blunt and essentially correct estimate of what happened is not complete. The dream of the Amur was destroyed not only by such "objective" factors as the shallowness of the river, the frighteningly adverse climate, the small number of inhabitants, and the almost total lack of any kind of infrastructure in the entire enormous area, but also, as the book indicates, by competing visions – whether that of the Pan-Slavs, pointing to Europe, or other visions pointing to Asia, but to Central Asia rather than the Far East. Reality could defeat dreams, but not stop people dreaming. And even destroyed visions could reappear in a mutated form. In a few fascinating pages in his "Conclusion" Bassin writes of the great communist project of the 1970s and 1980s, that of the BAM or the Baikal–Amur Mainline railway, which was built to run some 125–185 miles north of the Trans-Siberian and was, again, to open up an enormous territory for development and progress. "In precisely that same way that visions of the mid-nineteenth century proved to be so empty and misleading, so the great promise of the BAM railway gave way eventually to utter failure" (p. 281). What next?

Nicholas V. Riasanovsky
University of California, Berkeley

Acknowledgments

The interest in nationalism and ideology which underlies this study was initially stimulated by a series of brilliant undergraduate courses I took with George Mosse. It was a singular bit of good fortune to end up many years later as Professor Mosse's colleague at the University of Wisconsin, where I could continue to benefit from his intellectual insight and the very special pleasure of his company. The study itself originated out of my doctoral research at the University of California, and I would like to thank the individuals who helped and supported me at that time, most notably David Hooson, Nicholas Riasanovsky, Clarence Glacken, and James Parsons. Additionally, I should note my debt to Martin Malia, with whom I also studied and without whose magisterial work on Alexander Herzen I could not have conceived the present book. Professors Hooson and Riasanovsky have continued to provide valuable and much-appreciated advice and support down to the present, and I am particularly grateful to Professor Riasanovsky for agreeing to contribute a Foreword to the work in its final form. A very special note of thanks is due to James Gibson and John Stephan – both far more formidable experts on the Russian Far East than myself – for taking an early interest in my studies and encouraging me to present my research as a book. I might note that Jim Gibson bears the responsibility of directing my attention – over a heady cigar and a bottle of Portuguese rosé in a Toronto suburb – to the Amur in the first place. I am happy to say that he has now lived to regret it. My post-doctoral research on this project has been supported by the generous advice and assistance of numerous colleagues. In particular, I would like to thank Marc Raeff, Yi-Fu Tuan, and Abbott Gleason. Terry Martin helped with material on the Mennonites. Steven Marks and David Saunders both took the trouble to read the finished manuscript in its entirety and share their extensive knowledge of nineteenth-century Russian history with me. Their effort is enormously appreciated, and is (I hope) positively reflected through the many corrections and improvements they suggested.

Much of the material for this study was collected in the course of two extended sojourns in the USSR as a Fulbright Fellow on the IREX exchange. Of the many Soviet scholars with whom I have worked on these occasions and others, I would like to acknowledge in particular the assistance of the following: S. A. Kovalev, A. I. Alekseev, V. P. Esakov, B. P. Polevoi, A. P. Okladnikov, V. V. Vorob′ev , V. I. Bykov, and A. V. Postnikov and S. B. Lavrov. Intervening fellowships from the Institut für europäische Geschichte (Mainz), the Kennan Institute for Advanced Russian Studies, the Institute for Research in the Humanities at the University of Wisconsin, and the Remarque Institute for European Studies at New York University all provided an opportunity for me to develop my thinking, as did an extended affiliation as Research Associate with the (then) West German *Arbeitsgruppe* of the Commission for the History of Geography, International Geographical Union. I would also like to thank the staff of the following libraries: Doe Library, University of California; Lehman Library, Columbia University; Bobst Library, New York University; Memorial Library, University of Wisconsin; University Library, University of Illinois; Regenstein Library, University of Chicago; Library of Congress; Staatsbibliothek Preussischer Kulturbesitz, Berlin; Hauptbibliothek, Freie-Universität-Berlin; Slavic Collection of the University Library, Helsinki University; British Library; Gosudarstvennaia Biblioteka im. Lenina, Moscow; Gosudarstvennaia Biblioteka Akademii Nauk, St. Petersburg; Russkoe Geograficheskoe Obshchestvo, St. Petersburg; Gosudarstvennyi Kraevedcheskii Muzei, Irkutsk; University Library, Irkutskii Gosudarstvennyi Universitet.

I am no less grateful for the friendship and enouragement of many individuals who were not directly involved in the academic evolution of this book. First and foremost is Günter Gebhardt (and family), who offered an extended haven in Berlin–Lichterfelde on several occasions, where the important themes of this work were first thought through. If nothing else, I am happy that the publication of this book will finally provide some explanation for what actually happened to the notes I made on the back of the many computer cards that he so obligingly purloined for me from the Techological University. At the University of Wisconsin I would like particularly to thank Jost Hermand and Uli Schamiloglu, who in different ways helped me through some rather cloudy moments there before my move to London in 1995. My doctoral students Guntram Herb, Valentin Bogorov, Guven Sargin, and Donald Steele have been and remain the source of enormous stimulation, and together they have helped me learn at least as much as I may have taught them. Finally, I would like to thank my harried colleagues in the Department of Geography at University College London, in particular for their rather ironic good cheer as this project was drawing to a close and more broadly for offering a truly superb professional and intellectual environment in which to work. The possibility that my book could appear in the "Cambridge Studies in Historical Geography" series was first suggested by the late (and very much missed) Brian

Harley. I am grateful to Alan Baker for his patience in seeing the project through.

A special note of thanks is due to Anna Stawińska, for encouragement in the latter stages of this project and for helping me to recognize the (very) subtle glories of Manor House.

Mark Bassin
North London
May 1998

The Russian Far East (*c.* 1860)

Introduction

The work that follows is about three things: a geographical region, an historical episode involving this region, and an underlying enigma. The region in question is a massive chunk of territory at the southeastern-most extremity of the Russian landmass, where the Siberian frontier presses eastward to the Pacific ocean and south into Manchuria. The dominating natural feature of this area is the mighty Amur river. The Amur has its origins at the confluence of the Shilka and Argun rivers east of Lake Baikal and flows on for some 1,800 miles to empty into the Tatar straits opposite the northern tip of Sakhalin. With numerous tributaries feeding it from the north and south, the river commands a drainage basin of nearly three-quarters of a million square miles. It was not however this vast natural–geographical zone in its entirety that the Russians in the nineteenth century had in mind when they spoke variously about the *amurskii krai*, *amurskaia oblast'*, or *priamur'e*, and nor is it the scene of this study. The specifics varied considerably, as we will see, but for the most part they were referring to the river itself and the lands immediately along its northern bank. At a relatively late point, the Ussuri river – a major tributary which feeds into the Amur from the south – was included in this designation as well, along with all of the territory from the Ussuri east to the coast of the Tatar straits. It is this more restricted geographical zone that I will refer to in this study as the "Amur region."

The historical episode in question, played out over two decades in the middle of the nineteenth century, was Russia's acquisition of this territory from the Chinese empire. The Russians had originally entered and occupied the river valley in the mid-seventeenth century, but in a treaty signed with the Chinese in 1689, they acknowledged the river region to be the lawful patrimony of the Manchu dynasty. They duly disbanded their settlements and withdrew from the area, and over the next century-and-a-half the Amur remained largely forgotten. Beginning in the 1840s, however, signs of vigorous new Russian interest in the river and the region became apparent, both in the imperial capital and in Siberia itself. At first, this interest was articulated outside of the centers of official Russian policy, and indeed in rather fervid opposition to it, but by the middle of the 1850s the government had been won over as well. The grand culmination was the annexation of the Amur region by the Russians, secured through a number of treaties concluded with China in 1858 and 1860. The Amur and Ussuri rivers were established thereby as the international boundary between the two countries in the Far East, and they have remained as such down to the present day.

It is, however, the underlying enigma which is really at the center of this investigation. To speak of a new Russian "interest" in the river in the middle of the nineteenth century is something of an understatement, for what happened at the time was in fact a major redirection of the nation's attention to these remote territories in the Far East. For one brief historical moment, an obscure region which had not only been a virtual *terra incognita* for the Russians but moreover did not even figure as a part of their imperial dominions was able to attract the interest of the entire society, excite widespread enthusiasm, and even nourish the dreams of the country's most outstanding social and political visionaries. In the region itself, thinly populated up to that point by scattered indigenous groups and yet more scattered Manchurian officials, there was a flurry of activity as Russian military outposts were erected, commercial development initiated, and agricultural settlement on an ambitious scale undertaken. All of this gave rise to grandiose speculation and epochal plans regarding the brilliant and progressive future that appeared certain to occur. It was a shimmering prospect, the scope of which included not only the new Russian territories on the Pacific, but Siberia and more broadly Russia west of the Urals as well. Indeed, the remarkable significance which the Russians at the mid-century were prepared to attribute to the faraway and little-known Amur region transcended even the boundaries of the Russian empire, and a world-historical dimension was identified for it as well. An explorer who spent five long years on the Amur in the 1850s lectured a St. Petersburg audience at the end of the decade about how the region was attracting the attention of a Europe awestruck by its "incalculable importance and significance" for "politics and culture, for commerce and civilization." No less a figure than Alexander Herzen confirmed these sentiments from his London exile, declaring with characteristic grandiloquence in a letter to Giuseppe Mazzini that the Russian acquisition of the Amur represented "one of civilization's most important steps forward."[1] Then, abruptly, in the space of only a few years, all of this heady excitement came to a sudden end. The enthrallment was gone, the optimism evaporated, and the grand plans thwarted. The visionaries either gave up their dreams or found other regions upon which to project them, and the Amur – in spite of its new status as a formal part of the Russian empire – sank back into essentially the same obscurity and neglect

[1] [Radde], "Gustav Raddes Vorlesungen," p. 257; Gertsen, "Pis'mo k Dzhuzeppe Matstsini," p. 350.

that had surrounded it for the preceding century-and-a-half. To try and unravel this paradox, to determine just where what I will call the "Amur euphoria" came from, what it meant, and where it went – this is the principal task of the present work.

The first place we might think of turning for some insight into these questions is previous examinations of the topic. The secondary literature on the annexation of the Amur is small, but appreciable nonetheless. Most of it is the product of Russian scholarship (pre-revolutionary as well as Soviet[2]) but there are a few very good studies by Western scholars as well.[3] Yet while this literature taken as a whole gives us quite a full picture of the broad historical background, the main players involved, and the stream of events which culminated in the Russian acquisition of the region, nowhere is the enigmatic dimension of this process addressed – or indeed even recognized. There is, I believe, an entirely logical reason for this. Virtually all of the Russian works on the subject down to the present – with the notable but for our purposes insignificant exception of the revisionist historiography of Russian imperialism in the 1920s[4] – have been guided by an overriding interest in depicting the annexation of the Amur in the triple light of practical necessity, social progress, and historical justice. Toward this end, a teleology is set up, implicitly or explicitly, and the course of events it describes leads inexorably to a grand and effectively preordained conclusion, namely the incorporation of the Amur region as part of the Russian empire and Russia's achievement therewith of its "natural" boundaries on the Pacific.[5] In the historiography of the nineteenth century the reasons for this bias came from the exigencies of Russian nationalism and from the attempt to enhance the aura of glory around the figure of the main

[2] Nevel'skoi, *Podvigi*; Barsukov, *Graf . . . Amurskii*; Butsinskii, *Graf . . . Amurskii*; Efimov, *Gr. N. N. Murav'ev-Amurskii*; Sgibnev, "Vidy;" Shchukin, "Podvigi;" Romanov, "Prisoedinenie;" Zaborinskii, "Graf . . . Amurskii;" Shtein, *N. N. Murav'ev-Amurskii*; Kabanov, *Amurskii vopros*; Alekseev, *Amurskaia ekspeditsiia*; Sychevskii, "Russko-Kitaiskaia torgovliia."

[3] See especially Lin, "Amur Frontier Question;" Quested, *Expansion*; Mancall, "Major-General Ignatiev's Mission." John Stephan's panoramic survey of the history of Russian in the Far East offers an excellent if brief account of the background and events leading up to the Amur annexation. *Russian Far East*, pp. 26–33, 40–50.

[4] Writing in 1927, Kharnskii gave voice to this revisionism with the characterization of the Russian expulsion from the Amur after 1689 as a "well-deserved lesson" for its "pogroms" (pogromnye deianiia) against the indigenous peoples. The Russian advance on the Amur in the mid-nineteenth century was described as an imperialist "annexation" (anneksirovat'). *Kitai*, pp. 274, 322. For a similarly narrow view of the Russian presence, see Bakhrushin, *Kazaki*, *passim*.

[5] See, for example, Shtein, *N. N. Murav'ev-Amurskii*, p. 5; Kabanov, *Amurskii vopros*, pp. 15, 29–30. Russian historians have continued to assert this teleology down to the present day. In 1993, for example, A. V. Ignat'ev maintained that Russia's territorial expansion in the Far East was "fundamentally completed" when it "reached the natural maritime boundaries of the Pacific Ocean" (i.e. the Amur–Ussuri border) at the mid-century. Because the main subject of his essay is Russian activity in the Far East at the end of the nineteenth century, however, he is compelled to contradict himself in the very next line and extend the geographical scope of his teleology even further, to encompass the Russian move against Korea. It turns out that the Amur annexation left certain "assignments" unfulfilled, as he put it, one of which was the "establishment of ice-free ports"! "Foreign Policy," p. 251.

actor in the drama, Governor-General N. N. Murav′ev. In the Soviet period, the reasons relate again to the nationalist impulse but more immediately to the political problem that the legitimacy of the annexation and Russian domination in the region was seriously challenged by post-revolutionary China. Scholars in the West have been unfettered by such cumbersome agendas, it is true, but their accounts are nonetheless strongly influenced by the Russian work and in the final anaysis they have not gone very far beyond it, at least in their treatment of the Russian side.[6] Needless to say, if an analysis singlemindedly takes the legitimacy and historical necessity of the annexation as both its point of departure and its conclusion, there is not going to be much room in between for reflective appreciation of the ambivalences, ironies, or enigmas of the process.

There is, however, another factor which is perhaps even more important than an *a priori* political bias in explaining why aspects of the Amur annexation that seem to me so striking should have been overlooked. This relates to the manner in which the issue has been problematized. Specifically, most histories see the annexation exclusively in terms of its local origins and local implications. The relevant geographical context, accordingly, is restricted to adjoining areas of the Russian Far East and neighboring regions of East Asia; at most, it may be expanded to include Eastern Siberia. It is essentially within this arena that they seek the background for the Russian advance at the mid-century, and it is here that real significance of this advance is identified. There is, I would suggest, a rather considerable problem with this approach, which we can appreciate immediately in the fact that no accumulation of purely local considerations, however weighty, could possibly have motivated Herzen in London to make such a sweeping pronouncement about the importance of the Amur to the advance of world civilization. While not in the slightest dismissing the relevance of the far-eastern and Siberian dimensions in the present study – indeed, they will be emphasized throughout – it is not limited to them, and it will place the Amur squarely in much broader national context as well. My argument is that we can understand why the Amur was annexed, why this annexation occasioned such euphoria, and why this euphoria proved to be so ephemeral only if we appreciate the extent to which the thoroughly minor issue of a distant river on the Siberian–Manchurian frontier became intertwined with and, so to speak, energized by the truly major social and political issues of the day. One of the principal ambitions of this work is accordingly to demonstrate that an examination of Russian thinking about the Amur region not only informs us about Siberia and the Russian Far East but at the same time

[6] The most grevious example is that of Robert Kerner. See in particular "Russian Expansion," pp. 111–114. Two dissertations on the annexation of the Amur completed under his supervision, although far better informed than his own work, also repeat this interpretative bias: Stanton, "Foundations;" Sullivan, "Count N. N. Muraviev-Amurskii." On Kerner, see Rieber, "Historiography," pp. 27–31; Satsuma, " 'Scholarly Entrepreneur,' " *passim*. The more recent work of Quested and Mancall is particularly valuable in its careful incorporation of Chinese sources.

offers considerable insight into some of the most powerful impulses and pre-occupations driving Russian society as a whole in the nineteenth century, chief among them nationalism, social reform, and imperial expansion.

To say merely that the problem of the Amur region "became intertwined" with other issues of the day, however, risks oversimplifying what was in fact an extraordinarily complex process. The Amur, after all, *was* a local Siberian issue, at least to begin with, and an extremely obscure one at that. Prior to its becoming a *cause célèbre* in the 1850s, very few people in Russia's European capitals had as much as heard of it, let alone possessed an even approximate sense of where it was and what its practical significance might be. In order for the educated Russian public to embrace the cause of annexation, therefore, representations of the Amur had to be made available which would enable this public in some manner to envision and evaluate it. It was these representations or images which then served to give the region meaning and galvanize popular opinion around it, and it is through them that we can begin to make some sense of the enigma which was to characterize the entire experience. From this standpoint, therefore, this book is not so much a history *per se* of the Amur annexation as an interrogation of the images or geographical visions that accompanied this process and to a significant extent impelled it. The fact that so little was known about the area in no way impeeded the formation and proliferation of these visions; indeed, as I will repeatedly be emphasizing, it was largely by virtue of the Amur region's remoteness and its obscurity that it could provide such rich and yielding material for the Russians' imagination in the first place.

The study of geographical visions may seem an exotic pursuit, but there is a large literature devoted to it, some of which at least has been extremely influential. This includes numerous studies of images of Asia – the reader is likely to be familiar with Edward Said's bestselling *Orientalism*[7] – of Africa,[8] Australia,[9] the Americas,[10] and recently even the eastern half of Europe itself.[11] Closer to the subject of the present work is a small but vigorous literature specifically on Russian views of its own Orient.[12] Taken as a whole this is a diverse collection, but there are certain basic elements which all of these works have in common. They are all concerned with Western views of the

[7] Said, *Orientalism*; Honour, *Chinoiserie*; Steadman, *Myth*; Parry, *Delusions*; Greenberger, *British Image*; Winks and Rush, *Asia*; March, *Idea*; Smith, *European Vision*; Bishop, *Myth*.

[8] Curtin, *Image*. [9] Carter, *Road*.

[10] Henry Nash Smith's marvellous *Virgin Land* has been a steady source of inspiration for my own study from the outset. Also see O'Gorman, *Invention*; Echeverria, *Mirage*; Chinard, *L'Amérique*; Billington, *Land*; Honour, *New Golden Land*; Madsen, *Visions*; Greene, *Intellectual Construction*; Gerbi, *Dispute*. [11] Wolff, *Inventing Eastern Europe*.

[12] Layton, *Russian Literature*; *idem*, "Creation;" Slezkine, *Arctic Mirrors*; Diment and Slezkine, *Between Heaven and Hell*; Brower and Lazzerini, *Russia's Orient*; Schimmelpennick van der Oye, *Ex Oriente Lux*; Riasanovsky, "Russia and Asia;" *idem*, "Asia through Russian Eyes;" Becker, "Muslim East;" Hokanson, "Literary Imperialism;" Popkin, "Chekhov;" Bassin, "Expansion;" *idem*, "Russia between Europe and Asia;" *idem*, "Inventing Siberia," *idem*, "Russia and Asia."

regions in question, and beyond this they share an emphasis on the fact that there is at best only a very partial correspondence between these views and the actual material qualities of the regions depicted. Much more importantly, the former are to be seen instead as the product or creation of processes internal to the society which produces them. Although it usually arises out of some sort of knowledge about and contact with the regions they depict, a geographical vision is a cultural construct, and it is only by understanding it in this manner that we can appreciate its most basic analytical significance, namely that a society's picture of foreign peoples and places is above all an expression of its own domestic mentality. It informs us accordingly not so much about the object of representation as about the beliefs, hopes, prejudices, and frustrations of the group that authors it. This in turn suggests a corollary which is quite central to the theme of this book, namely that geographical regions are perceived and signified ideologically, in much the same way that social institutions and processes are signified – for example (in the case of Russia) serfdom or industrialization.

Historical geography has made a special contribution of its own to the study of regional images. The stimulus came largely from the seminal work of John Kirkland Wright, whose abiding interest in *terrae incognitae* was animated by an underlying concern with anthropological perception and signification of uncharted lands. To characterize what he had in mind Wright introduced the novel term "geosophy" or the study of the history of geographical knowledge.[13] Although the full scope of the term includes more or less the totality of subjective–psychological perceptions of the environment – on the individual as well as group level – the specific preoccupation of the present study with the attitudes of an educated social elite taken as a whole figures prominently. Wright is commonly celebrated as a harbinger of both the so-called behavioral geography of the 1960s as well as the humanistic geography which followed in the 1970s,[14] and while these unquestionably are important aspects of his influence, the degree to which his work and teachings on geosophy and regional perception fostered an important field of research in historical geography in their own right is perhaps rather underappreciated.[15] Hugh Prince has framed the orientation of this field of "historical geosophy" quite well, describing it as the concern with "past worlds, seen through the eyes of contemporaries, perceived according to their culturally acquired preferences, shaped in the images of their assumed worlds."[16] Despite its own par-

[13] Wright, "Introduction," pp. 6–7; *idem*, "*Terrae Incognitae*," pp. 82–88 and *passim*; [Billinge], "Geosophy," p. 138.

[14] Johnston, *Geography*, pp. 140, 142; Livingstone, *Geographical Tradition*, p. 336; Ley and Samuels, "Introduction," p. 8.

[15] See however Wright's impressive *Festschrift*: Lowenthal and Bowden, *Geographies*.

[16] Cited in Johnston, *Geography*, p.143. For an indication of how significant this perspective has been in influencing decades of work in historical geography, see the numerous entries for "historical geosophy" in Conzen *et al.*, *Scholar's Guide*.

ticular emphases and nuances, the present work is conceived very much as a part of this legacy.

Beyond the methodological and conceptual common ground of all this work the study of regional images diverges widely, and the examples considered in the present work have their special aspects as well. In particular, there are two characteristic qualities of Russian images of the Amur which are important to note at the outset. The first is that the specific object of the image varied in subtle but significant ways; indeed, it is possible to discern a range of more or less discrete objects. The "Amur" was understood variously to indicate the river by itself, the river and its valley, or the greater Amur–Ussuri region as already described. Moreover, in a manner that will become clear later in this work, the term could also be used in a much more general and amorphous sense, to refer to a location on the Pacific which was distinguished not by any contours or qualities internal to it but merely by its particular proximity to other Russian and non-Russian areas in the Far East. The second quality of these images relates to how they were formulated, and here again there was considerable variation. Some of the most colorful and clearly articulated representations of the Amur were the deliberate creation of Russian "promoters," enthusiasts who resembled in certain ways the prototype from the United States that played such a prominent role in the advance of the frontier across the North American continent. Indeed, one particularly enterprising American even made his way to the Amur, where he tirelessly preached a millenarian message of imminent commercial glory to a startled but appreciative Russian audience. His Russian counterparts sought to disseminate positive pictures of the region, often with the obvious ulterior motive of securing government support, encouraging migrants, attracting investment, and so on. These sorts of promotional images are the easiest to recognize and analyze, but their overall significance is limited. There were not very many of these individuals in the Far East, they made their appearance relatively late in the day, and their contribution was thus not an extended one. Much more important for this study was the protracted and – in contrast to the promoters – one might say sincere process of semantic accretion around the Amur, beginning in the 1830s, in the course of which a variety of significations and meanings were attached to the region as part of a much broader engagement of Russia's educated public with the social and political challenges of their day.

The specific themes and images associated with the Amur fall into two categories. The first of these includes what we might call place-specific images, in the sense that they highlighted a quality or group of qualities which were, if not always entirely unique to the region, then at least clearly characteristic of it. The region's Siberian, and more specifically its far-eastern Pacific location, was one such quality, as was the physical–geographical feature of the river itself, the agricultural lands which adjoined it, the region's natural resource endowment, and so on. Understandably, these images emphasized the local dimension of the Amur region's significance quite strongly – for example, its potential to serve as a base of agricultural production for food-hungry parts of Siberia and (in particular) for Russia's fur colonies of the North Pacific and Alaska. The view of the river as a navigable conduit connecting the oceanic coast with the continental recesses of the Transbaikal region was also an enduring prospect. This latter function could be intended as a solution of the local provisionment problem in the North Pacific, but it was usually framed rather more grandly, in the proposition that for a Siberia which as a whole was isolated and cut off from outside contact, the Amur represented a vital artery insuring a *vykhod k moriu*, or outlet to the sea.

At the same time, however, these place-specific images did not necessarily restrict the importance of the Amur to territories east of the Urals. The notion that the large-scale economic, industrial, and demographic development of Siberia possessed a special significance for Russian national development overall – a conviction which was to become extremely important in the late nineteenth and twentieth centuries – had its origins in the period and events examined in this study. This view grew out of older mercantilist attitudes from the seventeenth and eighteenth centuries toward Siberia as Russia's *zolotoe dno* (or gold mine), but was more immediately founded in the mood of future-oriented activism and the urge for national reconstruction that accompanied the emergence of an ideology of Russian nationalism in the early nineteenth century. To develop Siberia became a sort of imperative for the nationalists, for it would be an important means of developing and enhancing Russia as a whole. In a very different connection, Russia's ensconcement on the Pacific brought to light a fundamental tension between two differing geopolitical perspectives on the most appropriate course of Russia's future development. Should Russia consider itself first and foremost as a continental land power and focus its attention and energies on the landmass of northern Euro-Asia, or was it rather the world's oceans which offered the truly important arena for the country to enhance its stature among the other imperial powers and make good its international pretensions? The mutual incompatibility of these "continental" and "maritime" perspectives emerged gradually as the implications of Russia's new *vykhod k moriu* were worked through, and in the debates around the Amur in the 1850s we will see the incipient traces of a geopolitical discourse that was to become ever more articulated and emphatic as the century wore on.

More diverse and ultimately more revealing than these place-specific images were those for which the details of the Amur region's various natural–geographical qualities were not particularly significant. The degree of abstraction in these cases was far greater, and in them the region was converted into a geographical vision in a much more literal sense. The Amur became a sort of quasi-myth, the palpable realities of which were not only largely irrelevant but indeed positively obstructive, insofar as they could potentially restrict the extent to which it could be imbued with the desired kaleidoscope of meanings and significations. Rather than a material geographical place, the Amur essentially represented a concept, or better yet the shell of a concept which could be loaded with those preoccupations that happened to be uppermost in the mind at the moment. To be sure, the process of mythologizing was never complete, for at least one connection with the real world always remained, namely the fact that the Amur was a "foreign" region by virtue of its physical location outside of Russia's traditional historic space. As we will see, however, even this circumstance was nuanced and subject to rather different interpretations. Beyond this, the designation "Amur" could be used essentially as a metaphor, that is to say an ostensibly geographical zone which in fact was nothing more than an exotic name for those values, hopes, and expectations that had been invested in it. It was above all because the concept of the Amur could for a brief period in the mid-nineteenth century be semantically emptied and refilled with relative ease that it became such a useful and popular referent for the Russians at this time. Moreover, because whatever signification this might involve was not tied intrinsically to the region, it could be easily transferred elsewhere the moment that the image of the Amur was no longer realistically able to accommodate it. We will be able to follow this latter process at the conclusion of this study.

The most important of the preoccupations which shaped perceptions of the Amur region was the emergence and dynamic growth of Russian nationalism in the first half of the nineteenth century, with a climax of sorts in the aftermath of the Crimean War. The ideology of nationalism set out a comprehensive and multifaceted agenda, and as I will seek to demonstrate it was in terms of the objectives and priorities of this agenda that the Amur was "ideologized" and assigned a meaning at the time. For this reason, and only for this reason, the attention of Russia's educated public west of the Urals was directed to the Far East on a broad scale. Russian nationalism demanded a break with the stultifying conservatism of Nicholas I's regime, and the annexation of the Amur – a daring move on the international arena which the old tsar's government trenchantly resisted – seemed at long last to provide precisely that break. Russian nationalism demanded an activist program of national reform and reconstruction, which would demonstrate that the country's dynamism and its capacity for creative accomplishment were undiminished despite a protracted period of stagnation, and once again the Russian colonization of the Amur region appeared to provide a brilliant demonstration of precisely these qualities. Rather more ambitiously, Russian nationalism sought to provide a picture of what the reformed and regenerated Russia that compatriots ought to be striving for would look like, and as it happened this picture could in many respects be sketched out most satisfactorally on the unfamiliar and hence pliable canvas of remotest Siberia. In this connection, the geographical location outside of the traditional Russian pale was important, for this quality of absolute novelty was precisely what made it possible to project the vision of a revitalized Russian future upon it. At the height of the nation's excitement and anticipation, this vision led to the identification of the Amur region as Russia's very own New World or "America," and of the Amur river as Russia's very own Mississippi.

There was yet another way in which the advance in the Far East was accorded a significance in terms of the agenda of Russian nationalism. One of the most important aspects of the nationalist ideology articulated at the time was a powerful sense of universal mission, of having been selected to serve as an agent for bringing salvation and improving the welfare of other parts of the world. Stimulated by this confident conviction, the nationalists cast their eyes to the East. Rejecting the teachings of the Enlightenment and early Romanticism about the wisdom, sublimity, and perfection of the Orient, they instead discovered there just the opposite: a rich assortment of benighted peoples sorely in need of precisely the enlightenment and the multifarious benefits of Western Christian civilization that they were now rapidly realizing they wanted to provide. This attitude on the part of the Russians represented a significant link with the imperial mentality of other European states, where the ponderous notion of a "white man's burden" implied an essentially similar mission of salvation and enlightenment. The mere affirmation of a universal mission on Russia's part, however, was not enough. What was needed above all was the opportunity to realize it, by furnishing the restless and impatient energies of the nationalist activists with an arena upon which they could actually begin to demonstrate that they were indeed capable of fulfilling their newly-found responsibilities. Here again, the Amur region was the right place at the right time. For this particular purpose, the geographical identity of the region as a New World or a latter-day "America" was inappropriate, and its character as "Asiatic" was emphasized instead. Unsurprisingly, the Russians discovered there what they were looking for, namely a collection of indigenous peoples appearing for all the world to be in dire need of those blessings of civilization which they sought anxiously to bestow.

The importance of nationalist messianism may serve as a reminder that – *pace* the Russian and Soviet historiography noted above – the acquisition of the Amur was an act of political–territorial expansionism on the part of the Russian state. Rather than representing a dramatic *dénouement* which after centuries of thwarted effort finally secured or re-secured for Russia its natural and legitimate boundaries on the Pacific, it should rather be seen very differently as the beginning of the spectacular final phase of pre-revolutionary Russian imperialism. This period was subsequently to witness the incorporation of Russian Turkestan and, at the turn of the century, renewed expansionist activity in the Far East. Thus among other things the Amur epoch forms a chapter in the history of modern Russian – and, indeed, modern European – imperialism, and it would be appropriate to indicate the most important links with these questions which have informed my work. My argument that Russian thinking about the Far East was anchored firmly in the national sentiments of the day draws fundamentally on the perspective of what recent historiography of German imperialism terms the *Primat der Innenpolitik*. This interpretation identifies an intimate connection between domestic affairs and foreign policy, and moreover asserts the primacy of those impulses coming from the domestic scene over the international situation in influencing the pattern of political expansion.[17] There has been some resonance in the study of Russian imperialism to this approach, which must contend with the contrary view of Richard Pipes and others that "foreign policy and domestic politics in Russia were widely separated from each other."[18] Most significant in advancing an alternative perspective is the work of Dietrich Geyer, who has specifically indicated the relevance of what he calls "endogenous aspects" for the Russian advance to the Amur.[19] In a different regard, the connection between the drives for social and political reform on the one hand and political–territorial expansion on the other which is central to my story can be situated in the literature on so-called "social imperialism" in England, Germany, and elsewhere.[20] I may anticipate a point brought out later in this study, however, by indicating that there is no argument in the Russian Far East for the importance of economic factors in motivating the expansionist program, a interpretation which commonly figures in the social imperialism literature

Most of these studies, however, are preoccupied above all with the larger question of the "why" of imperial expansion, and do not engage very extensively with my own rather more limited concern – namely, the place of the imperial impulse within the nationalist doctrine that took shape in the 1830s and 1840s. This latter question, however, is one of the most critical of all for an understanding of Russian images and perceptions of the Amur region. For the most part, discussions of the imperial dimension in pre-revolutionary Russian nationalism have focused on the most visibly and consistently impe-

17 Mommsen, "Der moderne Imperialismus," *passim*.

18 Pipes, "Domestic Politics," pp. 145–150.

19 "Auch für den Erwerb des Amurgebietes und der fernöstlichen Küstenprovinz . . . ist der psychologische Zusammenhang mit der innerrussischen Lage evident." Geyer, "Russland," p. 343. Above all see Geyer's major work *Der russische Imperialismus*; on "endogenous aspects" see *idem*, "Modern Imperialism?," p. 50. For other treatments which seek to link domestic and foreign policy, see Ulam, "Nationalism;" McDonald, "Lever;" Hoffman and Flern, *Conduct*, pp. 31–90; Bialer, *Domestic Context*. It is incidentally extremely significant that virtually all of the examinations of foreign policy in post-Soviet Russia emphasize the close interconnection with domestic affairs. See for example Wallander, *Sources*; Skak, *From Empire*; Ra'anan and Martin, *Russia*; Buszynski, *Russian Foreign Policy*.

20 Semmel, *Imperialism*; *idem*, *Liberal Ideal*; Eley, "Defining Social Imperialism;" Wehler, "Bismarck's Imperialism;" *idem*, "Sozialimperialismus;" Kennedy, "German Colonial Expansion;" Gollwitzer, *Europe*, pp. 125ff; Ascher, "National Solidarity."

rialistic nationalists, namely the Pan-Slav movement.[21] While the Pan-Slavs are indeed central to my study in a number of ways, I take a rather different approach to the question of nationalism and imperialism. To begin with, I will argue that an imperial vision was not a later addition or elaboration from the period after 1860 (when the Pan-Slav doctrine became fully articulated) but rather figured as an essential part of Russian nationalism from the outset – that is to say, as the latter took recognizable shape in the 1830s and 1840s. This point calls into question an assumption common to the literature on European nationalism in general, namely that the practical contrasts between nation-states and empires as political entities corresponded to an doctrinal contradiction and incompatability between the ideologies of nationalism and imperialism. Scholars as divergent as Hannah Arendt and Benedict Anderson have emphasized this point, the latter with an evocative metaphor indicating the impossibility of ideologically stretching the "short tight skin" of the nation over the "large body" of the empire.[22] While a nationalism with clear imperialist overtones may well have existed, it is therefore suggested, its appearance post-dated that of nationalism proper – usually located in the final quarter of the nineteenth century – and it is furthermore distinguished from the latter as being something qualitatively different: a "new kind of imperialist nationalism," an "emotive nationalism," or a "perverse (missratener) nationalism," as Anthony Smith, Wolfgang Mommsen, and Hans-Ulrich Wehler, respectively, have put it.[23]

My point is not to deny the obvious fact that nationalism in various countries underwent transmutations of this sort (although the extent to which nationalism can be said to have ever been entirely free from extra-national hegemonic inclinations has been seriously questioned[24]). What I would suggest is that from the very beginning there was no such contradiction or incompatability in Russia. Here nationalists, however Russophilic and – more importantly – however implacably opposed to the dynastic and autocratic state, at once

[21] This particular aspect is examined most comprehensively in the work of Hans Kohn, which has been fundamental for my own thinking in a number of regards. See especially his *Pan-Slavism*; "Messianism;" "Dostoevsky;" "Russia: Permanent Mission." Also see Petrovich, *Emergence*; Bohatec, *Der Imperialismusgedanke*; Hunczak, "Pan-Slavism."

[22] Arendt, *Origins*, pp. 126–126ff; Anderson, *Imagined Communities*, pp. 86 (quote), 109–110. Paradoxically, Arendt goes on to provide considerable evidence to the contrary (not, incidentally, the only paradox in her brilliant work).

[23] Smith, *Nationalism*, pp. 9–10; Mommsen, *Das Zeitalter*, pp. 15 (quote), 16–27; Wehler, "Einleitung," pp. 13 (quote), 14–15. Also see Hayes, *Generation*, pp. 216–220ff; Alter, *Nationalism*, pp. 43–44, 46–47. "Imperialist nationalism" coincided chronologically with the emergence of so-called "integral nationalism," and the latter term often subsumes the former. "By the 1880s, nationalism and imperialism became intertwined in a way which had never been the case before." Mommsen, "Varieties," p. 217; Hayes, *Historical Evolution*, pp. 164–231; Lemberg, *Nationalismus*, I, pp. 195–225.

[24] Reinhold Niebuhr, for example, traces what he calls "national imperialism" back to the sixteenth century. *Structure*, pp. 201–216. Also see Shafer, *Nationalism*, pp. 149–150, 179–180; Lichtheim, *Imperialism*, pp. 81–99.

embraced the entirety of their unmistakably multinational empire, and did so with singular devotion. Beyond this, they virtually unanimously endorsed the desirability and even necessity of further political–territorial expansion into non-Russian areas as an important part of their program of national advancement and renewal. This enthusiasm was characteristic, as we will see, not only for reactionary chauvinists but for tolerant liberals and even for proto-socialist radicals. Moreover, not only did an unmistakable expansionist element have a presence in nineteenth-century Russian nationalism from the moment of its first articulation, this element was an inalienable part of it, which derived from the same general rationale and spoke to the same needs. Effectively, nationalism and imperial vision were joined in a common project and could not be divorced. This was moreover a condition which endured. While fully acknowledging the fundamental changes in ideology and reality that the intervening century-and-a-half have brought, I would nonetheless suggest that the enormous difficulty in disengaging nation from empire in Russia's post-Soviet consciousness of the 1990s is not entirely unrelated to their original conflation examined in this book.

It is also important to appreciate that in a practical sense there was a distinct amorphousness in this expansionist urge. This condition can be understood only in view of the sources and the broader framework of Russian nationalism as a whole. As had been the case in Germany some decades earlier, nationalism in Russia arose out of the confrontation with the West, as Russians sought to rationalize and master the overwhelming sense of inadequacy with which this confrontation left them. Within the ideological framework of Russian nationalism, the prospect of political–territorial expansion was intended not so much to satisfy an appetite for control of foreign lands and peoples as to secure evidence of positive or even superior national qualities which could then serve to raise Russia's stature *vis-à-vis* the West. The imperialist project thus assumed a significance for the national psychology as what Adam Ulam has called a "mechanism of compensation for backwardness," and its real concern was accordingly not with the object of conquest and incorporation but rather with Russia itself.[25] Indeed, because the foreign region in question possessed a mere instrumental and conditional value, its specific location was not necessarily of much importance in the first instance.

The fact remained, of course, that there could be no expansion without these foreign regions, and so a practical program was necessary which, by providing a geographical object and suitably inspiring rationale, could enable the nationalists to translate their imperial inclination into action. By far the most successful of these programs was Pan-Slavism, "the most popular cause ever taken up by Russian imperialism" as Emanuel Sarkisyanz reminds us, which

[25] Ulam, "Nationalism," p. 44. Also see Simon, "Russischer und sowjetischer Expansionismus," p. 103; Geyer, "Russland als Problem," pp. 344–345.

directed the attention of the nationalists to the relatively familiar Slavic regions of Eastern Europe, the Balkans, and the Near East.[26] It was not, however, the only such program. In something of a parallel fashion, Russian nationalism also included a clear and ultimately sophisticated vision of disseminating Russian domination to the East, across Asia. If the rationale for the latter was not exactly identical with that for Pan-Slavism, then at least the two shared a variety of important elements, for which reason Russian expansion in Asia throughout the nineteenth century – beginning in the 1850s with the annexation of the Amur – was commonly supported with great enthusiasm by the Pan-Slavs.[27] At the same time, there were significant differences, and by the end of the century the so-called *Vostochniki* would be urging Russia's advance to the East in terms of a rationale and practical program that had little to do with Pan-Slavism.[28] The common roots of both movements in Russian nationalism, however, were evident.

In the final analysis, experience in the Far East at the mid-century was to demonstrate that there were limits to the extent to which the imagination of the Russian public west of the Urals could manipulate the image and the meaning of the Amur region. Indeed, through the process of manipulation and appropriation the prospect of the Amur actually came to present a peculiar challenge of its own to Russian nationalism. There had been enormous satisfaction at the ease with which the Amur could be filled with the dreams for the future of a reformed Russia, but after the initial excitement had died down a question was inevitably bound to arise. Was Russia's new acquisition on the Manchurian frontier, with its monsoon climate and sub-tropical vegetation, really Russia after all? And if it wasn't, then how legitimate was it to try and locate such a profound national significance in it? Characteristically, the gravity of this problem had nothing to do with the Amur region itself, which as events quickly showed could be written off and forgotten at essentially no cost. What it brought to light, rather, was a particularly vulnerable aspect of Russia's view of itself as a nation, namely the fact that there was no clear and commonly accepted notion of exactly, or even approximately, what its geographical contours were. The question "where is Russia?," in other words, was neither elementary nor self-evident, but instead one which had no commonly accepted answer. Because this was the case, we will note throughout this study an implicit ambivalence regarding the Amur which was expressed in a variety of forms. At one moment or for one person, the region could be the most progressive and genuine part of Mother Russia, while at

[26] Sarkisyanz, "Russian Imperialism," p. 64.

[27] See Geyer, *Der russische Imperialismus*, p. 77, who notes this point but dates it only from the 1870s.

[28] There is no full study of the *Vostochniki*, although Schimmelpennick van der Oye's recent dissertation *Ex Oriente Lux* devotes considerable attention to them. For other discussions, see Malozemoff, *Russian Far Eastern Policy*, pp. 41–50; Marks, *Road*, pp. 136–138; Hauner, *What is Asia*, pp. 56–60; Geyer, *Der russische Imperialismus*, p. 146.

another moment or for someone else it was merely a remote and thoroughly un-Russian part of the empire. Needless to say, very few Russians at the time took much note of this vacillation, and they were equally unperturbed by the more general lack of clarity regarding the extent of their specifically national space. As I have pointed out, nation and empire could coexist in what can almost be called a symbiotic relationship, and for most people there was simply no point in trying to disengage them. We will see in conclusion, however, that the dilemma was indeed sensed by a few particularly perceptive individuals, for whom such a coexistence was unacceptable, and who consequently tried to draw a clear boundary between nation and empire. As far as I am aware, this was the first such attempt in Russian national thinking, and there is more than a little irony in the fact that the Amur region, which had served to focus their attention on this problem in the first place, was to be excluded from the realm of "genuine" Russia in the process.

The present work, therefore, is the history of a geographical vision, which will seek to shed light both on the fate of a region and on the self-conception of the society which envisioned it. As I have set it out and pursued it, the problem has always seemed to me to be quintessentially geographical, and the work of historical geographers described above certainly confirms this. At the same time, some readers will find that it makes more sense to approach the work as intellectual history, and indeed both my acknowledgments as well as my bibliography make quite clear my enormous debt to intellectual historians. The fact is, however, that this study crawls along the borders between the fields. It does so, I would like to point out, not (or not only) by virtue of professional schizophrenia on my part or a cagey reluctance to acknowledge and submit to the discipline of academic boundaries. More than this, it is bound up in the very nature of the subject I have engaged. Intellectual history provides a way into the mind of imperial Russia, a framework for identifying the images of the Amur, and a methodology for problematizing and studying them. But geography provides an appreciation of the Amur region as a real existing place, with physical characteristics, settlement patterns, some sort of level of development and potential for future development. At the most basic level, *Imperial Visions* grows out of and is driven by the interplay and the tension between the two.

PART I

1

Early visions and divinations

"A region beautiful and bountiful"

The Russians first appeared in the Amur river valley in the 1630s, in the course of their movement east across Siberia. This movement had begun some 50 years earlier, with the penetration across the Urals by Yermak's cossack band and their victory over the Tatar prince Kuchum. From the outset, the Russians were pulled eastward by a single and simple goal: the quest for the fabulous fur wealth of the Siberian taiga.[1] Furs played a critical role in the finances of the early Russian state, serving not only as the most important item of barter with foreign countries but as a major commodity of domestic exchange as well. Indeed, furs represented one of the most significant sources of mercantile capital for the Muscovite economy, and were used in much the same way and for the same purposes as were the gold and silver of the New World by the Iberian empires.[2] The high value of Siberian pelts ensured that they would be hunted intensively, and as the fur-bearing population of one locale was exhausted the *promyshlenniki* or fur traders pressed further east, seeking out new reserves. In this manner, furs may well be said to have drawn the Russians across the north Asian landmass, and indeed they did so with remarkable rapidity. Most accounts date Yermak's initial crossing of the Urals to 1582, and the Pacific coast was reached by Ivan Moskvitin in 1639. In primitive log boats, on sled and by foot, well over 3,000 miles of some of the roughest terrain on the face of the globe had been traversed in the space of only 57 years.[3]

The main line of this eastward advance, logically enough, followed the

[1] On the Siberian fur trade, see Pavlov, *Promyslovaia kolonizatsiia*; Fisher, *Russian Fur Trade* ; Gibson, *Feeding*.

[2] Baikalov, "Conquest," p. 561; Fisher, *Russian Fur Trade*, pp. 17, 119–120; Gibson, "Significance," p. 443; Bassin, "Expansion," pp. 11–15.

[3] On Russian expansion to the Pacific, see Lantzeff and Pierce, *Eastward*; Foust, "Russian Expansion."

trapping grounds for the very best fox, ermine, and sable pelts, and consequently remained deep in the Siberian taiga, well to the north of the Amur region itself. The Russians set up their *ostrogi* or forts at Yeniseisk in 1619, at Bratsk on the Angara in 1631, and at Yakutsk on the Lena in 1632. It was cossacks from the latter outpost who sailed upstream along the eastern tributaries of the Lena river and descended down the southern slope of the Stanovoi range to the valley of the Amur river in the 1630s. They were drawn south in this fashion not so much by the prospect of furs but rather in response to one of the most serious crises in the logistics of the fur trade, namely the problem of provisionment. From the beginning, the Russians in Siberia were plagued by the lack of a reliable supply of foodstuffs, and this problem intensified in direct proportion to their eastward advance. In Eastern Siberia, where limited numbers combined with particularly harsh physical–geographical conditions to render agriculture impossible for all practical purposes, the problem of securing an adequate food base was acute; indeed, the search for provisionment often rivaled that for furs as an impetus for exploration and occupation.[4] Thus, stories related to the Russians by the indigenous peoples about the mild climate of the Amur valley, about the fish which abounded in the river and the golden grain which was said to ripen in abundance along its banks attracted considerable attention, and in 1643 the cossack Vasilii Poiarkov was ordered to survey these unknown territories. Poiarkov returned to Yakutsk three years later with glowing accounts of flourishing agricultural tribes cultivating no less than six varieties of grain, and where along with cattle, horses, cows, rams, and pigs abundant stocks of sable, lynx, and fox were to be found.[5] He completed this enticing picture with the confident assurance that the entire river valley could be easily subjugated by as little as 300 men,[6] and in 1649 Yerofei Khabarov was dispatched with a band of cossacks to do just that.

In this manner, the Amur region was imbued from the very outset with an allure of an entirely special nature for the Russians in Siberia. This was an aura which Khabarov, for his part, made every attempt to enhance. In the report on his mission to the river which he sent to Moscow in 1650, he asserted that the "great river Amur" contained "more kaluga, sturgeon, and other fish than the Volga," and that its banks were lined with dense forests in which roamed "sables a-plenty and all sorts of [fur-bearing] animals." Most importantly, he assured his superiors that the region could without any question support a bountiful agriculture, which would supply grain to Yakutsk and other parts of Siberia. "And just imagine, lord," he concluded, "that this land of Dauria will be more profitable than the Lena . . . and in contrast to all of Siberia will be a

[4] Gibson, *Feeding*, *passim*; Shchapov, "Istoriko-geograficheskoe raspredelenie," pp. 197–198; Kharnskii, *Kitai*, p. 269. [5] *Dopolneniia k aktam*, III, Doc. 12, pp. 51, 57.
[6] Miller, "Istoriia," p. 13; Romanov, "Prisoedinenie," 4, pp. 182–183.

region beautiful and bountiful."[7] At the same time that Khabarov was tantilizing Moscow with these seductive accounts, rumors about this glorious new realm to the south had began to spread among the Russian population of Eastern Siberia itself. The hyperbolic character of these early images of the region was conveyed by the German historian Gerhard Müller, who in the 1750s published the first historical account of Russia's early experiences on the Amur. Müller noted how Khabarov and his men referred to the Amur as a "inexhaustible source of wealth," and he paraphrased their evocative depictions:

> There was an abundance of gold, silver, the best sables, cattle, grain, and orchard fruits to be found there [on the Amur]. The local peoples wore no other type of clothing besides gold, brocade, and silk, and even the Russian cossacks themselves [who had been there] possessed these sorts of clothes and showed off in them, so that no one would doubt their stories. In short, the lands along the Amur were praised as a *New Canaan and Paradise* (raiskaia zemlia) in Siberia. Everyone wanted to take part in [settling] them.

The effect that the prospect of such an El Dorado would have had on the hungry and discontented Russians in less favored parts of Siberia can readily be imagined, and our chronicler went on to recount how many men "left their homes, their wives and children, imagining that life on the Amur would be incomparably better."[8] In the decades that followed Khabarov's expedition, cossacks and *promyshlenniki* streamed to the Amur in substantial numbers, and a number of settlements were established along its northern bank.

It seemed too good to be true, and indeed it was, for the actual conditions in the Amur region did not begin to correspond to these effusive depictions. The climate did in fact differ somewhat from that in other parts of Siberia, but it remained in all events far from mild. The quality of the local furs, moreover, was decidedly mediocre in comparison to those available in the heart of the taiga further north, and accounts of rich deposits of precious metals proved apocryphal. Most disappointing, however, was the fact that whatever agriculture practiced by the scattered and miniscule indigenous groups whom the Russians encountered was on a subsistence level, and produced no more food than they needed themselves in order to survive.[9] Moreover, the expectation

[7] *Dopolneniia k aktam*, III, Doc. 72, pp. 260–261; Miller, "Istoriia," pp. 15–29; Bassin, "Expansion," p. 13. In the seventeenth century, Dauria was a common designation for the Amur region; subsequently it was used more rarely and usually in specific reference to the lands of the upper Amur, where most of the Russian settlement prior to 1689 had been concentrated. See [Gakman], *Kratkoe zemleopisanie*, p. 22n.

[8] Miller, "Istoriia," pp. 101–103, 201 (emphasis in original quote).

[9] Iakovleva, *Pervyi russko-kitaiskii dogovor*, p. 28. Subsequent accounts have explained the striking exaggeration of Khabarov's early depictions of the Amur as a carefully calculated attempt to attract manpower necessary for his conquest of the region. Miller, "Istoriia," p. 103; Romanov, "Prisoedinenie," 4, p. 187.

that in the absence of a native agricultural base the Russians themselves could develop the large-scale production of agricultural foodstuffs for export to other parts of Siberia was effectively thwarted by the very nature of the society which took shape there. The extreme remoteness of the area, together with the fact that many of those who came to the Amur did so in part to escape the local authorities in other parts of Siberia,[10] made it impossible for the government to effect the organization necessary for such an undertaking. Consequently, the initial optimistic expectation that the Amur region would become Eastern Siberia's *zhitnitsa* or breadbasket and solve its problems of provisionment came quickly to naught. Indeed, the new Russian settlements along the Amur were already experiencing severe food shortages of their own in the 1650s.[11] Ultimately, the Amur valley proved to be an El Dorado only insofar as it enjoyed the primitive anarchy of a freebooters' camp.[12]

It was not, however, problems of provisionment which ultimately were to determine the fate of the Russians in these remote territories. At the very moment that they had begun to establish their presence on the Amur, the Manchus to the south were in the final stages of consolidating their victory after decades of struggle to establish their domination over China. Their ascendance in the 1640s insured that greater attention would begin to be paid to the Amur valley, for the region formed the northern boundary of their Manchurian homeland, and the Amur figured as a sacred river in Manchu mythology. The territories along it on both sides were considered to be rightfully part of China's imperial domains, the native peoples inhabiting them were regarded as imperial subjects, and the new Russian presence was seen as nothing less than an unlawful intrusion. The Russians, it may be pointed out, were unaware of Manchu pretensions to these lands as they began to occupy them,[13] but they discovered quickly enough, and in the early 1650s armed clashes were already taking place between the two powers. Hostilities increased in number and intensity over the following two decades, to the point that by the 1680s a number of Russian settlements were razed to the ground.[14] Given the region's more favorable proximity to the political center of the Chinese state, the latter was in a position to assemble a military presence there which the Russians could not hope to match. These persistent hostilities on the frontier between the two empires were finally resolved by negotiations held at the town of Nerchinsk in August 1689. The Chinese mustered a massive display of arms and men for the occasion, and were easily able to impose their

[10] Koz'min, "M. B. Zagoskin," pp. 161–162.

[11] Miller, "Istoriia," pp. 120, 126; Iakovleva, *Pervyi russko-kitaiskii dogovor*, p. 53; Gibson, "Russia on the Pacific," p. 25.

[12] Lantzeff, *Siberia*, p. 82. This condition did not prevent the government from making use of the region as a place of banishment. The most illustrious exile from this period was none other than Archpriest Avvakum, who helped found the town of Nerchinsk in 1658. Holl, "Avvakum," *passim*; Kurts, *Russko-kitaiskii snosheniia*, p. 17.

[13] Mancall, *Russia*, p. 21.

[14] Miller, "Istoriia," pp. 29, 108–109, 206–212.

own conditions on the agreement which emerged out of it. By the letter of the Treaty of Nerchinsk, Russia relinquished all claims to the territories along the Amur and the native peoples living there, and agreed to withdraw its settlements and cease all its activities in the region. A boundary running well to the north of the river was agreed upon in principle; because however the topographical information at the disposal of either side was scanty and highly imprecise, a specific demarcation could not be made.[15]

Thus after nearly four decades, the Russians were expelled from the Amur valley under circumstances which subsequently would make it possible to claim that they had been forcibly and unjustly evicted from the region entirely against their will. The Chinese, it would be asserted, had in effect stolen territories which their northern neighbors had rightfully claimed and settled, and which moreover were valued very highly. There is some degree of accuracy in this argument, to be sure, but it is worth noting at this point that in certain respects it is misleading. The importance of the Amur lands to the Russian state had been measured in terms of the same mercantile concerns that underlay Russian interest in Siberia overall, and thus once it became clear that the river valley would neither supply furs and precious metals nor contribute significantly to the logistical support of other areas which could, the region simply lost its value, at least in the eyes of the government in Moscow. Indeed, it even became something of a liability, for in the second half of the seventeenth century, as we will see presently, the Russians were keenly interested in fostering commercial relations with China, and the incessant territorial disputes over this remote, unknown, and apparently worthless region threatened to undermine Russia's elaborate overtures toward this end. The pronounced displeasure of the Chinese regarding the situation on the Amur was stated quite clearly in a missive to the tsar in 1657, in response to which Aleksei Mikhailovich went so far as to plead forgiveness for the ignorance of his cossacks who had ventured into this region "not knowing that the Daurian lands are part of your Dominion."[16] In the same spirit, a Russian diplomatic mision to China headed by one Sentkul Ablin emphasized the government's preparedness to halt the activities of the cossacks in the Amur valley for the sake of a commercial agreement between the two countries.[17] Finally, the various sets of instructions sent three decades later from Moscow to Fedor Golovin, Russia's chief negotiator at Nerchinsk, made it quite clear that the Russians were, if not exactly anxious then in any event entirely willing to sacrifice their territorial claims to the Amur valley, if by so doing they

[15] Miller, "Istoriia," pp. 291–323.

[16] *Russko-kitaiskie otnosheniia*, I, p. 229. With this, the Russian tsar explictly admitted that the Amur valley was Chinese, and not Russian territory, a fact that the editors of this Soviet collection have tried to obscure with the improbable clarification that he was actually referring to lands along not the Amur but rather the Sungari river, a Manchurian tributary to the south. *Russko-kitaiskie otnosheniia*, I, p. 554. Also see Kappeler, "Die Anfänge," p. 32; Bassin, "Expansion," pp. 14, 20n.

[17] Savvin, *Vzaimootnosheniia*, pp. 25–26.

could facilitate progress toward a formal trade agreement with the Chinese.[18] This is precisely what Golovin accomplished, and thus the Treaty of Nerchinsk was not denounced as the unlawful theft of the "Russian" river Amur but rather was welcomed as a diplomatic success of considerable significance.[19]

"This river can . . . be useful for Kamchatka and [Russian] America"

The immediate effect of the Russian departure from the Amur valley in the southeastern corner of Siberia was the re-direction of the attention to the fur trade to the north.[20] The exploration and occupation of the peninsula of Kamchatka began in the 1690s, almost immediately after the Russo-Chinese accord, and this set the stage for Russia's further penetration east over the waters of the North Pacific, the chain of the Aleutian islands, and finally onto the North American continent itself. Throughout the eighteenth and early nineteenth centuries, Russian settlements in these regions, and to a lesser extent along the continental coast of the Sea of Okhotsk, became the principal centers of the fur trade and hence of Russian activity overall in the Far East. This geographical shift involved a change not only in the types of furs obtained, as the sable and fox of the Siberian taiga were replaced by sea otter pelts harvested in the waters of the North Pacific, but also in the destination for sale, from largely European to largely Chinese markets. It did nothing, however, to alleviate the chronic logistical problems of provisionment. Indeed, if anything these problems became even more acute, as now – despite a local supply of foodstuffs pressed from the indigenous population – an even greater proportion of the food for the fur trade had to be transported from afar.[21] Beginning in the 1730s, this was accomplished by means of a tortuous overland passage across rugged mountainous terrain from Yakutsk to the coastal settlement of Okhotsk on the Sea of Okhotsk, and further by boat to the port of Petropavlovsk at the southern tip of Kamchatka.

This arrangement proved so thoroughly unsatisfactory, however – by virtue of both the arduous trek involved and the abysmal qualities of the port facility itself at Okhotsk[22] – that various other methods of supply had to be

[18] Bantysh-Kamenskii, *Diplomaticheskoe sobranie*, pp. 50–57.

[19] Iakovleva, *Pervyi russko-kitaiskii dogovor*, pp. 202–204; Foust, *Muscovite*, p. 5–7; Mancall, "Kiakhta Trade," p. 23; Stephan, *Russian Far East*, pp. 31–32.

[20] Aleskeev, "Amurskaia Ekspeditsiia," p. 76; Mancall, Russia, p. 10; Gibson, "Russian occupancy," pp. 64–7.

[21] Sgibnev, "Vidy," p. 563; Gibson, "European Dependence," pp. 364–376; *idem*, "Russian Expansion," p. 135.

[22] For first-hand accounts of the various problems with Okhotsk, see Shemelin, *Zhurnal*, II, pp. 202–204, 210; *Materialy dlia istorii russkikh zaselenii*, pp. 119–20. Also see Sgibnev, "Vidy," pp. 566, 580–581; Kabanov, *Amurskii vopros*, p. 42.

sought. One solution, decided upon in the 1780s but not actually implemented until the early years of the nineteenth century, was to support the fur trade with provisions shipped halfway around the world from Russia's Baltic port at Kronstadt.[23] Under the direction of Russia's most celebrated admirals, including Ivan Kruzenshtern and Ferdinand Vrangel′, the celebrated world voyages that resulted added illustrious pages to the history of the country's naval exploits, but the overall success of the venture was mixed at best. Yet more dramatic was the the attempt to supply food more locally, from agricultural bases situated in more temperate locales in the Pacific basin. The interest in establishing such a base was one of the factors stimulating Russian efforts in the 1810s to secure a sphere of influence in the Hawaiian islands, in which endeavor they lost out to the Americans.[24] Ironically, they had better luck on the North American continent itself, where in 1821 the Russian agricultural settlement of Fort Ross was founded on Bodega Bay in northern California, with the express purpose of providing grain for Russian Alaska.[25]

None of these measures were successful in relieving the critical logistic pressure on the Russian presence in the North Pacific. Because of this, the vision of the Amur river valley – long abandoned by the Russians and forgotten by most – continued to exercise a special appeal for many of those involved in the fur trade and the general administration of the Russian Far East. The specific quality of the vision, however, had been transformed from what we have seen in the seventeenth century, and now corresponded to the new circumstances and geographical situation of the fur trade. Rather than emphasizing the potential for the settlement and agricultural development of the river valley itself, the Amur was now recognized to be a natural and indeed the only water route connecting the interior of Eastern Siberia to the Pacific ocean. Foodstuffs from the relatively rich agricultural regions both to the west and east of Lake Baikal, it was fancied, could easily be shipped down the river to its mouth and then northwards across the Sea of Okhotsk to the fur colonies, thereby replacing the cumbersome overland route from Yakutsk. In this altered guise, the Amur thus once again appeared as a sort of panacea which could offer vital logistical support for the far-flung fur trade.[26] The prospect of the Amur as a supply route for Kamchatka and Russian America was first outlined in 1743 in a recommendation to the government from Aleksei Chirikov, a lieutenant who served under Vitus Bering in the celebrated Kamchatka expeditions, and it was taken up repeatedly in reports and mem-

[23] "Nachalo nashikh krugosvetnykh plavanii," pp. 501–502.

[24] On this ill-fated venture, see Zavalishin, *Zapiski*, II, pp. 22–23; Shteingel′, "Zapiski Barona V . . . a I . . . a Sh . . . a," p. 71; Mazour, "Doctor Yegor Scheffer," *passim*; Okun′, "Politika," pp. 34–39; Pierce, *Russia's Hawaiian Adventure*, pp. 1–33.

[25] Okun′, *Russian–American Company*, pp. 118–152.

[26] Kabanov, *Amurskii vopros*, p. 28; Lin, "Amur Frontier Question," pp. 6–7; Gibson, "Russia on the Pacific," p. 20.

oranda by various individuals throughout the eighteenth century.[27] Fedor Shemelin, an officer in the first world voyage carrying provisions from the Baltic to Kamchatka, stressed the importance of this potential function of the Amur in his report in the 1810s.

> [T]his river can however be useful for Kamchatka and [Russian] America, where the manufacture and trade of the Russian–American Company is now being established. The convenience with which it is possible by means of this water route to supply items of every sort – whatever their size and weight – at the least possible cost offers not only the easiest alternative, but enormous advantages as compared to the difficult and inadequate . . . route from Yakutsk to the port of Okhotsk.[28]

Paul I's ceremonious establishment in 1799 of the Russian–American Company for the purposes of administering the fur trade was an indication of a heightened level of government interest in the region. In the decades that followed, various projects were advanced for securing and expanding Russia's maritime empire in the North Pacific. As part of these, the prospect of the Amur as a link to Russia's fur colonies took on a enhanced significance. In the 1810s, for example, one Peter Dobell – a charismatic Irish–American from Pennsylvania active in the Pacific trade, who became a Russian subject and gained an ear in high governmental circles – urged the Russians to develop Kamchatka as a base from which they could expand their commercial activities in the Far East. In his evocative descriptions, the proposition appeared virtually irresistible. "There is not a place nor a harbor on the globe," he claimed with regard to Kamchatka, "which has such an advantageous position for commerce and which is surrounded at such close distances by fertile and populous countries, abounding in the most splendid products of art and nature." To drive his message home for a Russian audience which delighted in the stories of James Fennimore Cooper and which was apparently more comfortable about its geographical knowledge of the frontier in North American than with its own remote Pacific reaches, Dobell made his points about the Russian Far East by means of comparisons with the New World: in terms of its natural environment, he assured his readers, Kamchatka compared favorably with the Ohio Valley.[29] Dobell specifically indicated the

[27] Sgibnev, "Vidy," pp. 564–565, 580; Middendorf, *Puteshestvie*, I, p. 170; Romanov, "Prisoedinenie," p. 338; Aleskeev, "Amurskaia Ekspeditsiia," p. 73; Kabanov, *Amurskii vopros*, pp. 38, 34; Lensen, *Russian Push*, p. 62.

[28] Shemelin, *Zhurnal*, II, pp. 206–207 (quote), 208–210; Kirilov, "Zametki," p. 85; Sgibnev, "Vidy," p. 695; Middendorf, *Puteshestvie*, I, p. 170n; Romanov, "Prisoedinenie," pp. 370–371.

[29] [Dobell], "O Petropavlovskom porte," p. 55; Parry, "Yankee Whalers," p. 37. The government listened carefully to Dobell, but remained unconvinced. To his claim to be able to import tea more cheaply from China by boat and thus make it accessible to even poor Russians, the Minister of Finance is reported to have retorted that poor Russians had no need for tea, as they already have their vodka! [Balasoglo], "Vostochnaia Sibir′," p. 162. Also see Sgibnev, "Vidy," p. 645.

importance of the Amur River as part of this North Pacific complex, and with his reference to Ohio initiated a practice of cross-continental comparisons that decades later was to be developed into something of a fine art, as will be seen presently.[30]

At the same time, a number of individuals close to the leadership of the Russian–American Company, intoxicated by the success which they perceived the Hudson Bay Company and the East India Company to be enjoying in enhancing and spreading Britain's imperial glory across the globe, were busy developing their own schemes for promoting Russian political expansion in the Far East. It would be both possible and highly advantageous, they suggested, for Russia to extend its domination beyond Alaska, down the Pacific coast of North America to California. Among the advocates of this plan were a number of future Decembrists, notably Kondratii Ryleev, Baron Vladimir Shteingel′,[31] and a young and particularly enthusiastic midshipman by the name of Dmitrii Irinarkhovich Zavalishin. Zavalishin, who actually had visited California on a supply mission to Kamchatka in 1823,[32] was completely absorbed with this prospect and presented Alexander I with an elaborate plan for a campaign – a veritable Russian "crusade," as he rather ambitiously described it later – to liberate Mexico from Spanish domination and annex California to Russia. In his view, control of California would enable Russia to extend its Pacific dominions to include the Amur River, Sakhalin, and the Hawaiian islands.[33] None of these magnificently inflated dreams of a trans-oceanic Russian empire in the North Pacific survived the early 1820s, and in 1824 Russia signed a convention with the United States in which it agreed not to establish any new settlements south of 54°40′.[34] Nonetheless, the conviction persisted that Russia needed to assert and enhance its political and commercial presence on the Pacific, and in a very different form it was to reemerge in the following decades as attention in the Far East shifted away from the fur colonies to territories to the south.

Although down to the 1840s the significance of the Amur river was thus understood primarily in terms of the link it could provide with the Russian settlements in the Sea of Okhotsk and the North Pacific,[35] other important functions were associated with it as well. The original hope that the river valley itself might become a center of agricultural production for export faded, although it did not disappear entirely and was to be easily reanimated, as we

[30] Bolkhovitinov, "Vydvizhenie," pp. 266, 268–269.

[31] Shteingel′, "Zapiski Barona V . . . a I . . . a Sh . . . a," p. 71; Boden, *Amerikabild*, pp. 82–85.

[32] Pasetskii, *Geograficheskie issledovaniia*, p. 73.

[33] Zavalishin, "Kaliforniia," *passim* (English: "California"); Mazour, "Dmitry Zavalishin," *passim*; Okun′, "Politika," pp. 34–41; Shatrova, *Dekabrist D. I. Zavalishin*, pp. 26, 32–33; Bolkhovitinov, "Vydvizhenie," p. 269; Obolenskii, "Vospominanie," p. 315; Azadovskii, "Putevye pis′ma," p. 210.

[34] Kushner, *Conflict*, pp. 59–60; Bolkhovitinov, "Sale," p. 197.

[35] Romanov, "Prisoedinenie," pp. 370–371.

will see presently.[36] More importantly, the Amur began to be envisioned as an artery for trade which could link European Russia and Siberia to the commercial arenas of east Asia and the south Pacific. In a rudimentary form, this notion was actually extremely old, having originated in the seventeenth century out of interest in finding a transit route leading across the north Asian continent to India and to Cathay. Such a route conceivably could have been either overland across western Siberia and central Asia, or by sea following a northeast passage across the Arctic into the Pacific.[37] Convinced that the perennial ice floes of the eastern Arctic would never allow a boat to pass all the way to the Pacific, the Croatian priest Yurii Krizhanich, exiled to Siberia in the 1660s and 1670s, appears to have been the first to suggest a combined overland–ocean route. In an extensive manuscript about Siberia composed in the early 1680s, he proposed following a network of rivers from European Russia across Siberia to the Pacific, and thence southward to the shores of China and beyond. Although his geographical information was fragmentary and often inaccurate, Krizhanich clearly specified the Amur as a vital link in this network.[38]

In the late seventeenth century, the only real interest in such a passage across Russia to the East came from foreign commercial agents – Krizhanich, for example, composed his manuscript at the request of a emissary from the King of Denmark.[39] In the wake of the Petrine reforms, however, the Russians quickly became interested as well. News of his proposed route spread remarkably quickly, and already in 1716 two trading houses in St. Petersburg requested permission to follow this "short route to Japan and East India," again naming the Amur.[40] In a memorial of 1730 to Tsarevna Anna Ivanovna, after his first expedition to the North Pacific, the Dane Bering emphasized the potential usefulness of the river for Russian trade with Japan,[41] and subsequent observers developed the point extravagantly. At the turn of the century, the hydrographer Gavril Sarychev expressed the conviction that if Russia were still in possession of the Amur river, it not only would be able to conduct trade on the Pacific with "incomparably greater advantage than other European powers," but would "without doubt control (vladet′) the entire Pacific ocean."

36 [Spasskii, G.], "Istoriia," ch. 16, p. 271; Kornilov, *Zamechaniia*, pp. 35–36. As Peter Kropotkin noted in the 1850s, dreams about the potential agricultural bounty which the Amur ("where the vine grows wild") could supply were kept alive particularly among the local population in Eastern Siberia, who passed them on from generation to generation. *Memoirs*, p. 204.

37 Bassin, "Expansion," p. 5.

38 [Krizhanich], "Historia," pp. 214–215; *idem*, *Politika*, pp. 386, 395; Middendorf, *Puteshestvie*, I, p. 169n. An earlier and less complete translation of Krizhanich's original Latin manuscript was made by Grigorii Spasskii in 1822: [Krizhanich], "Povestvovanie."

39 Krizhanich, "Historia," p. 161.

40 *Das veränderte Russland*, p. 219; Sgibnev, "Vidy," p. 319; Middendorf, *Puteshestvie*, II, pp. 169–170n; Lensen, *Russian Push*, p. 36; *Russkie ekspeditsii*, p. 22.

41 Berkh, *Pervoe morskoe puteshestvie*, p. 106; *Russkie ekspeditsii*, p. 97; Sgibnev, "Vidy," pp. 532–535; Lensen, *Russian Push*, p. 45.

Some years later, a Siberian official endorsed Sarychev's point with a reference to the "indescribable advantages" of trade with Japan, China, and India that could be Russia's, if only it could utilize the Amur river.[42] In 1810, a statistical survey of Siberia made an elliptical reference to the fact that Russian trade had "lost a great deal" due to the sacrifice of the Amur.[43]

It is important to note that, to the extent that the Amur was envisioned as an artery for communications rather than as an agricultural zone, Russian pretensions on the river were not necessarily conceived in terms of formal territorial acquisition. All that was really needed was that the power which controlled the region – in other words, the Chinese – grant privileges of navigation to Russian ships. Bering's lieutenant Chirikov, for example, drew this distinction quite clearly in 1743, when he requested the Russian government to secure the agreement of the Chinese to allow Russian boats to use the Amur and perhaps to establish a small settlement on its mouth, "without any infraction of our [already-existing] treaties with them and without disrupting our commerce with them."[44] At the same time, however, voices were occasionally raised calling for military campaigns against China, in order to bring the Amur valley once again under Russia's political control. The historian Gerhard Müller, who carried out extensive research in Siberian archives and wrote at length about the early history of the Russians on the Amur, had very definite feelings on the issue and in 1763 submitted a proposal for a war with China "to avenge . . . the many offences we have endured from them," the most grevious of which was the loss of the Amur.[45] In a similar spirit, Mikhail Lomonosov felt strongly enough about the issue to include the following thought in a ode composed for the coronation of Tsarevna Elizabeth Petrovna in 1747:

> We will praise your gift to the heavens
> We will erect a marker of your munificence
> Where the sun rises, and where the Amur
> Winds in its green banks,
> Desiring to taken from the Manchurian
> And be returned once again to your dominion.[46]

In the early nineteenth century, the Amur valley was occasionally mentioned in the various projects for the reform of Russia that the Decembrists prepared prior to their insurrection, and several insisted on the need to include it within the boundaries of the new state. Pavel Pestel's rather muddled picture of the

[42] Sarychev, *Puteshestvie*, I, p. 147; Kornilov, *Zamechaniia*, p. 36; Middendorf, *Puteshestvie*, II, p. 170n. Also see Shemelin, *Zhurnal*, II, pp. 205–206; Kabanov, *Amurskii vopros*, pp. 33, 45; Lensen, *Russian Push*, pp. 40, 62, 45, 98–100.

[43] *Statisticheskoe obozrenie Sibiri*, pp. 26–27.

[44] Quoted in Sgibnev, "Vidy," p. 564; Bartol'd, *Istoriia*, p. 412; Kabanov, *Amurskii vopros*, pp. 28, 33.

[45] [Müller], "Rassuzhdenie," pp. 378–379; Maier, "Gerhard Friedrich Müller's Memorandum," *passim*.

[46] Lomonosov, "Oda," p. 204,

geography of northeast Asia, for example, did not prevent him from calling for Russia to acquire that "part of Mongolia [sic], so that the entire course of the Amur river, which begins with Lake Dalaia [sic], will belong to Russia."[47] Successive governor-generals of Eastern Siberia certainly harbored this dream of reacquiring the Amur, until one of them – the subject of the present study – was finally successful, and a variety of schemes for campaigns against China were discussed.[48] Nevertheless, until the 1840s sentiments such as these tended to be the exception rather than the rule. For the most part, those who spoke about the importance of the Amur understood that China remained a formidable foe, and that while it was perhaps not an optimal arrangement, the most the Russians nevertheless could reasonably hope for was the acquiescence of their neighbors to the south in letting them "share" in the use of their river. At the same time, this would be entirely sufficient for the purposes of supplying Russia's settlements on the North Pacific, or alternatively fostering commerce with east Asia.

Up to this point, we have examined the viewpoints of individuals. It remains to consider the attitude of the Russian government itself in regard to the territories it sacrificed in the Far East. The position of Peter the Great, the first tsar for whom the Amur was foreign territory, remains debatable. There is no question that he was extremely anxious to develop trade with Asia,[49] and he was no less anxious to expand Russia's commercial and political presence on the Pacific. At the same time, however, his focus remained squarely on the North Pacific, as his plans for the Bering expedition would appear to demonstrate,[50] and the assertion made repeatedly in the mid-nineteenth century that he considered the Amur estuary, along with those of the Neva and the Don, to be among the three "most important" geopolitical points in the whole of the empire is almost certainly apocryphal.[51] He was interested in Siberia and is reported at one point to have expressed the desire to travel "to the land of the Tungus at the Wall of China,"[52] but at no point indicated any displeasure whatsoever with provisions of the Treaty of Nerchinsk, which had after all been signed at the beginning of his reign. Throughout the eighteenth century, there were sporadic expressions of interest in the Amur on the part of the

[47] Pestel, *Russkaia Pravda*, pp. 16–17; Svatikov, *Rossiia*, pp. 17–19; Kabanov, *Amurskii vopros*, p. 81. Also see Mamonov, "Iz bumag," p. 146; Semevskii, *Politicheskaia i obshchestvennaia idei*, pp. 386, 394–395.

[48] Sgibnev, "Vidy," p. 674; Veniukov, *Starye i novye dogovory*, p. 21; Middendorf, *Puteshestvie*, I, 164–165n.

[49] [Zubov], "Obshchee obozrenie," p. 100; Gagemeister, *O rasprostranenii*, p. 23; Maikov, *Rasskazy*, p. 49.

[50] Two scholars have recently argued that Peter's real interest in the expedition was not "scientific" – i.e. the determination of the relationship of the Asian and American continents – but rather to prepare the way for Russian expansion south along the North American coast. Polevoi, "Glavnaia zadacha"; Fisher, *Bering's Voyages*.

[51] Romanov, "Prisoedinenie," pp. 332–333; Sgibnev, "Vidy," p. 319; Kabanov, *Amurskii vopros*, p. 26.

[52] *Des veränderten Russland*, p. 124.

government in Moscow. In 1756, following the advice of Chirikov, an official entourage was dispatched to China to negotiate for Russian navigational rights on the Amur, but the mission was unsuccessful. Catherine II included the exploration of the Amur estuary in the instructions for what was intended to be the first world voyage from Kronstadt to the North Pacific in 1786, but this mission was cancelled before it could begin. Finally, in 1805 a special embassy headed by Gavril Golovkin was sent to China to negotiate, among other things, for Russian navigational rights on the Amur for the purposes of supplying Kamchatka. Golovkin's entourage was stopped on the Mongolian border and turned back to Russia.[53] This was to be the last government-sponsored attempt to redefine Russia's position on the Amur until the 1850s, for the government of Nicholas I most definitely had its reasons for adhering strictly to the existing status quo in its relations with China.

"Our . . . friendly relations with China"

During the reign of Nicholas I (1825–1855), foreign affairs in Russia were concerned almost exclusively with European relations, specifically with the problems of preserving a post-Napoleonic order increasingly threatened by the rise of popular nationalism and revolutionary sentiment. The only arenas outside of Europe which attracted significant attention were to the immediate south, most notably the Caucasus mountains, across which Russia advanced in a slow but unrelenting process of penetration throughout the first half of the century. Beyond this, Russian interests collided with those of the Ottoman empire over a variety of issues, and indeed it was a war on this front which would ultimately bring about the collapse of his administration and the system by which he ruled the country for three decades. On the remaining non-European frontiers of the empire, relations with foreign powers were not entirely neglected, but until the 1850s were assigned an importance that was clearly subordinate. This was particularly true of the Far East. Despite the establishment of the Russian–American Company at the turn of the century, by the time Nicholas I ascended to the throne the fur trade had begun its final decline. Russian settlements on Kamchatka, the Okhotsk coast, and Russian America were stagnating, and St. Petersburg did not feel moved to ameliorate, or indeed even to address, this situation.[54] The only issues in these remote eastern reaches of the empire which stirred any official interest at all involved Russia's relations with its neighbor to the south. Thus insofar as it is possible to speak at all of a "far-eastern policy" under Nicholas I, it reduced in effect to relations with China, and it was entirely within this context that the significance of the Amur river was understood.

[53] Sgibnev, "Vidy," pp. 580, 626–627; Middendorf, *Puteshestvie*, I, pp. 164n, 170n; Romanov, "Prisoedinenie," pp. 343–344; Clubb, *China*, p. 67; Kabanov, *Amurskii vopros*, pp. 34–35, 45–46; Lensen, *Russian Push*, pp. 62, 98.

[54] Romanov, "Prisoedinenie," p. 355.

Russia's policy toward China can best be understood within the larger framework of relations between the Western powers and China which had been evolving since the sixteenth century. Since this early date, European governments and commercial interests had made concerted efforts at the economic and political penetration of China, only to be turned back by the consistent and no-less-concerted efforts on the part of the Chinese to resist such incursions. This situation in the Middle Kingdom stood in marked and – from the Europeans' standpoint – rather galling contrast to other parts of Asia, where they had been far more successful in their attempts not only to acquire trading privileges but also to establish territorial footholds. The Spanish were long secured in the Philippines, the Dutch in Indonesia, and the British in Singapore, Indochina, the Indian sub-continent, and elsewhere.[55] In China, however, it was the Chinese who continued to dictate unilaterally the terms of commercial and political relations with the West, and the conditions they set were most unsatisfactory. Foreign traders were not permitted to reside within China, and from the middle of the eighteenth century were further constrained to conduct all of their business at a single point: the port city of Canton along the southern coast. Even here their transactions were limited to certain times of the year, and they were required to deal exclusively with appointed representatives of China's merchant guilds. Portugal was the only country to have been granted, in the 1550s, something resembling a territorial enclave in the form of the port of Macao south of Canton, and it was here that traders from other nations, most notably the British, were constrained to retire for those months when activities in their "factories" in Canton were prohibited.[56]

Commercial relations between the Russians and the Chinese both resembled and diverged from the pattern of interaction with the nations of Western Europe. The most obvious difference was that the exchange of goods between the two countries was not maritime, but overland. Initially trade had been conducted by caravans from Siberia across Mongolia to the Chinese capital, but a treaty between the two countries concluded in 1727 designated the town of Kiakhta, south of Lake Baikal on Russia's frontier with Mongolia, as the entrepôt for commerce between them.[57] By the middle of the century, Kiakhta had completely eclipsed the caravan trade, and for this reason the problem of access to and official status in oceanic ports was for the most part not critical for the Russians.[58] Indeed, even though it was perceived as a major issue in the 1850s, as we will see, it would become genuinely significant only toward the end

[55] Hudson, "Far East," pp. 685–686.

[56] Fairbank, *Trade*, pp. 23–53; McAleavy, *Modern History*, pp. 36–41; Wakeman, *Fall*, p. 129.

[57] Mancall, "Kiakhta Trade," p. 25; Palmer, *Memoir*, pp. 37–38; Romanov, *Poslednie sobytiia*, p. 28.

[58] On the few abortive Russian efforts to conduct maritime commerce with China, see Clubb, *China*, p. 73; Quested, *Expansion*, pp. 33, 40; Sgibnev, "Vidy," pp. 596–597; Sladkovskii, *History*, p. 61. See Shchekatov, *Kartina*, II, p. 210 for a depiction of the route by which goods made their way from Kiakhta to European Russia.

of the nineteenth century, when the completion of the Suez canal made convenient oceanic connections between Russia's Black Sea ports and the Far East possible. Also setting the Russians apart from the other European powers was the highly significant fact that they were the only ones whose commercial privileges were officially codified in a treaty. This document stipulated that trade at Kiakhta was to be conducted exclusively on a barter basis, meaning that goods could be traded only for goods, and cash or credit dealings were prohibited. Down to the early nineteenth century, the Russians traded Siberian furs for Chinese cotton fabrics and silks. After this point Chinese tea became the main commodity desired by the Russians, who increasingly traded manufactured items, primarily woolen and cotton fabrics, as the already dwindling market for furs was now being further eroded by competition from the Americans.[59]

The situation of Russian commerce in China fundamentally resembled that of the Western nations, however, in that it was the Russians who were the party interested in fostering it. The tariffs deriving from the commercial activity at Kiakhta had increased steadily throughout the early decades of the nineteenth century, and represented a modest but significant source of revenue for the state.[60] Beyond this, Chinese goods were always popular and in high demand, both in Siberia as well as the rest of the country. Yet the trade remained subject entirely to the whim and caprice of the Chinese officials, who could interfere with it as they liked, or even stop it entirely. Indeed, in the course of the eighteenth century there were several instances when Kiakhta was simply closed to commerce for years at a time. Thus the Russians shared much of the same frustration experienced by the Europeans, and the China trade remained for them an alluring but elusive prospect. They were well aware of the extent to which they were at the mercy of the Chinese, and the Russian government sought at all costs to avoid issues of any sort which might displease or antagonize the Chinese and thus further jeopardize this already highly fragile trade. The essence of this policy was spelled out clearly by the director of the Asiatic Department of the Foreign Ministry in 1833. Russian policy toward China, he explained, was animated by dual concerns, political and commercial.

> The former consists of the preservation and development of friendly ties with China, as a state with which we share such a long border; the latter consists of the broadening and development of our trade relations with the Chinese for the benefit of Russian industry and mutual advantage.[61]

This preoccupation with maintaining "friendly ties" with China set the tone for official attitudes in regard to the question of the Amur river. The ministers

[59] Mancall, "Kiakhta Trade," pp. 28–31; Foust, "Russian Expansion," p. 478–479; Sladkovskii, *History*, pp. 73–74; Ritchie, "Asiatic Department," p. 44; Clubb, *China*, p. 61.

[60] Sladkovskii, *Istoriia*, pp. 199–200; *idem*, *History*, p. 61; Clubb, *China*, p. 71; Mancall, "Kiakhta Trade," p. 27.

[61] Quoted in Bunakov, "Iz istorii," p. 97; Clubb, *China*, p. 69; Gillard, *Struggle*, pp. 8–9.

of Nicholas I well appreciated that in whatever form they might raise the issue – as a maximalist demand for the return of the territory altogether or in a more moderate form seeking only navigational rights – it would not please the Chinese, and would result most certainly in yet another rupture in the already precarious Kiakhta trade. Moreover, from a practical standpoint it seemed pointless to advance any claims in a region in which the existing balance of forces seemed no less disadvantageous to the Russians then it had been in the 1680s, when they had been compelled to relinquish the river.[62] In view of these considerations, the Russian government under Nicholas I followed a policy which was direct and unequivocal. With the Treaty of Nerchinsk, it was maintained, the Russians had in good faith recognized Chinese authority over the river and the native peoples living along it, and Russia was morally committed to respect this. Activities of any sort on this river would represent nothing less than an infraction against the existing international order, and were most strictly forbidden.[63] The Minister of Finance Yegor Kankrin spelled this position out clearly in 1832, in explaining his objections to a plan submitted by Colonel Mikhail Ladyzhenskii for a reconnaissance expedition to the Amur. "Our activities in Siberia should be directed simply at supporting and maintaining the friendly relations with China necessary for the development of the Kiakhta trade. Undertakings of this sort from our side may do much damage to these relations and for this reason cannnot be tolerated."[64]

Although official resistance to such proposals for Russian activity on the Amur was motivated primarily by the concern not to disturb the Chinese unduly, it was at the same time buttressed by a very different factor, namely the ignorance and misconceptions reigning in the imperial capital about the physical geography of the regions in question. The great expanse of land beyond the Yablonovy mountains stretching east to the Tatar straits was a *terra incognita* in the fullest sense of the term, for the Russians no less than for the rest of Western civilization. Its orography was not known, and thus the exact border with China, which according to the Treaty of Nerchinsk depended on the configuration of the Stanovoi range of mountains, was entirely unclear. There was moreover considerable confusion surrounding the geography of the Amur river itself, for the river had never been charted and consequently its course was not known with even approximate accuracy. The most critical area of misinformation, however, related not so much to the river as to the characteristics of its estuary on the Tatar straits and the relationship of Sakhalin to the continent. In 1787, the French explorer Lapérouse sailed into the Tatar straits northwest of Hokkaido. He noted that the passage narrowed markedly as he proceeded northwards, and although he was unable to

[62] Sgibnev, "Vidy," p. 675; Romanov, "Prisoedinenie," p. 366.
[63] Gibson, "Russia on the Pacific," p. 22; Romanov, "Prisoedinenie," p. 355.
[64] Quoted in Stanton, "Foundations," p. 99n; Middendorf, *Puteshestvie*, I, p. 164n; Semenov, "Amur," p. 186.

reach its uppermost extent he concluded that Sakhalin was in all probability connected by an isthmus to the mainland and that the mouth of the Amur would be located north of this, making the river inaccessible from the south.[65] Ten years later, in 1797, the English explorer William Broughton was able to penetrate somewhat further north than Lapérouse had, but he too remained "fully convinced" of the inaccessibility of the Amur from the south and the resulting impossibility of a passage through the Tatar straits north into the Sea of Okhotsk.[66]

It took yet another decade before the Russians themselves explored these these regions on the Pacific coast. In the summer of 1807, the ship *Nadezhda* under the command of Ivan Kruzenshtern sailed from Kamchatka and descended south along the Okhotsk coast, approaching the Tatar straits for the first time from the north. Kruzenshtern, who carried the accounts of his French and English predecessors along with him and was familiar with their conclusions, determined in a similar fashion that the Amur could not be reached from the south because Sakhalin was connected to the continent by a land-bridge south of the river's estuary.[67] If the conclusions of these three explorers were indeed true, then the usefulness of the Amur as a link to the Pacific basin would be severely reduced, for in fact the river would provide access only to the Sea of Okhotsk. The significance of even this access was undermined, however, by doubts that were raised as to just how navigable the river was in its lower reaches. All of the explorers' accounts suggested that the estuary was far too shallow and obstructed by sand-bars to permit passage to any ocean-going vessel.[68] The speculative element in this unencouraging geographical picture was apparent even at the time, to be sure, and by no means all experts in Russia supported it.[69] It was, however, accepted unquestioningly by Nicholas I's ministers in St. Petersburg, and it served to lock them even more firmly in their resistance to permitting Russian activity on the river. The

[65] Lapérouse, *Voyage*, pp. 275–276; Sgibnev, "Vidy," p. 581; Romanov, "Prisoedinenie," pp. 341–342.

[66] Broughton, *Voyage*, p. 302; Romanov, "Prisoedinenie," p. 342–343.

[67] von Krusenstern, *Voyage*, II, pp. 176–182; Sgibnev, "Vidy," pp. 595–596; Romanov, "Prisoedinenie," pp. 352–354; Kabanov, *Amurskii vopros*, p. 42; Alekseev, *Amurskaia ekspeditsiia*, pp. 6–8.

[68] See the editor's comments in [Krizhanich] "Povestvovanie," p. 244n.

[69] Romanov, "Prisoedinenie," pp. 312–314. Indeed, a variety of Russian atlases and geographies throughout the eighteenth and nineteenth centuries showed Sakhalin quite clearly to be an island. See *Atlas rossiiskoi . . .*, map 19; [Gakman, Iogann and Fedor Iankovich], *Vseobshchee zemleopisanie*, II, p. 120; Vil′brekht and Maksimovich, *Obshchii uchebnyi atlas*, maps 19–20; Akhmatov, *Atlas istoricheskii*, II, no. 54; Proskuriakov, *Podrobnyi uchebnyi atlas*, map XIII. On the basis of all this, Alekseev argues that Russians up to the eighteenth century had an accurate picture of the configuration of these territories, and that the erroneous notion as to Sakhalin's peninsular nature came only as Russians began to rely on geographical information coming from Western Europe. Alekseev, "Amurskaia Ekspeditsiia," pp. 69, 78–79; Alekseev, *Russkie geograficheskie issledovaniia*, p. 90. For Russian depictions of Sakhalin as an peninsula, see *Karmannoi atlas*, map facing last page of text; Semivskii, *Noveishie, liubopytnye i dostvernye povestvovaniia*, map after p. 230; Shemelin, *Zhurnal*, II, p. 201; Kornilov, *Zamechaniia*, map appended to text.

significance of the Amur had after all been presented largely in terms of improving communications with the Sea of Okhotsk and the Pacific, functions that were effectively precluded by the river's assumed non-navigability and the lack of an outlet to the south.

This then was the state of affairs in regard to the Amur valley in the early years of Nicholas I's reign. A certain sense of regret over the loss of the river had never entirely disappeared after the Russians abandoned it in 1689, but those who still felt this way a century-and-a-half-later were a small and select group. As a remote, unknown, and wild region located in a foreign country, the Amur region was bathed in a fog of obscurity, and it is safe to say that very few Russians in the early nineteenth century had ever as much as heard of it, much less possessed any practical sense of what its significance might be. The government, moreover, was quite determined that this status quo should be preserved at all costs. Yet it was not to be preserved, for considerations both foreign and domestic. Externally, Britain's victory over China in 1842 in the so-called "Opium Wars" signaled the beginning of the transformation of relations in the Far East, not only within the Middle Kingdom but indeed across the entire Pacific basin. Internally, Russian educated society found itself in an increasingly agitated state of intellectual and ideological ferment as Nicholas I's reign wore on. New preoccupations with issues of national identity and destiny, combined with the spreading conviction of the need for far-reaching national reform, were all to transform the way Russians viewed themselves and their country. As a result of all these changes, the far-eastern mists began to dissipate, and the vision of the Amur was to be fundamentally transformed as well.

2

National identity and world mission

"The East is not the West"

The first stirrings of a Russian nationalist movement are to be sought in the early decades of the nineteenth century. The overriding issue that stimulated the Russian intelligentsia at this time to confront and wrestle with the thorny problems of national identity was the question of Russia's "Europeanness" and its general relationship to the West. To be sure, this relationship had become a concern for the Russians much earlier, indeed in the immediate aftermath of the Petrine transformations. Throughout the eighteenth century, the comparison and contrast between Russia and the West was a regular preoccupation for Russian intellectuals, as can be seen in the writings of notables such as Nikolai Novikov, Denis Fonvizin, and numerous others.[1] What served to make the turn of the century into a watershed, therefore, was not the problem itself but rather the nature of the conclusions about the Europe–Russia contrast that began to be drawn. Despite whatever critique of European society that Russians might have entertained in the eighteenth century, they nonetheless remained steadfastly committed to the universalism of the Enlightenment and the Age of Reason. Russian society may well have differed from that of the West, in other words, but these differences could still be measured and evaluated within the common context of a single set of values and principles that were equally valid for and shared by all civilized societies. After 1800, however, under the intense cultural influence of Romanticism on the one hand and the political experience of the French Revolution and the ensuing Napoleonic invasion on the other, this perspective was gradually but inexorably radicalized into a doctrine of national uniqueness and exclusive national virtue.[2]

The exponents of this incipient nationalism were varied. An early and rather extreme representative was the admiral and subsequently Minister of Education Alexander Shishkov, who in 1803–1804 published two philological

[1] Walicki, *History*, pp. 17–18; 34; Greenfeld, *Nationalism*, pp. 189–274; Rogger, *National Consciousness*, pp. 70–84. [2] Rogger, *National Consciousness*, pp. 1–7.

treatises advancing the theory that Russian was the most basic of all human languages and that all other languages ultimately derived from it. Shishkov bewailed the pollution of Russian that resulted from the incorporation of foreign words, and insisted that the latter be purged in order that the purity of the langugage be protected.[3] Nikolai Karamzin's richly patriotic *History of the Russian State* (1816–1829), was a major contribution to the developing appreciation of Russia's past that was an essential element of the new nationalism, and it continued to influence nationalist sentiments throughout the century. Karamzin's work was followed by Nikolai Polevoi's *History of the Russian People* (1829–1833), a veritable panegyric to the manifold glories of Russia's national achievement. These individuals by no means spoke with a single voice; indeed, their emphases and interpretations varied widely and even conflicted, as was clearly apparent in the vitriolic polemics that Karamzin carried on both with Shishkov as well as Polevoi.[4] What unified them all *de facto* was their common conviction that Russia as a culture and society, animated as it was by a unique national ethos, differed fundamentally from all others. Rather than seek to understand Russia in terms of some vaguely postulated collection of universal principles and values, therefore, it was necessary instead to scritinize the Russian ethos itself, for such scrutiny alone would enable Russia to appreciate the unique character, needs, and destiny of the country.[5]

It was an unmistakable tribute to the growing appeal of nationalist sentiments in the early nineteenth century that tsar Nicholas I himself endorsed a version of "nationality" as a leading inspirational principle throughout his reign. This endorsement was codified by his Minister of Education Sergei Uvarov, who in 1833 proposed the formula "Autocracy, Orthodoxy, and Nationality (narodnost′)" as a sort of ideological trinity inspiring all aspects of governmental philosophy and policy. It is important to note, however, that "nationality" as refracted through the prism of this tsar's idiosyncratic inclinations and priorities came to stand for something rather special.[6] For Nicholas I, the most important of the three principles represented in Uvarov's trinity was without question that of autocracy. Uvarov had underscored the precedence of this principle from the outset with his characterization of autocracy as the "main condition of [Russia's] political existence" and the "cornerstone" of its imperial greatness,[7] and Nicholas I's understanding of what this meant was blunt and straightforward. Quite simply, autocracy represented the absolute and unconditional subservience of all segments and layers of Russian society to the word and the will of the monarch. "I cannot permit," he declared a year after his accession to the throne, "that one single

[3] Ivanov, "Shishkov," pp. 287–288; Flynn, "Shishkov," pp. 1–3.

[4] On Polevoi see Saunders, "Historians," pp. 55–57.

[5] Riasanovsky, *Russia and the West*, pp. 1–9.

[6] Saunders, "Historians," p. 58.

[7] Uvarov, *Desiatiletie*, p. 3.

person [in the realm] should dare to defy my wishes the moment he has been made exactly aware of them."[8] In view of his obsessive preoccupation with rank, hierarchy, and unquestioning obedience, the tsar's notorious fascination and delight in all things military becomes readily understandable, and he left ample indications of his desire to reorganize Russian society quite literally along military lines. Early in his reign, for example, decrees were issued requiring engineers, professors, students, and eventually even the nobility to wear uniforms of a specified cut and color,[9] and in 1835 he promulgated a statute which spoke in explicit terms of introducing the order of military service, the principle of rank, and the strictest "precision in the fulfillment of even the simplest orders" into all institutions of higher education.[10] "Il est toujours l'homme qui veut être obéi," observed the Marquis de Custine tartly, "d'autres ont voulu être aimés."[11]

The principal concern that animated 30 years of iron rule by this uncompromising individual was the steadfast and unconditional maintenance of the status quo, in regard both to domestic as well as foreign affairs.[12] Internally, the efforts of Nicholas I's administration were directed toward the preservation and enhancement of Russia's traditional forms of social organization, most notably the medieval system of serfdom, which was viewed as a sort of institutional foundation supporting the rest of society. Beyond this, social mobility in the country was to be frozen as much as possible and education reduced to a bare minimum. What remained of open or public intellectual activity was subject to the strictest government observation, for which purposes the censorial and surveillance apparatus of the state was dramatically expanded.[13] In regard to foreign affairs, the policies of Nicholas I were guided above all by his dedicated conviction that the principles of divine right and monarchial legitimacy were no less sacred outside of Russia than within it, an orientation that in practice translated into an effort to maintain the European post-Napoleonic order established in Vienna in 1815. This was no simple undertaking, for it involved resisting the increasingly irresistible wave of liberal and nationalist sentiment that was washing across the whole of Europe and was ultimately to erupt in the insurrections of 1848. Nicholas I demonstrated his unfailing preparedness to intervene militarily in those situations where he preceived a threat to legitimacy and the old order, a practice which in the long run served to isolate Russia quite completely and to win for the tsar the infamous designation as "the *Gendarme* of Europe."

It was thus into an ideological perspective that was emphatically, even quintessentially conservative that Nicholas I absorbed the concept of *narodnost'*,

[8] Cited in de Grunwald, *Tsar Nicholas I*, p. 79. [9] de Grunwald, *Tsar Nicholas I*, p. 80.
[10] Presniakov, *Emperor Nicholas I*, p. 29. [11] de Custine, *La Russie*, I, p. 174.
[12] Comprehensive examinations of governmental principles and practice under Nicholas I can be found in Riasanovsky, *Nicholas I*; Presniakov, *Emperor Nicholas I*; Lincoln, *Nicholas I*; Strakhovsky, *L'Empereur Nicholas I*. [13] Monas, *Third Section*, *passim*.

and the significance he attached to it was nuanced accordingly. Writing in the 1920s, the historian Presniakov characterized his attitude very well:

> The word "nationality" was understood to mean official patriotism, unconditional admiration for governmental Russia, for its military strength and police power, for Russia in its official aspect . . . It also meant admiration for a Russia decorated in the official style, hypocritically confident of its power, of the incorruptibility of its ways, and intentionally closing its eyes to enormous public and state needs.[14]

Predictably, rather than in any way celebrating or even acknowledging the significance of the *narod* or the Russian people as an autonomous entity, nationality in its Nicholaian version viewed them ideally as a passive and inert mass, whose devotion to the Russian homeland was most properly expressed in their unconditional personal devotion to the autocrat. As Uvarov unabashedly set forth in 1834, "here [in Russia] the Tsar loves the Fatherland in the person of the people whom he rules as a Father . . .; and the people are unable to separate Fatherland from Tsar and see in Him their happiness, strength, and glory."[15] The official insistence on the identification of Russia in all of its manifestations with the person of the tsar induced one sychophant, in a fit of what could only have been delirious enthusiasm, to propose that the country actually be renamed "Nikolaevia."[16]

Clearly, rather than having anything in common with those nationalist impulses that inspired writers such as Karamzin or Polevoi, "nationality" in Nicholas I's system instead ominously echoed the traditional absolutist spirit of a Louis XIV.[17] The tsar's intention had from the outset had been to strap an ideological straitjacket onto all of thinking Russia, and he was able to claim no little success in this mission of inhibiting and intimidating the intellectual and spiritual life of the time. The goal of entirely eradicating all traces of independent thought, however, was not achieved, and neither the vigilant scrutiny of his army of censors nor the ruthlessness of his secret police were ultimately able to prevent the emergence and proliferation of oppositionist sentiment. To a large extent, this opposition was expressed through the affirmation of the principles of a populist and "genuine," as opposed to a coopted and "Official" Russian nationalism.[18] Such sentiments were evident already in the 1830s in the writing of the Slavophiles, and they went on to develop with particular intensity in the following decade. The period of the 1840s – the "remarkable

[14] Presniakov, *Emperor Nicholas I*, p. 44.

[15] Cited in Thaden, *Conservative Nationalism*, p. 20.

[16] Riasanovsky, *Nicholas I*, p. 139n. For a fuller consideration of Nicholas I and the national question in Russia, see Koyré, *Philosophie*, pp. 194–207.

[17] It is extremely indicative that in 1847 Uvarov, in the face of proliferating nationalist sentiments in Russia's western borderlands, was prepared to drop *narodnost′* from the original trinity altogether, leaving intact only orthodoxy and autocracy. P. B., "Ob ukraino-slavianskom obshchestve," p. 348; Saunders, "Historians," p. 61.

[18] Riasanovsky, *Nicholas I*, pp. 137–139, and Thaden, *Conservative Nationalism*, pp. 23–24.

decade," as one contemporary recalled it – was in many ways the most liberal and relaxed of the entire reign, and it witnessed a brilliant flowering of intellectual and cultural life.[19] As an important part of this, nationalist sentiments were espoused on an unprecedentedly broad scale. They served the immediate function not only of organizing and giving a new depth of meaning to the accumulating resentment against Nicholas I's system, but beyond this offered at least some rudiments toward a vision of what a viable and healthy alternative might look like.

Nationalist movements were taking shape throughout Europe in the first half of the nineteenth century, and the emergence of a more-or-less clearly defined nationalist ideology in Russia during the 1830s and 1840s was very much part of this larger process. Indeed, the articulation of nationalist sentiments in other countries, notably in Germany, provided a direct stimulus and an important source of inspiration for the Russians themselves.[20] Common to all was the vision of the "people" – the German *Volk* or the Russian *narod* – which was identified as the country's most important resource, the repository of its precious traditions and historical experience, and the source of its vital strength and energy. An imperative was therefore felt by all nationalists to better the economic and social conditions of the masses, to improve public education and raise literacy, and to involve them directly in all aspects of the civic life of the homeland. In Russia, the *conditio sine qua non* for all this was the abolition of serfdom, and this issue became the main rallying point of the nationalist camp.[21] This circumstance pointed in turn to another element which Russian nationalism shared in common with other European countries – namely, its character as a progressive movement for social change, for reform that was "democratic" (if only vaguely so), and for national renovation. In Russia, indeed, the nationalists envisaged the abolition of serfdom as but the first step of an elaborate program of reforms and innovations which would transform the country into a more egalitarian and just society, reestablish its integrity, and breathe back into it the creative dynamism that had been lost through decades of Nicholas I's stultifying conservatism.[22] Inspired by all of these lofty goals, the nationalists were possessed of an absolutely irrepressible activism, an impatience to create and to actually *do* something constructive in order to set their country in motion toward the new heights they envisioned for it.

Despite its affinities with nationalist movements elsewhere in Europe, however, the Russians nonetheless "borrowed their concepts from Europe to

[19] Annenkov, *Extraordinary Decade*, *passim*; Malia, *Alexander Herzen*, pp. 279–280.

[20] Kohn, *Pan-Slavism*, pp. 222–223; Kohn, *Twentieth Century*, pp. 22–24ff; Malia, *Alexander Herzen*, pp. 72ff; Berlin, "Birth," p. 124.

[21] Riasanovsky, *Nicholas I*, pp. 139–140; Petrovich, *Emergence*, p. 56.

[22] Presniakov, *Emperor Nicholas I*, pp. 12–13; Seton-Watson, "Russian Nationalism," p. 19.

idealize and mobilize Russia against Europe," and their sympathies were emphatically anti-Western.[23] The issue of Russia's relationship to Europe, as noted above, had been fundamental to the entire nationalist project from the outset, and it was the special tension that characterized this relationship which more than anything else served to give Russian nationalist ideology its particular tone and character. Russia, it was widely felt, had for over a century offered the very best of its energies and resources to the West, and in return had received only scorn and contempt. The Russians were increasingly conscious of their status – not only political, but social and cultural as well – as junior members of the European community, and this disagreeable realization raised the existential question of Russia's true identity with a unprecedented urgency. In responding to it, the nationalists of the 1830s and 1840s pressed the scepticism of earlier decades yet farther. Russia's Europeanness, they asserted, was illusory: it was a charade and a thin veneer clumsily construed to cover a chasm of fundamental cultural and social difference that separated it from the West. In fact, Russia was a world apart, a unique society animated by a constellation of values and beliefs that could not be fit within or even reconciled to those which inspired Western Europe. "The East is not the West," lectured Mikhail Pogodin, professor of Russian history at Moscow University and an ardent nationalist,

> we have a different climate from the West . . . a different temperament, character, different blood, a different physiognomy, a different outlook, a different cast of mind, different beliefs, hopes, desires . . . [We have] different conditions, a different history. Everything is different.

He did not neglect, moreover, to cast these differences in a light favorable to Russia: "Suspicion and fear reign in the West, while among us there is only trust."[24] And in a celebrated verse, the poet Fedor Tiutchev gave expression to the essential uniqueness of the Russian ethos by stressing its inscrutability:

> Russia can't be understood with the mind,
> It can't be measured with a common yardstick.
> It has a special stature
> In Russia you can only believe.[25]

The conclusion seemed inescapable that it was time for a radical break in Russia's constricting ties with the West, ties that continued to be maintained all too carefully by Europe's "defender and custodian," that is, Nicholas I.[26] It was necessary for Russia to turn inward in the single-minded pursuit of what was to become a veritible keyword for the entire epoch: *samopoznanie*, or a combination of self-knowledge, self-understanding, and self-awareness. In other words, Russia had to learn to recognize and cultivate its own native

[23] Kohn, *Pan-Slavism*, pp. 108–109.
[24] Quoted in Lincoln, *Nicholas I*, p. 251.
[25] Tiutchev, *Stikhotvoreniia*, p. 195.
[26] von Rauch, "J. Ph. Fallmeyer," p. 192.

resources and energies, and must seek out its national destiny along an independent path.[27]

As it turned out, this break with the West and quest for an independent path was rather more easily called for than accomplished. The multifarious bonds that tied Russia to Europe were after all the product of over a century of deliberate endeavor, and could not simply be dissolved by an act of will. Indeed, for no one was this more true than for the very educated elite who filled the ranks of the nationalist movement. However sincerely and passionately they may have come to disappreciate Europe and all it stood for, the fact remained nevertheless that they had been raised and educated in a milieu that was expressly European, were imbued with many of its values, and not uncommonly even spoke its languages quite as well as their own. The radical rejection of this entire heritage, therefore, created a rather serious dilemma, for if Russia was no longer to draw its inspiration and guidance from European models, it would have somehow to generate its own. The nationalists were therefore constrained to identify an array of values and positive qualities which were genuinely native and which could act to counterbalance and replace those they had purged. These values and qualities, then, would represent what one historian has termed "seeds of Russia's future," giving substance to the claim of Russia's national preeminence and at the same time serving as the basis upon which the country's glorious future would be constructed.[28]

In the event, the Russians responded energetically to this challenge, and their emerging nationalist ideology was rapidly embellished with an array of unique qualities which they located in different aspects of the Russian ethos. There was little order or logic to this process, and different individuals and tendencies emphasized different and even conflicting elements of Russia's national personality. The traditionalist Slavophiles, for example, tended to be backward-looking, and in stressing the notion of *sobornost'* – a special sense of social collectivity – as a uniquely Russian (or Slavic) attribute, they idealized the organicism and harmony of those early periods of Russian history before Peter the Great ruptured this tradition with his criminal imposition of European norms.[29] They paid great attention to the social institution of the *obshchina* or rural peasant commune, in which they suggested that the traditional collectivist principle of the pre-Petrine era had been preserved and protected from the corrosive influence of Western ideas and morals. Political radicals, most notably Alexander Herzen, also emphasized the importance of the peasant commune in setting Russia apart from Europe, but toward a rather different end. They reasoned that, because of its strongly collectivist basis, the existence of the *obshchina* in effect served to move the country at large closer

[27] Petrovich, *Emergence*, p. 46; Kohn, "Dostoevsky," pp. 505ff; Presniakov, *Emperor Nicholas I*, p. 58. [28] Malia, *Alexander Herzen*, p. 142. [29] Lincoln, *Nicholas I*, pp. 265–266.

than even the advanced West to the progressive socialist order of the future.[30] And while some nationalists argued that Russian civilization was older than that of Western Europe, or at least for various reasons purer,[31] Herzen insisted upon precisely the opposite. He suggested that the relative youth of Russian culture and its lack of substantial contribution to world civilization – considerations which a bare generation earlier had induced a shudder of nervous discomfort among educated Russians and had provided the basis for Peter Chaadaev's vituperative denunciations of his homeland – represented a positive national virtue. Because of Russia's youth, he reasoned, it would be able to avoid the mistakes and undesirable paths of development taken by the "advanced" but deeply flawed West.[32] Finally, Russian nationalists came to be animated by the earnest belief that their country was the bearer of a special calling, a mission whose goal was nothing less than the redemption of the rest of the world, or at least some significant part of it. And although once again the specifics of this mission were conceived in a variety of very different ways, the simple circumstance of having been so chosen conveniently served the function of a national virtue equally well for all who accepted it.

As a movement for social reform and national rejuvenation, Russian nationalism represented the virtual antithesis of Nicholas I's reactionary Official Nationality. It was therefore natural that, as conscious – if not yet overt – political opposition to his reign intensified and spread in the 1840s and early 1850s, it would adopt the ideological framework of nationalism. Moreover, in the same way that this opposition was eventually to become nearly universal, so nationalist ideals came to represent a common *Zeitgeist* of sorts that inspired an entire generation of Russian educated society. This point is important to appreciate. The period under consideration was a turbulent one in Russian intellectual and political life, which witnessed intense conflicts between a variety of different tendencies and movements: the Slavophiles, for example, against the Westerners, or the moderate liberals in the imperial bureaucracy against radicals such as Herzen or the utopian socialists of the Petrashevskii circle. The differences between these individuals and groupings were real enough, and their various visions of Russia's future – as well as the means by which this future was to be attained – diverged widely. Nevertheless, on a deeper level it can be argued that to a significant extent they were all animated and inspired by a common belief and faith in the ideals of Russian nationalism described above. Reflecting back on the Slavophile–Westernizer dispute, for example, Herzen himself was most emphatic on this

[30] Gertsen, "La Russie," p. 187. [31] [Magnitskii], "Sud′ba," pp. 392–399.
[32] Gertsen, "La Russie," pp. 186–187; Iskander, "Kontsy," p. 1298; Gerschenkron, "Problem," p. 31. This sense of youthfulness *vis-à-vis* Europe had been espoused to much the same effect earlier by nationalists in Germany, although the view of Russia as a *tabula rasa* – and the advantages thereof – had been entertained by Russians in the eighteenth century. Berlin, "Birth," p. 120; Schapiro, *Rationalism*, p. 33.

point, insisting quite passionately that the nationalist devotion the two groups shared between them over-shadowed all differences.

We shared a common love . . . From early childhood, a single strong, instinctive, physiologically passionate feeling was imprinted on us as it was on them . . . the feeling of limitless love for the Russian people, for the Russian way of life, for the Russian cast of mind, [a love which] encompassed all of our existence. And, like Janus or the two-headed eagle, we looked in different directions, but at the same time a *single heart was beating within us.*[33]

Indeed, even the great literary critic and ultra-Westernizer Vissarion Belinskii could freely agree with much the Slavophiles said in critique of the fetish of Europeanism in Russian national life. He did not regret the fact that Peter the Great had thrust Russia aggressively into the European milieu, as they most emphatically did, but in 1846 he nonetheless confirmed his belief that "Russia has exhausted and lived out" this transformation, that Peter's reforms "had done for [Russia] everything they could and should have done," and therefore that "the time had come for Russia to develop autonomously (samobytno), out of itself."[34] That the participants themselves perceived a common ground of sorts between perspectives which otherwise clashed so strongly and in so many regards is to be explained by the particular nature of the nationalist appeal. In Russia, as elsewhere in Europe, this appeal was founded precisely on the possibility it extended to *all* members of the national community for self-fulfillment and union through common participation in an unfolding national destiny, regardless of more specific political or ideological divergences. All Russians, therefore, could share in the exhilaration that this alluring and invigorating vision had to offer.

"Who is closer to Asia than us?"

Messianic thinking in Russian intellectual and spiritual history claims a rich tradition, which can be traced back at least to the 1510s when the Pskov abbot Filofei first set forth the familiar vision of Muscovy as a "Third Rome" in a missive to Tsar Vasilii III.[35] Nevertheless, the particular messianic sensibility that took shape in the 1830s and 1840s was strongly colored, if not indeed determined, by the ideological emphases and needs of contemporaneous Russian nationalism. The belief that Russia was the chosen bearer of a special mission for the deliverance of Europe and the world, a notion embraced with

[33] Cited in Vetrinskii, "Sorokovye gody," pp. 100–101 (emphasis in original). Greenfeld has stressed this point most recently: *Nationalism*, p. 265. Also see Gerschenkron, "Problem," pp. 24–25; Riasanovsky, *Russia and the West*, pp. 87–90; Offord, *Portraits*, p. 22; Malia, *Alexander Herzen*, pp. 294–296.

[34] Belinskii, "Vzgliad," pp. 17–20; Offord, *Portraits*, p. 22; Schapiro, *Rationalism*, p. 63.

[35] Kirillov, *Tretii Rim*, pp. 5–7, 27.

remarkable intensity during this period, was a product of the nationalist quest for positive qualities and virtues which could both demonstrate Russia's national exclusivity and superiority as well as illuminate an illustrious path into the future. As was true in regard to their nationalist sentiments in general, messianism was characteristic of similar movements in France, Germany, Italy, and Poland. In Russia, however, these convictions developed with a passion and endurance that was singular.[36] Outwardly in all directions – to the west, south, and east – Russians saw themselves confronted with societies that stood in dire need of redemption and salvation of some sort, from either external oppression or internal decay, and upon their chosen shoulders devolved this exalted responsibility. "We ought not to forget," instructed the youthful Siberian historian Afanasii Shchapov, "that the Russian nationality exists not only for itself exclusively, but for all of the European peoples, and for all of humanity, for the West as well as . . . for the East."[37] The very special status which this conferred was described by one nationalist in 1833:

> The ordinary person lives as external circumstances arrange his life, while a person of fate, a person *called by Providence* constantly and unswervingly, with his entire life and under all circumstances, realizes one idea which has been pre-determined for him. God Himself has shown him the way! Such is Russia.[38]

As with virtually all aspects of its nationalist ideology, the specific character of Russia's mission of salvation could be identified in a variety of different ways. Some looked to the distant past, and believed along with Alexander Pushkin that the hardships of the *mongol′skoe igo*, or Mongol yoke, which blighted Russia's early development had in fact represented the fulfillment of a mission to protect Europe and Christianity from despoilation by the infidel hordes of Asia.[39] Yet while the appeal of such a thoroughly positive evocation of ancient Russia's tribulations was unquestionable, the remote and essentially indirect influence on the fate of Europe that it offered was hardly sufficient to fill out and inspire the dynamic vision of Russia's predominant role among the peoples of the earth that the nationalists were now trying zealously to create. For most of them, it was clear that Russia's mission – its *schastie* or "happi-

[36] Kohn, "Dostoevsky," pp. 501–504.

[37] Shchapov, "Novaia era," p. 4; von Rauch, "J. Ph. Fallmeyer," p. 188.

[38] [Magnitskii], "Sud′ba," p. 404 (emphasis in original). On messianism during this period, see Malia, *Alexander Herzen*, pp. 288–289; Hunczak, "Pan-Slavism," p. 86.

[39] Pushkin expressed this thought in an indignant draft of a response to Peter Chaadaev's first "Philosophical Letter," in which the latter had written that Russia was unable to claim a single historical accomplishment of any significance.

> [N]ous avons eu notre mission à nous. C'est la Russie, c'est son immense étendue qui a absorbé la conquête Mongole. Les tartares n'ont pas osé franchir nos frontières occidentales, et nous baisser à dos. Ils se sont retiré vers leurs déserts, et la civilisation Chretienne a été sauvée.

Letter to Chaadaev (19 October 1836), p. 387. Also see Vernadskii, "Pushkin." p. 64; Karpovich, "Pushkin," pp. 186–188; Sarkisyanz, "Russian Attitudes," p. 245; Schapiro, *Rationalism*, p. 45. The poet apparently never sent this reply to Chaadaev.

ness," as Herzen put it[40] – was something that remained to be fulfilled: a prospect or vision of imminent accomplishment that both challenged and inspired their entire sense of the future. "The West is perishing!" exclaimed the Romantic poet Prince Vladimir Odoevskii in the early 1840s, and in the conclusion to his cycle *Russian Nights* he breathlessly depicted the glorious coming destiny that this circumstance conveniently made available to his homeland:

> Providence itself . . . nurses a nation, which will have the task of showing anew the path from which [the rest of] mankind has strayed, and [this nation] then will occupy the first place among all nations. But only a young and innocent nation is worthy of this great deed, only in it, or by means of it, is the rebirth of a new world possible, [a world which] encompasses all the spheres of intelligence and social life . . .
>
> [W]e are placed on the border of two worlds: the past and the future; we are young and fresh; we are not accessories to the crimes of old Europe . . . Great is our calling and difficult is our task! It is we who must breathe new life into everything. We must inscribe our spirit onto the history of the human mind, as our name is entered on the panels of victory. A different and higher victory – the victory of science, art, and faith – awaits us on the ruins of decrepit Europe.[41]

Odoevskii's fervor was entirely typical for the period. Among other things, it left little doubt that Russia's mission was not something which would be achieved passively, through simple good intentions, or even by setting a positive example of the preferred direction for civilization's future development for all to appreciate. Rather, the mission of salvation was necessarily activist, and could be pursued only through deliberate and self-conscious intervention into the fate of those peoples who were to be saved. Russia must cease to live as "external circumstances" arranged its life, dumbly responding to events taking place around it, and undertake instead to initiate actions and then direct them toward that higher goal which it determined for itself. This interventionist imperative supplied nationalist energies with an alluring channel for constructive activity, and one which was to take them – in spirit, if not always in practice – far beyond Russia's own national borders. Furthermore, although the most important result of this intervention was ultimately to be the moral and spiritual revival of the societies affected, the nationalists did not fail to appreciate that their efforts at salvation would of necessity take on a political dimension as well. Indeed, distinctions of this sort were hardly made at all, and the prospect of the extension and enhancement of Russian political influence beyond its national boundaries seemed an entirely natural, even organic part of the noble historical task conferred upon them by Providence. In this way, "national messianism thus becomes the cradle of an unbridled imperialism: the nation, the chosen vehicle of God's designs, sees in its political triumph

[40] Quoted in Vetrinskii, "Sorokovye gody," p. 101; Berlin, "Birth," p. 126.
[41] Odoevskii, *Russkie nochi*, pp. 341–344.

[beyond its borders] the march of God in history."[42] This is by no means to say that messianism in Russia was always by its very nature expansionist in this manner, for it could also have a more purely religious focus which lacked a political dimension. Indeed, the contrast between the two was apparent even in the period under discussion.[43] Ultimately, however, the messianism that Russian nationalism absorbed in the early nineteenth century had indeed succumbed to the "temptation of imperialism" that Nikolai Berdiaev argued was always an option for it, and it accordingly carried an articulated imperial vision.[44]

Through this messianic impulse, therefore, an active desire for the export of national, and ultimately political influence became interwoven into the very fabric of Russian nationalist thought in the 1840s and was fully rationalized in terms of its overall ideological structure. This desire, in turn, formed one of the most important sources of nationalist opposition to Official Nationality. Nicholas I's foreign policy, as noted above, was devoted to the maintenance of the international status quo. Intervention beyond Russia's boundaries was countenanced only toward the end of resisting movements for change, and this policy was strictly followed. Indeed, rather than support even the revolts of coreligionists in the Balkans against Ottoman domination, Nicholas I angrily denounced such "shocking, blamable, even criminal" insubordination against "their legitimate sovereign," and in 1848 actually dispatched some 45,000 Russian troops to crush uprisings of Christian Orthodox Rumanians in Wallachia and Moldavia.[45] Political or territorial aggrandizement of Russia for its own sake was quite simply out of the question.[46] In stark contrast, the nationalists were emphatic in their call for an assertive foreign policy with the single ultimate goal of promoting the country's national interests. For them,

Russia expanded to become Slavdom, Russian destiny advancing to the Elbe, Vienna, and Constantinople. Indeed, the entire world was to be recast in response to this call of fate, through blood and iron if necessary. The messianic Russian future called for an adventurous, aggressive, even revolutionary, foreign policy which represented the very opposite of Nicholas I and his government.[47]

It is clear from these latter observations that such a policy would in the first instance lead Russia to the West, and this period accordingly witnessed the initial articulation of the sentiments which at a somewhat later point would be formalized into Russian Pan-Slavism. In fact, the original inspiration for Pan-Slavism had come not from the Russians at all but from the West Slavs, who

[42] Kohn, "Messianism," p. 13. [43] von Rauch, "J. Ph. Fallmeyer," p. 158.
[44] Berdiaev, *Russian Idea*, pp. 9 (quote), 195–196.
[45] Lincoln, *Nicholas I*, pp. 118 (quote), 312–313.
[46] Gillard, *Struggle*, pp. 39, 64–65, 104–105; Sarkisyanz, "Russian Imperialism," p. 54; Fuller, *Strategy*, pp. 220–221. For a dissenting view, see Presniakov, *Emperor Nicholas I*, p. 62.
[47] Riasanovsky, *Nicholas I*, pp. 137–138 (quote), 164–165.

argued that broad cultural, linguistic, and historical affinities served to bond the various Slavic peoples together, and they promoted the union of these peoples into a single all-Slavic political entity. These teachings were quickly taken up by the Russian nationalists, who organized them in terms of their own priorities into a program for the messianic Russian salvation of their Slavic brethren and domination of their future union. The focus, logically enough, was on eastern and southeastern Europe, where the nationalists felt that Russia ought to intervene in order to defend the rights of the oppressed Slavic nationalities. "What is [Russia's] historical significance?" demanded the poet Tiutchev of Pogodin.

> It is precisely that Russia, as the sole independent representative of the entire [Slavic] tribe, has been pre-destined to resurrect this independence for the entire tribe. This historical law was for Russia a existential condition, outside of which there is no historical life for even her.[48]

Outside of Europe proper, the Pan-Slavs felt it to be Russia's responsibility to conduct a crusade against the Ottoman empire to the south, in order to "liberate" Constantinople from the rule of the infidels and resurrect it as the capital of the future Slavic union. It is critical to note, however, that these preoccupations with Russia's western and southern flank by no means exhausted the geographical sweep of the nationalists' messianic vision. At the same time that these Pan-Slav concerns were being articulated, in other words, and in something of a parallel fashion, Russian nationalists were busy at work setting the conceptual foundations for a program of activity that would lead them to the East, to Asia.

The late eighteenth and early nineteenth centuries witnessed what one specialist has termed an "Oriental Renaissance" in Western Europe, that is a notable revival of interest in the peoples and cultures of Asia.[49] The rediscovery of the Orient at this time generated a great deal of excitement among Europe's intellectual elite, and although on the whole not much was known or understood about these lands and peoples, the incomplete picture which the Europeans possessed provided exotic material for their imaginations. In Asia, it was fancied, all of civilization's most exalted qualities – high moral principles, enduring veneration of tradition, and intellectual enlightenment – were all realized with a sublime perfection which their own societies, tormented as they had been by discord, bloody revolution, and war, could not match. The scope of this fascination with the East assumed truly remarkable dimensions, and it figured as an important stimulus, among other things, for the incipient Romantic movement. In 1800 Friedrich Schlegel affirmed that Asia was the most exalted source of inspiration for Romanticism ("Im Orient müssen wir

[48] Letter to Mikhail Pogodin (11 October 1855), p. 422. On Russian Pan-Slavism in this early period, see especially Kohn, *Pan-Slavism*; Petrovich, *Emergence*; Hunczak, "Pan-Slavism."
[49] Schwab, *Oriental Renaissance*, p. 9.

das höchste Romantische suchen"), Herder refered to the Orient as the "soil of God," and Victor Hugo spoke of it as a "mer de poésie," saturated with transcendental meaning.[50] China in particular stood out for admiration, and in France the popular fashion of *chinoiserie* idolized not only Chinese letters, but Chinese art, ceramics, and clothing as well.[51]

Russia's educated elite fully shared in this awe and veneration for the Orient, and even sported its own version of *chinoiserie*, called by its Russian name *kitaishchina*.[52] In literary treatments of the late eighteenth and early nineteenth centuries, China was depicted positively, as the land of "the wise Confucius, the good Emperor, the morally just individual, and the scholar," and was often used didactically as an example to demonstrate how an absolutist state ideally ought to be run.[53] During the reign of Nicholas I, Oriental studies in Russia were strongly promoted by the Minister of Education Uvarov, who reaffirmed the view that it was in Asia where Europe must seek "les bases du grand édifice de la civilisation humaine." Uvarov (who incidentally counted among his ancestors Fedor Golovin, the diplomat who negotiated the Treaty of Nerchinsk with China[54]) believed that Europe owed a tremendous intellectual debt to the Orient, and suggested that the study of Asia could provide spiritual relief and rejuvenation for those heirs of the French revolution and the Napoleonic wars, "fatigués des sanglants excès commis au nom de l'esprit humain."[55] He accordingly prepared an elaborate project for the establishment of an Oriental Academy in St. Petersburg, a plan which won the enthusiastic endorsement of no less a figure than Johann Wolfgang von Goethe, who shared the Minister of Education's fascination with the East.[56] It is interesting to note that Uvarov accorded the development of Oriental studies a significance within the larger project of Europeanization of Russia, and suggested that it would work to enhance Russia's standing among European nations.[57]

Beginning in the 1830s, however, and under the influence of the new ideas

[50] Hugo, "Les Orientales," p. 332; Schwab, *Oriental Renaissance*, pp. 13–14. On the image of the Orient at this time in Europe, also see Said, *Orientalism*; Steadman, *Myth*; Honour, *Chinoiserie*.

[51] Kiernan, *Lords*, p. 152.

[52] Maggs, *China*, *passim*; Alekseev, "Pushkin," p. 115; Billington, *Icon*, pp. 294–295; Schwab, *Oriental Renaissance*, pp. 449–552.

[53] Maggs, *China*, pp. 216, 289–291. This thought, for example, lay behind the translation into Russian of *The Book of Court Teachings and Moral-Didactic Thought of the Manchurian and Chinese Khan K'ang-Hsi*, which depicted the emperor K'ang-Hsi (1661–1722) as just such a model ruler, a veritible paragon of sagacity and justice. Some decades later, however, this very Manchu emperor – during whose reign the Treaty of Nerchinsk was concluded – was to become the Chinese arch-villain responsible for stealing the Amur from the Russians!

[54] Whittaker, "Impact," p. 509.

[55] Ouvaroff. "Projet," pp. 4, 23. On Uvarov and Orientalism in Russia during the reign of Nicholas I, see Riasanovsky, "Russia and Asia," p. 171; *idem*, "Asia," p. 11; Whittaker, "Impact," pp. 513–514; Bartol'd, *Istoriia*, pp. 232ff; Frye, "Oriental Studies," pp. 39–42.

[56] Schmid, "Goethe," pp. 138–143; Riasanovsky, "Russia and Asia," pp. 173–175.

[57] Ouvaroff. "Projet," p. 9; Uvarov, *Desiatiletie*, pp. 26–27; Layton, *Russian Literature*, pp. 75–76.

about Russia's identity and destiny we have been discussing, a very different image of Asia was taking shape. Asia's heritage as the hearth of civilization was not disputed, but the new perspective placed its emphasis instead on the state of decay into which this once-flourishing cultural realm had degenerated. Fascination and veneration gave way with surprising ease to condescension and disgust, and rather than spiritual enlightenment, lofty aesthetic accomplishment, and virtuous and wise social principles, the East now came to epitomize precisely the opposite: social degeneracy, and intellectual and spiritual inanition. The one word which came to be used ubiquitously in Russia to characterize Asia in general and China in particular – *nepodvizhnost′*, that is, stagnation or immobility – betrayed the uncompromisingly critical stance which the new attitude implied. While hyper-conservative proponents of Official Nationality such as Uvarov were entirely sincere in their admiration of the Chinese ability to enjoy "leur suprême bonheur dans la plus parfaite immobilité," no quality served to make the Orient more thoroughly reprehensible in the eyes of the restless nationalists, obsessed as they were with remaking and improving their own society through constructive activity.[58]

The lack of movement and progress of any sort which the Russians now began to perceive in Asia was simply appaling, and it served to transform the Orient into a contemptible model of social, political, and cultural lassitude. Much of the country's educated public could agree wholeheartedly with Peter Chaadaev's conclusion that Chinese civilization represented "only some dust left for us to look at," preserved by Providence as a grim lesson of the depths to which humanity was capable of sinking.[59] This attitude became a standard component for Russia's nationalist ideology, and those who shared it made up for their commonly palpable ignorance about the regions in question with the unrestrained intensity of their convictions. Vissarion Belinskii, to cite one particularly notable example, wrote a devastating review of a book which noted the existence of positive qualities in Chinese society, and if the celebrated critic was at all aware of the fact that the author was one of Russia's most knowledgable experts on China who knew far more about it than Belinskii himself, he did not allow this circumstance to deter him. "Up to the present, from time immemorial," he declared with confidence, Asians "have been rotting in moral stagnation, and repose in undistubed slumber on the lap of Mother Nature."[60] To be sure, Belinskii and authors of similar denunciations might well have had

[58] Ouvaroff. "Projet," p. 6 (quote); Riasanovsky, "Russia and Asia," p. 181; *idem*, "Asia," p. 16. This aversion to Oriental immobility was common throughout Europe at the time, for as Kiernan notes "a philosophy of standing still was not one that the age of Progress could sympathize with." *Lords*, p. 154. [59] McNally, *Major Works*, pp. 144, 151.

[60] [Belinskii], Review, pp. 44 (quote), 45, 49. For a similar criticism of the same book, which faults the author for "excusing the most scandalous shortcomings" in Chinese society and for failing to "introduce a European perspective," see Kovalevskii, *Puteshestvie*, I, p. 40. This latter critic was an outspoken nationalist and Pan-Slav, who in his capacity as an official in the Ministry of Foreign Affairs played a important role in shaping a forward policy in the Far East.

more than China in mind, for critically-minded Russians not infrequently used the Orient as a metaphor for Russia itself, hoping in this way to elude the watchful eye of the censor and air their discontent in print.[61] Nevertheless, imagery of this sort could be effective only because the negative associations with Asia were so pronounced to begin with.

The prospect of an entire vast continent wasting away in languid somnolence proved absolutely irresistible to the nationalists. Here was an expansive, beckoning arena where they would have full freedom to assume their messianic duties, an arena which moreover lay directly at their own back door and over which they could therefore claim a priority against other prospective civilizers from the West. That Asia belonged rightfully to them appeared unquestionable; indeed, given Russia's geographical location and particular historical experience, the legitimacy of its claim to exclusive proprietary rights over this world region could not have seemed more self-evident. The manner in which Alexander Balasoglo – a particularly well educated and politically progressive member of Mikhail Petrashevskii's radical *kruzhok* – expressed this conviction in the early 1840s was characteristic:

> The East belongs to us [Russians] unalterably, naturally, historically, voluntarily. It was bought with the blood of Russia already in the pre-historic struggles of the Slavs with the Finnish and Turkic tribes, it was suffered for at the hand of Asia in the form of the Mongol yoke, it has been welded to Russia by her cossacks, and it has been earned from Europe by [the Russian] resistance to the Turks.

As we will see presently, the prospect of the Amur was to provide an opportunity for the *petrashevtsy*, and Balasoglo in particular, to articulate these sentiments much more specifically.[62]

Establishing the principle of Russia's legitimate authority over this region was only the beginning, however, for the more critical point was the precise manner in which Russia could realize its mission of salvation and enlightenment there. Once again, a number of alternatives were available. The uncomplicated prospect of conquest and imperial domination in Asia was powerfully attractive, in this period as always, particularly in view of the nationalist imperative for Russia to assert itself through demonstrations of its national vigor and might. A telling, if somewhat obscure expression of this sentiment can be found in the conclusion to a play written in 1845 by Nikolai Polevoi about Yermak's conquest of Siberia. At the play's end the cossack hero, mor-

[61] In a letter to Herzen, the anarchist Mikhail Bakunin made this non-geographical and metaphorical sense of the term "Asia" explicit: "In a moral, social–political sense, Asia begins wherever tyranny and force dominate. If this is so, then are we not now in Asia, or rather, does not Asia reign throughout the entire Russian empire?" Bakunin, "Russkim . . . druziam," p. 1027. Also see Shchegolev, *Dekabristy*, III, p. 19; Offord, *Portraits*, p. 40; Sarkisyanz, "Russian Attitudes," p. 246; Dulov, "Neizvestnye stat'i," pp. 163–164.

[62] Desnitskii, *Delo Petrashevtsev*, II, p. 44. See below, pp. 90–94.

tally smitten in battle on the banks of the Tobol river, delivers the following inspired soliloquy with his final breath:

> My eyes see clearly our future destiny!
> Across Siberia, to distant seas
> Our fellow Russian is travelling, and will there find
> A new world . . .
> One eagle's wing has touched
> The diamond mountains of rich India.
> The other is resting on the floes of ocean ice.
> Waves of gold are flowing from the mines and sands of Siberia.
> The Bashkir, the Persian, the Mongol, the Indian, and the Chinaman
> Will bring us their tribute. O, how clear and bright
> Is your future destiny, Mother Russia!
> As bright as the Russian mind. It is burning
> Like a candle before an icon, with an immortal light![63]

Leaving the audience to digest this pregnant image, Polevoi's stage instructions direct the final curtain to descend slowly as the dying hero "falls silent in ecstasy." From the standpoint of the historical record, the suggestion that such a grandiose vision actually inspired Yermak's foray across the Urals in the late sixteenth century was simply ludicrous.[64] As an indication of the national mood of the mid-1840s when it was written and performed, however, it was much more meaningful.

Even more popular and effective than this sort of unself-conscious chauvinism, however, was the attempt to present Russia's mission in Asia in altruistic and even philanthropic terms. From this perspective, Russia was seen as the conveyer of enlightenment and civilization to the ossified societies of the East, in effect as God's chosen emissary to carry out the noble and proud task of rescuing these peoples from the decay and stagnation benighting their present existence. This was an awareness of a sort of "white man's burden," differing significantly from that of other European countries in that it did not rely upon an elaborate racialist ideology,[65] but sharing nonetheless the same appreciation of the yawning gap between "backward" and "advanced" societies, as well as the unquestioning confidence in the salutary benefits which were theirs to bestow as representatives of the latter. To be sure, the notion of Russia as a *prosvetitel'* or enlightener in Asia was by no means the exclusive property of the nationalists, and even official governmental organs freely indulged in self-congratulatory depictions of the "crude and unenlightened" tribes of Siberia who paid the yearly *iasak* or fur tribute to Moscow "eagerly" in grateful recognition of the munificent patronage accorded by the

[63] Polevoi, *Ermak Timofeich*, p. 144. [64] Bassin, "Expansion," pp. 11–12 and *passim*.
[65] Sarkisyanz, "Russian Imperialism," p. 73.

Russian tsar.[66] The new messianism, however, was decidedly more ambitious. Above all, it was aggressively expansive, and actively sought out new geographical arenas not just within but beyond Russia's imperial boundaries upon which the country could exercise its civilizing activities. In this role as civilizer, the nationalists acquired yet another national quality with which to embellish their vision of Russian virtue and of a glorious future to come, one which would at least match them with the West, if not indeed put them ahead of it.

In a lecture delivered in 1840 entitled "On Russia's Relation to the East," Vasilii Grigor′ev offered a revealing expression of this new messianic mentality. Grigor′ev was one of Russia's most prominent specialists of the day on the history of Asia, and he was the first to offer a university course on the subject. Along with this, he was a fervent nationalist. He identified the essential clue to Russia's messianic destiny in a sort of geographical teleology, according to which Russia had been situated quite deliberately as an intermediary between West and East. Placed on the borders of two worlds, physically containing in itself half of one and half of the other, Russia was obviously "fated by destiny itself to have a great influence on the fortunes of each." It was, however, with the nature of this influence on the East that he was primarily concerned:

> I do not know if there can be on earth a higher, more noble calling for a people and for a state than the calling of Russia in regard to the tribes of Asia: to preserve them, set their lives in order, and enlighten them. We are summoned to protect these peoples from the destructive influence of Nature itself, hunger, cold, and sickness . . . We are summoned to put these peoples' lives in order, having taught these rude children of the forests and deserts to acknowledge the beneficent power of the laws [of civiliation] . . . We are summoned to enlighten these peoples with religion and science.

Russia's unique proximity to the East, furthermore, indicated that such a mission was pre-eminently one for Russia alone. "Who," he demanded, "is closer to Asia than us?"

> Which of the European peoples has preserved in itself more of the Asiatic element than the Slavs? . . . Yes! if the science and civic life of Europe must speak to Asia through the mouth of one of its peoples, then it will of course be us. . . . Is it not obvious that Providence preserved the peoples of Asia as if intentionally from all foreign influences, so that we would find them in an entirely undisturbed condition, and therefore more able and more inclined to accept those gifts which we will bring it?[67]

In these thoughts, Grigor′ev has summed up the main points behind the Russian messianistic mission to the East. It was a task which had been conceived by God, for which reason He placed Russia in its particular geograph-

[66] These comments appeared in a statistical handbook on Siberia published by the Ministry of Internal Affairs. *Statisticheskoe obozrenie Sibiri*, pp. 96–97.

[67] Grigor′ev, *Ob otnoshenii*, pp. 4, 7–9.

ical location between Europe and Asia.[68] Moreover, the salvation of Asia was primarily the responsibility of Russia alone, which was unquestionably better suited to carry it out than Europe.

We see here the entire ideological mechanism of Russian nationalism at work. Through the conception of a national mission – and a divine mission at that – Russia acquired its noble and worthy destiny, which in turn would enable it not only to achieve a comfortable parity with the West but even to surpass it. It should be emphasized, moreover, that this particular vision of a Russian mission to the East and of Asia as an open arena upon which Russians could exercise their civilizing activities enjoyed an appeal that was virtually universal among the educated public. There was as yet no trace of the stigma that was to become attached to the notion of imperial domination by the end of the century, and it consequently was espoused with equal enthusiasm by conservatives such as Grigor′ev and Pogodin, moderate liberal reformers, and political radicals.[69] On the pages of Herzen's *Kolokol*, for example, there was mention of the "irresistible historical fate of Russia, the world mission of which consists in the transfer of enlightenment from Europe to Asia." In the same spirit, an early Russian disciple of Karl Marx wrote to his mentor in 1850 explaining how the Slavs could include themselves in the ranks of the "European civilizers" by acting as the "carriers of creative ideas" to the peoples of Central Asia[70] – a point, incidentally, to which the recent authors of the *Communist Manifesto* could warmly respond.[71] The universal appeal of this prospect indicates better than anything the pervasiveness of nationalist sentiments at the mid-century and the degree to which they transcended the standard boundaries of political opinion to become a truly collective spirit of the age.

This new view of Asia made it possible for the original nationalist rejection

[68] The notion that geography or location provides the essential clue to Russia's impending destiny has always been a popular one, and deserves a fuller study. Echoing Grigor′ev, the liberal geographer and statistician Konstantin Arsen′ev made the following point in a geography text in 1848:

> In a *political respect*, Russia's location is unique; not a single other state has such immediate influence on dry land. Bordering upon or neighboring with the most important powers of Europe and Asia, Russia by virtue of this fact alone should . . . strongly influence the fate of many peoples.

Arsen′ev, *Statisticheskii ocherk*, p. 23 (emphasis in original).

[69] Treadgold, "Russia," p. 541.

[70] [Dolgorukov], "Pis′mo," p. 1935; letter of N. I. Sazonov to Marx (2 May 1850), cited in Riazanov, *Ocherki*, p. 378.

[71] In a letter to Marx the following year, Engels extolled the "genuinely progressive" effect of Russia on the East and noted that the influence of the former, "despite all of its Slavic dirt" (slawische Schmutz), is nonetheless "civilizing for the Black and Caspian seas and Central Asia, for the Bashkirs and the Tatars." In particular, Engels felt that the Russification of the colonial subjects of the empire – a practice which a more enlightened age was later to appreciate as an exercise in national chauvinism and intolerance – represented instead a positive indication of Russian cultural strength. "Selbst die Juden," he pointed out with apparent satisfaction "bekommen dort slawische Backenknochen." Letter (23 May 1851), pp. 266–267.

of Europe to be reformulated into something rather more subtle, namely a geographical choice between West and East. One historian of Siberia claimed that Peter the Great had intended his turn to the West to last only for a limited period, after which "it would be time for us to redirect our attention back to our own East and think only about Russia, courteously turning away from Europe."[72] In effect, a spiritual and moral turn away from Europe could be supported and facilitated by a corresponding geographical redirection of energies and attention to the East. Grigor′ev used this logic in a proposal he submitted in 1837 to the Senate of St. Petersburg university for the establishment of a chair for the History of the Orient. The domination of Western ideas and education in Russia, he wrote, was categorically evil, and threatened Russian nationality itself with absorption. Decisive and far-reaching measures would be required to counteract this pernicious influence. There was no better way to go about this than by fostering Oriental studies in Russia, an undertaking which would serve "as a counterweight to the preponderance of Western principles that oppress our national development." "The best means to resist the influence of the West," he concluded, "is to rely upon the study of the East."[73] We may be sure that Grigor′ev's intertwining of Oriental studies with the fostering of healthy Russian national development was more than a little self-serving – the chair, after all, would have been his – but this need not call in question the sincerity with which he presented this argument. As we will see, this attitude grew stronger after 1850, especially in the aftermath of the Crimean War, and it was to have considerable influence on the evolution of views toward the Russian Far East.

A final point remains to be made in this discussion of changing Russian views of Asia. No matter how vociferous the nationalist denunciations of the "decrepit" civilizations of Western Europe might have been, or how dire the warnings about the pernicious influence of Western ideas which spread cancer-like within Russia itself, the Russian nationalists never entirely rejected the West nor disassociated themselves completely from it. Indeed, this fact becomes clear precisely through those attitudes toward the East which we have been examining. When confronting Europe directly, Russian nationalists loudly insisted upon their separate identity, but when these same individuals turned to Asia the calculus was subtly but unmistakably inverted, and they universally saw themselves acting as European agents. The enlightenment and civilization which it – Russia – was to transmit to the East was unquestionably *European* enlightenment and civilization.[74] Indeed, a number of nationalists,

[72] Nebol′sin, *Pokorenie*, p. 146.

[73] Veselovskii, *Vasilii Vasil′evich Grigor′ev*, pp. 33 (quote), 231. Grigor′ev's proposal was not acted upon until 1863, ironically as part of a program to bring Russian universities up to the standards of universities in Western Europe.

[74] Riasanovsky, "Asia," pp. 6–9, 14–17; Sarkisyanz, "Russian Imperialism," p. 66.

among them Herzen and Balasoglo, characterized Russia's role precisely as that of a *posrednitsiia*, or intermediary between a civilized Occident and a backward, stagnant Orient.[75] That such a cosmopolitan perspective did not particularly resonate with either Russian nationalism's explicitly anti-Western stance or, yet more strikingly, the vision of Russia as the redeemer of a fallen Europe, only speaks to the very fundamental ambivalences with which this ideology was fraught. We will have occasion to consider these ambivalences in much greater detail later in this study. For the moment, it is important to appreciate that the Russians advanced to the East as self-appointed representatives of the West. The true dimensions of this paradox become apparent with the fact that, for even the most fervent nationalists, Russia's mission in the East often represented a way of proving Russia's moral worth, toward the ultimate goal of full membership in the European family of nations. Russia's great destiny as Asia's saviour and enlightener thus turned into nothing more that a means of securing what could with disarming frankness be referred to as its *vkhodnoi bilet* or "entry ticket" into Europe, a coupon that would be of absolutely irreproachable validity.[76] This notion was to be a significant element in those images which came to be associated with Russian activity on the Amur.

"What does this boundless space portend?"

The turning inward that the nationalists urged in their quest for *samopoznanie* and native Russian virtues was intended in the very broadest of senses. Thus, along with the "discovery" of positive moral qualities such as sobornost′ and of social institutions such as the *obshchina* which such introspection yielded, the nationalists made a rather more literally geographical discovery at the same time. They were struck with a new appreciation of the fact that their own country, sprawling as it did from the Baltic to the Pacific, was a vast and grand imperial universe unto itself. The Slavophiles, for their part, had concluded that only states which were physically large had any claim to self-determination and national greatness, and who would not agree that none was as magnificently immense as Russia?[77] Russia covered fully one-sixth of the earth's land surface (a fraction which Russians have remained fond of repeating down to the present day) and it represented an extraordinarily diverse bazaar of peoples and habitats, which by any measure could only be seen as truly unique. An indication of just how enthusiastically Russians could respond to this particular aspect of their imperial grandeur can be seen in the following depiction by the historian Pogodin in 1838.

[75] Desnitskii, *Delo Petrashevtsev*, II, pp. 41–43; Gertsen, "Pis′mo iz provintsii," p. 132; Malia, *Alexander Herzen*, p. 143.

[76] Sarkisyanz, "Russian Attitudes," p. 245; *idem*, "Russian Imperialism," pp. 48–49.

[77] Petrovich, *Emergence*, p. 59.

Russia! what a marvelous phenomenon on the world scene. Russia – a space ten thousand *versty* [from west to east], on a straight line from the river that is virtually at the center of Europe across all of Asia and the Pacific to the distant lands of America! A space of five thousand *versty* [from north to south], from Persia . . . to the edge of the inhabited world – to the North Pole. What state can equal it? Its half? How many states can equal its twentieth, its fiftieth part? . . .

Russia – a state which includes all soils and climates, from the hottest to the coldest, from the sweltering environs of Yerevan to icy Lapland; a state which abounds in all products necessary for the maintenance, comfort, and pleasure of life . . ., an entire world, self-sufficient, independent, absolute. Many of its products are ones which individually have served for centuries as the source of wealth for entire large states, while Russia has them all grouped together.

The sheer abundance of Russia's resources, Pogodin reasoned, put it in a class entirely of its own and endowed it with strengths that the West simply could not match. "Of gold and silver, which are practically exhausted in Europe," he continued,

we have mountains . . . Bread – we can feed all of Europe in a hungry year. Timber: we can rebuild her if – God forbid – she burns down. Flax, hemp, leather: we shall dress and shoe her . . . Is there anything that we lack, is there anything that we cannot produce ourselves? That we cannot supply to others?[78]

Rejected – so they felt – by the West, and desiring in any event to disassociate themselves from it, the Russians found comfort, reassurance, and new depths of meaning in the vast expanses of territory which stretched from the central provinces of European Russia to the Black and Caspian Seas, across the Caucasus and the Kazakh steppe to the Tien-Shan mountains, the Altai, and far across Siberia to the Pacific and Alaska. On the most basic level, the simple, almost unimaginable immensity of the Russian realm could by itself be judged a positive quality, entirely autonomous and emphatically non-European. "The mind grows dumb when confronted with your expanse," mused the writer Nikolai Gogol′ in the early 1840s in the concluding passages of his *Dead Souls*:

what does this boundless space portend? Is is not here, is it not in you that a limitless thought will be born, because you yourself are limitless? Is this not the place for a Russian knight, here where he has room to spread out and stride about?[79]

The Orientalist Grigor′ev developed essentially the same thought. "It is enough to look only at the awesome extent of the earth's surface covered with the name Russia," he wrote,

for the idea of its grand fate involuntarily to dominate one's thoughts. Neither the kingdom of the Macedonian hero, nor Rome in the flower of its might, nor the

[78] Pogodin, "Pis′mo," pp. 2–3; Barsukov, *Zhizn′*, V, pp. 166–167.
[79] Gogol′, *Mertvye dushi*, p. 237.

Caliphate in the springtime of the Arabian outpour ever occupied such an enormous territory.

Providence had given Russia this space intentionally, "so that there would be somewhere for Russian activity to spread, [and] so that its broad knightly chest would have something to breathe in."[80] The elemental power which this vision of physical immensity and grandeur exerted on the Russian imagination at the time is well conveyed in one of Alexander Pushkin's most famous poems. Responding irately to European critics of Russia's suppression of an insurrection in Poland in the early 1830s, the poet evoked precisely this geographical image and presented it in the form of a challenge. "What do you think of us?" he jeered at the *klivetniki* or slanderers of Russia:

> Has the Russian become unused to victories?
> Or are we too few? Or
> won't the Russian land
> From Perm′ to Tavrida
> From the cold Finnish cliffs to the fiery Colchis
> From the shaken Kremlin
> To the walls of motionless China
> Arise, its steel bristle flashing?[81]

The nationalist preoccupation with the immensity and diversity of Russia's imperial domains led many educated Russians to a new appreciation of the country's remote provincial reaches. Indeed, such a redirection of attention was actively encouraged, and the manifest geographical ignorance of most Russians about their homeland was bewailed. On the pages of Pogodin's journal *Moskvitianin*, for example, one commentator vented his ire upon those of his compatriots who "are more interested in foreign lands and are not aware of the diversity of our homeland." The author called for Russians to stop "rushing abroad to marvel at the wonders found there" and to direct their attention instead to regions within the imperial boundaries.[82] Such exortations did indeed produce results, one of the most notable of which was the unprec-

[80] Grigor′ev, *Ob otnoshenii*, p. 4; Petrovich, *Emergence*, p. 90.

[81] Pushkin, "Klevitnikam," pp. 209–210. On Pushkin as a "poet of Russian imperialism" see Berdiaev, *Origins*, pp. 24, 78–79.

[82] Zenzinov, "Pis′mo" (1844), p. 234; also see Layton, *Russian Literature*, p. 15. Compare Zenzinov's complaint to the nationalist summons issued in 1824 by Prince Odoevskii in the first issue of Kiukhel′beker's short-lived literary almanac *Mnemosyne*. The most important goal of the new movement, Odoevskii insisted, was to

> direct the attention of Russian readers to little-known subjects in Russia, at the very least to compel people to talk about them, . . . [and] finally to show that not all [native Russian] topics have been exhausted, that we, searching for trifles in foreign countries that we might study, forget about the treasures that are close to us [in Russia].

Quoted in Sakulin, *Iz istorii*, I, pt. 1, p. 110; Rammelmeyer, "V. F. Odoevskij," p. ix. Although the "treasures" that Odoevskii has in mind are those of literature and philosophy, the precise parallel with Zenzinov's complaints about Russian tourism abroad and the lack of geographical appreciation of the homeland is remarkable.

edented interest that now began to be directed to the territories stretching east of the Ural mountains across northern Eurasia to the Pacific. Throughout the second quarter of the nineteenth century, the nationalists developed new images of Siberia that contrasted strongly with the more traditional and conservative views propounded by the representatives of Official Nationality. This perceptual transformation of the Russian east[83] is of considerable importance to our study, for it comprised an vital element of the overall framework within which the issue of the Amur river was resurrected and resignified at the mid-century.

As we have noted above, since the sixteenth century Siberia had played the role of a mercantile colony for the Russian state. In the heyday of the fur trade in the seventeenth and eighteenth centuries, the exploitation of this resource provided fabulous profits for Muscovite coffers. Lomonosov made his celebrated pronouncement that "Siberia will foster the growth of Russian imperial grandeur," and the region was popularly regarded as Russia's *zolotoe dno*, or gold mine.[84] By the beginning of the nineteenth century, however, the fur trade had entered a period of definitive decline, as changing styles of dress eliminated international markets and the protracted over-harvesting of the fur-bearing population depleted its numbers precipitously. With this exhaustion of Siberia's value as a source of mercantile wealth, it was not immediately clear how this vast region could continue to remain useful for the state, and thus representatives of official Russia were increasingly inclined to see it unfavorably, as a frozen and useless wasteland. This attitude became particularly pronounced after 1825. In the eyes of the state, the only use which this region retained was as a place of exile, a "deep net" – as Nicholas I's foreign minister Count Karl Nesselrode ungenerously characterized it – that was suitable for convicts and the other dregs of Russian society, but only as long as it remained isolated and undeveloped.[85] Indeed, even as conscientious and well intentioned a statesman as Mikhail Speranksii, who prepared a careful reform of the Siberian administrative structure in 1822 and was by no means unfavorably disposed toward the region, came to the not altogether encouraging conclusion that Siberia was "not a place for life and for more advanced civil organization."[86] A similarly negative attitude was conveyed rather more evocatively by someone who stood well outside of government circles. The poet P.V. Shumakher spent only one year in government service in Siberia in the 1830s, but was nonetheless moved to commemorate the experience in the following gloomy tones:

83 For a fuller discussion, see Bassin, "Inventing Siberia," *passim*.
84 Lomonosov, "Kratkoe opisanie," p. 498; [Soimonov], "Drevniaia poslovitsa," *passim*; Svatikov, *Rossiia*, p. 6; Iadrintsev, "Sibir′ pered sudom," p. 23.
85 Cited in Barsukov, *Graf . . . Amurskii*, I, p. 670; Kuznetsov, "Sibirskaia programma," pp. 17–18.
86 Quoted in Potanin, "Zametki," p. 192. On Speranskii's reforms see Raeff, *Siberia, passim*.

O you, bitterness, cruel stepmother, Siberia,
Your snowy steppes have spread out far and wide:
Unfree, unfriendly, deserted, hostile,
Unappealing, inhospitable, and cold.[87]

"The very name [of Siberia] is enough to terrorize a Russian," commented an explorer of the Arctic in 1830, "who sees there only inexorable separation from his homeland, and a vast dungeon, inescapable and eternal."[88]

In the 1830s, however, Russian nationalists beganto formulate a perspective on Siberia which was quite different. In their quest for elements in the Russian ethos that could embellish their exalted sense of national pride and accomplishment, they discovered in Siberia a vast repository of useful material. In the chronicles of Siberian history, for example, they could identify undeniable evidence of the Russian capacity for independent action and creative accomplishment. The daring exploits of the cossacks and *promyshlenniki*, or fur traders, who conquered Siberia from the remnants of the Golden Horde and then displayed extraordinary endurance in traversing and exploiting these wild domains represented a model of positive historical achievement, and one which moreover provided a reassuring counterbalance to the accomplishments of the European *conquistadores* in the New World. In the figures of such cossack heroes as the great Yermak, Yerofei Khabarov, and Vladimir Atlasov, the Russians could point with pride to their own native Cortez or Pizzaro, and proceeded to do precisely that.[89] "We did not discover America," wrote Pogodin to the young tsarevich Alexander in 1837, "but we opened up a third of Asia, . . . and does [this] not enhance the discovery of Columbus?"[90]

An indication of the concerns which motivated this new interest in Siberia can be seen in the attempts to produce a new historiography for the region to replace that inherited from the eighteenth century. Written for the most part by eighteenth-century German scholars such as Gerhard Müller and Johann Fischer, the older histories had portrayed Siberia's cossack conquerers indifferently and even negatively as the obstreperous leaders of lawless brigand bands interested only in their own gain. Indeed, even Karamzin had characterized them as a "band of drifters, moved by a base hunger for profit and the . . . love of glory."[91] All of this offered precious little encouragement for the new nationalist enthusiasm. The first specialized works on the subject by Russian scholars began to appear in the 1830s and 1840s, and they left no question about the nationalist motives that inspired their research.[92] One his-

[87] Shumakher "Pesnia katorzhnogo," in his *Stikhi*, p. 123 [88] Gedenshtrom, *Otryvki*, p. 4.
[89] Nebol'sin, *Pokorenie*, pp.138–140; Mirzoev, *Istoriografiia*, p. 223; Becker, "Contributions," pp. 331–353. [90] Barsukov, *Zhizn'*, V, p. 171. [91] Karamzin, *Istoriia*, IX, p. 232.
[92] E.g. Slovtsov, *Istoricheskoe obozrenie*; Nebol'sin, *Pokorenie*.

torian introduced his work by stressing his overriding desire to educate every Russian who felt "the need to become acquainted with lands that, although remote, are nevertheless native, [and to know them] better than regions that, although neighboring, are foreign and with which we do not and cannot possibly have any blood bonds, any common interests, or national sympathy."[93] In these new historical monographs, the cossack conquerers of Siberia were depicted in a manner rather more appropriate to nationalist sensibilities, as courageous and deeply patriotic sons of Russia whose deeds were animated by the overriding desire to enhance the glory of the fatherland. In addition to these scholarly works, the Russian reading public was showered during this period with an array of historical novels, poems, and theatrical productions based on the theme of Yermak's foray across the Urals similar in spirit to Polevoi's drama cited above. Indeed, the appeal of Siberia's historical legacy proved infectious for no less a literary giant than Alexander Pushkin himself. Already in the 1820s, the poet expressed the intention of composing an epic poem about Yermak, and at the time of his death was collecting material on the Russian Far East, apparently in preparation for a work about Vladimir Atlasov, the conquerer of Kamchatka.[94]

In their advance across Siberia the Russians had obviously conquered peoples as well as territory, and the nationalists of the 1830s and 1840s did not pass up the opportunity to identify these conquests as noble achievements in Russia's historical record. Toward this end, Siberia's indigenous inhabitants were generally depicted as depraved pagan savages, and the cossack heroes correspondingly cast in the appealing role of Christian crusaders and bearers of civilization and spiritual enlightenment, who had easily been the equal of anything the West could offer. "Russians for some reason become uncomfortable when they hear all around themselves the high-flown tales about different foreign heroes who conquered unknown lands," the historian Nebol'sin observed regretfully in his study of Yermak's campaign, especially when "they hear the new designation of 'the civilizer' bestowed upon rich England." Yet "these same Russians [should] know that our Russia, no less than England, can lay a claim to the title of 'civilizer,' that we as well had our Cortez and our Pizzaro."[95] And, in Polevoi's play about Yermak, the protagonist's victory in a wrestling match with a Tatar prince occasioned the following reflective aside:

> As the pagan idol fell, so will fall
> The infidel before the Orthodox faith,

[93] Nebol′sin, *Pokorenie*, p. 2.

[94] Pushkin, Letter to N. I. Gnedich, p. 100; *idem*, "Zametki." On Pushkin and Siberia see Selinov, "Proshloe," pp. 152–154; Bassin, "Inventing Siberia," p. 781n.

[95] Nebol′sin, *Pokorenie*, p. 2. Indeed, later in this work the author went so far as to reject this parallel between Yermak and Cortez, arguing that the deeds of the former so surpassed those of the Spaniard that any comparison was implicitly denigrating for the Russians. *Pokorenie*, pp. 138–140.

And divine grace and light will shine
Over the Siberian realm, which hitherto
Has stagnated in the darkness of idolatry.[96]

In its capacity as the benevolent civilizer of Asiatic Siberia, the nationalists could locate yet another example of Russia's noble and humanitarian national achievement.

For the most radical of the nationalists, those who were absolutely uncompromising in their rejection of the status quo under Nicholas I and categorical in their insistence upon sweeping revolutionary change, Siberia offered a very special sort of appeal. The region could be envisioned as a positive alternative to Russia west of the Urals, and even a source of its future regneration and reform. This perspective is clearly apparent in the thinking of Alexander Herzen, one of the most brilliant and engaging intellectuals of the period and at the same time one of its most prominent dissidents. Herzen was exiled to the town of Viatka in 1835 for his activities in a student group critical of the regime, and he spent a total of five years in exile, in Viatka and later in Vladimir.[97] At this early point, Herzen was still working out his own particular nationalist perspective, and his eyes were open wide for qualities unique to Russia which could help guarantee a great destiny for the country. We may be certain that he did not initially suspect he would find anything of this sort in Siberia, for at the outset of his exile he shared the general negative view expressed in the verse by Shumakher. On his way to Viatka he was moved at one railroad station to jot down the verses of Dante:

Through me you enter the woeful city
Through me you enter eternal Grief

with the observation that they were "equally well adapted for the road to Siberia as to the Gates of Hell."[98]

There was an unmistakable element of melodrama in Herzen's reference to the "road to Siberia," for the town of Viatka, in the upper basin of the Kama River, was situated several hundred miles west of the Urals, and Herzen thus never actually saw Siberia. This did not, however, interfere with his determination to locate a positive alternative to Offical Russia, and what he heard about Siberia now at relatively close range suggested that it might offer precisely what he was looking for.[99] His view of it, accordingly, changed markedly. His former prejudice against the region as a desolate land of exile began to dissipate, and he quickly developed a positive, even enthusiastic appreciation for it. Above all else, this critic of autocratic tyranny and despotism valued the qualities of freedom and egalitarianism which he per-

[96] Polevoi, *Ermak Timofeich*, p. 93. [97] Malia, *Alexander Herzen*, pp. 135–136.
[98] Gertsen, *Byloe i dumy*, VIII, p. 219. As we will see, Dante's negative imagery was a popular medium for the Russians to express their dim views of Siberia.
[99] Malia, *Alexander Herzen*, pp. 143–144.

ceived in Russian society in Siberia. He was impressed by the absence of a landed nobility and urban aristocracy, and marveled at the outspoken sense of independence on the part of the common Siberians, a characteristic rarely encountered in European Russia. He described with great satisfaction the general disdain for official representatives of tsarist authority – the *chinovnik* or governmental bureaucrat, and the *gendarme* – who in Siberia were appropriately regarded by the local population "more like an occupying foreign garrison installed by a conquering power" than the respectable and legitimate guardians of the social order. The Siberian peasant, unburdened by the oppressive weight of serfdom, compared in Herzen's estimation entirely favorably to his Great-Russian counterpart, in terms of physical well-being as well as intelligence. The great distances in Siberia, and the associated isolation of rural settlements, left the Siberian much more self-reliant, resourceful, and – of no little importance – prepared when necessary to demonstrate his resistance. Herzen noted the relatively infrequent contact of the average Siberian with the church, owing again to the great distances and isolation, and saw in this a decidedly beneficial effect, for it has "left his mind freer from superstition" than was the case with the peasantry in European Russia.[100]

It was not only by virtue of its unique social order that Siberia was appealing for Herzen's nationalist sensibilities. Like many of his contemporaries, he was immensely impressed by the historical record of the Russian occupation of these lands, which he depicted as a catalog of creative accomplishment in response to the extraordinarily difficult challenge of mastering a wild and inhospitable natural environment. Just like the other expansionist Western nations of his day – for example, the British and the Americans – the Russians as well had worked to advance civilization through their activities in "taming" and making productive the primeval, untouched expanses of Eurasia. Herzen's emphasis here, it should be noted, was far more on "civilizing" wild nature than a heathen population. "A handful of cossacks and several hundred homeless peasants," he observed, "crossed oceans of ice and snow at their own risk, and everywhere these exhausted bands settled, the frozens steppes which had been forgotten by nature bubbled with life, and fields became covered with crops and herds, from Perm′ to the Pacific."[101] In a letter to Giuseppe Mazzini, Herzen stressed this point with considerable emphasis, and tried to convince his Italian comrade of the appreciable quality of Russia's historical achievement precisely through the example of the colonization of Siberia. Russia's colonists beyond the Urals, he wrote, were in every respect like settlers on the North American frontier, and just like them the Russians brought these empty territories into the pale of modern civilization

[100] Gertsen, *Byloe i dumy*, VIII, pp. 256–257.
[101] Iskander, "Rossiia" (1859), p. 259; Kucherov, "Alexander Herzen's Parallel," p. 36.

by building cities, hospitals, and schools, and by introducing modern commercial activity.[102] Depicted from this perspective, the occupation and development of Siberia served admirably as evidence of the Russian capacity for independent national accomplishment.

Underlying all of these positive manifestations of Siberian society and history, Herzen felt he saw a deeper principle at work. On reflection, it seemed obvious to him that it was Siberia's very backwardness, together with its lack of advanced development and refinement, that had conditioned the qualities which so impressed him. When compared to the cosmopolitan enlightenment of Europe, including even European Russia west of the Urals, Siberia remained undeniably primordial, with an absence of accumulated historical traditions and patterns, an elemental rawness and an outspoken lack of cultivation. All of these qualities were powerfully attractive to Herzen. He described his reactions in an enthusiastic letter written back to friends in Moscow shortly after his arrival in 1835.

> what is Siberia? – here is a land you do not know at all. I filled my lungs the icy air of the Ural mountains; its breath is cold, *but fresh and healthy*. Do you realize that Siberia is an entirely new country, an America *sui generis*, precisely for the reason that it is a land without aristocratic origins, a land which is the daughter of the cossack bandits which doesn't remember its ancestry, in which people are renewed, shutting their eyes on their entire past existence . . .? Here all are exiles and all are equal . . . Back there [in European Russia] life is more enjoyable, and there is enlightenment, but the more important points are: freshness and newness.[103]

This, then, was something very different from the reigning view of Siberia. In stark contrast to European Russia, which had been stifled and spoiled by the oppressive weight of Nicholas I's Official Nationality, Siberia was characterized by thoroughly positive qualities, as a fresh and developing society. In effect, Herzen located in this exotic, remote region at least some of the very qualities which he felt to be sorely lacking west of the Urals. As we will see later in this study, he was subsequently to generalize these factors of youth and pastlessness first identified by Siberia into qualities characteristic for Russia as a whole. What is most striking about Herzen's ideas are not his conclusions concerning Russia's advantage as a *tabula rasa* – a point which was being made already in the eighteenth century by Russian neophytes of the Enlightenment and frequently repeated thereafter[104] – but rather his approach to this notion through the example of Siberia. This latter perspective was entirely novel, and very much a product of the period we are discussing.

Although Herzen was not entirely alone in viewing Siberia as a respository of unspoiled national qualities that could regenerate or even replace Russia west of the Urals,[105] the notion remained quite radical, and it would be mis-

[102] Gertsen, "Pis'mo k Dzhuzeppe Matstsini," p. 350.
[103] Letter to N. I Sazonov (18 July 1835), pp. 45–46 (emphasis in original).
[104] Schapiro, *Rationalism*, pp 33, 42. [105] See below, p. 93.

taken to assume that it was widely shared among even the ranks of committed nationalists. What they all did have in common, on a more basic level, was the determination to seek meaning and inspiration within the boundaries of the Russian realm itself, a determination that led to the sort of geographical reconceptualization of the empire we have just examined. In the first instance, this process yielded a new pride in Russia's imperial expanses as a whole; more specifically, it led to a positive reevalation of the significance of Russia's vast but remote eastern reaches to the nation at large. The precise nature of Siberia's significance may have varied, but all nationalists shared some sort of heightened appreciation of its potential contribution to Russia's coming destiny. Indeed, this new fascination with the Russian east reached even into the Winter Palace itself, where in 1837 the tsar's son Alexander – who as we will see was not untouched by the anticipation of a reformed and revivified Russia – included Siberia in the venue of an official tour of the empire. Although his visit was limited to a hurried passage through a few towns on Siberia's westernmost fringes, it secured for him nonetheless the distinction of being the very first Romanov ever to venture across the Ural mountains.[106] It was within the framework of these entirely novel perceptions of the Russian East that the evolution of views on the Amur issue in the 1840s and 1850s was to take place.

"Asia, Europe: influence on the entire world!"

The process of rethinking Russia's relationship to Europe and Asia came to a feverish climax with the outbreak of the Crimean War in 1853. The joint Anglo-French declaration of hostilities delivered an elemental shock – psychological no less than military – which jolted Russia to its very foundations and which the regime of Nicholas I was ultimately not to survive. For Russia's educated public, the onset of the war made tragically obvious what had been sensed inwardly for a long time, namely that Europe stood in a united front against Russia, that it considered Russia in no way to be a fraternal member, and indeed identified its own interests in clear opposition to it. At the same time, however, this public could not over the long term find within itself the spirit to rally behind a regime that, through a quarter-century of reactionary bureaucratic stagnation, had left the country weak and unable to resist the onslaught from the Western powers. The dimensions of the desperation felt at the time were truly existential, and many saw no alternative but to accept the eventuality of military defeat, with all hopes pinned on a far-reaching national rejuvenation that would take place after a peace had been made. This prospect of national rejuvenation, therefore, served as a sort of psycho-

106 [Rastorguev], *Poseshchenie*, *passim*; Tatishchev, *Imperator Aleksandr II*, I, pp. 79–80; Pereira, *Tsar-Liberator*, pp. 17–18.

logical palliative, and in elaborating on it the nationalists drew fully on the vision of Asia as the true and meaningful arena for Russia's future attention and activity.[107]

"Asia" in this case proved to be a rather amorphous geographical concept. On one level, the coming turn to the East would imply a redirection of national interest inward, onto Russia's own imperial domains which contained more than their share of Asiatic territories and peoples. In this spirit, an army officer noted the following thoughts in his diary as the war wore on:

> If we are fated to become a kingdom of the East, then perhaps the real theater of Russia's genuine and solid might is located there. In Europe we have for a long time had nothing to do. It is hostile to us, and there was little that it should have taken from us or we from it. St. Petersburg as a capital did unspeakable, incalculable damage to us . . . From Moscow we laid our hand on the East, on what was, at that time, the Tatar Volga, the Urals, Siberia, entire kingdoms: the best, inalienable, and a guarantee of our unshakable might. Nizhnii Novgorod is the focus of internal industry and Russia's bazaar with Asia, and also has traditions dear to the people, a center of intelligent, strong beautiful Russian stock!

St. Petersburg had brought Russia into contact with Europe, with the result that "more than anything we feared and were ashamed to be Russians! We were the monkeys of Europe!"[108]

Even more enthralling, however, and perhaps more effectively soothing for the country's wounded sensibilities, was the prospect that Russia might assert itself in that Asia that lay beyond its borders. It was really in this arena that Russia could find just rewards for its labors, and could achieve or regain the global significance it had lost through a quarter-century of simpering acquiesence to Europe and what was certain to be an ignominious military defeat. An aging Aleksei Khomiakov wrote regretfully to Mikhail Pogodin in the early days of the war that Russia, in contradiction to its "natural instincts," had for so long been preoccupied with Europe. Its "true advantages," on the other hand, "summoned the country to intensified activities in the East, which could have become ours very easily."[109] Pogodin himself more than agreed, and took the opportunity to demonstrate once again his rhetorical brilliance at transforming frustration and chagrin into optimistic and aggressive zeal:

> Leaving Europe in peace in the expectation of better circumstances, we should turn all of our attention to Asia, which we have let fall almost entirely from our sight, although it is actually for the most part pre-destined for us . . . Let the European peoples live as they know how and arrange themselves in their own countries as they wish, while half of Asia – China, Japan, Tibet, Bukhara, Khiva, Persia – belongs to us if we want. And

[107] Riasanovsky, *Nicholas I*, pp. 165–166.
[108] The officer was P. Kh. Grabbe. Diary entry of 28 January 1854, quoted in Barsukov, Zhizn′, XIII, p. 37.
[109] Quoted in Barsukov, *Zhizn′*, XIII, p. 16; emphasis in original.

perhaps we should spread our dominions, in order to disperse the European element across Asia, and [let] Japheth tower above his brothers.

Not content with this statement of principle, however, the historian went on to outline a future program of Russian activity in strokes that were positively dizzying.

Lay new roads into Asia or search out old ones, develop communications, if only in the tracks indicated by Alexander the Great and Napoleon, set up caravans, girdle Asiatic Russia with railroads, send steamships along all of its rivers and lakes, connect it with European Russia, . . . send European goods . . . to Asia, import Asian goods from the richest countries such as China and Japan for ourselves and Europe, and you will increase happiness and abundance across the entire globe.

Asia, Europe: influence on the entire world! What a magnaminous future for Russia![110]

In this way, the prospect of Russian activities in Asia and the notion of a mission of salvation were intensified in the early 1850s and given the tone of a fateful imperative, in response to the extreme deterioration of relations with Europe. This state of affairs was critical in helping set the stage for the ensuing transformations in Russia's perceptions of, as well as its policy toward, the Far East.

[110] Pogodin, "O russkoi politike," pp. 242–244; Riasanovsky, "Russia and Asia," pp. 178–180.

3

The rediscovery of the Amur

"My ardent love for Russia . . . will serve as my excuse"

One of the clearest indications of the obscurity surrounding the Amur region in the early decades of the nineteenth century was the fact that practically nothing about the river appeared in print. What was published at the time was located for the most part in a highly specialized literature and directed to a select audience which had some practical reason to be concerned with the Russian Far East. The picture of the region that emerged out of these accounts was unalluring and undistinguished, and moreover was one which carefully accorded with the official position of the government. The most extensive of these treatments – Grigorii Spasskii's 1824 essay "Historical and Statistical Notes about Places along the River Amur" – offered a good example of this latter point, for although the author made it clear that Russia had relinquished the region to China in the late seventeenth century, he did not comment on the injustice of the Treaty of Nerchinsk and did not even allude to the issue of reacquisition.[1] Elsewhere Spasskii did discuss the value of the Amur in a manner that had rather more contemporary overtones, by indicating the region's potential importance as a supply base for Russia's North Pacific settlements, but again he did not call for the reacquisition of the Amur, even in muted tones or by implication.[2] Writing about the Russian experience on the Amur in the seventeenth century, the naval historian Vasilii Berkh not only did not discuss the significance of the river or its loss but even offered a mixed evaluation of the exploits of Vasilii Poiarkov, the first cossack conquerer in the region. Berkh stressed the cossack's interest in enhancing his own personal glory, and spoke explicitly about his mistreatment of the native peoples he encountered, including rumored incidents of cannibalism.[3] The

[1] G. S., "Istoricheskie i statisticheskie zapiski." [2] [Spasskii], "Istoriia," ch. 4, pp. 270–271.

[3] Berkh, "Otkrytie," pp. 50–51, 55. The theme of cannibalism on the Amur was to reemerge during the Russian occupation in the nineteenth century, this time in reference to a contemporary incident. See below, p. 246.

author of a lengthy book about Eastern Siberia appearing in 1828, which featured an extensive discussion of commerce with China at Kiakhta, did not even find it necessary or desirable to mention the Amur at all.[4] Indeed, the only reference to the issue of reacquisition in this early periodical literature appears to be that made by the future Decembrist Gavrill Baten′kov (not coincidentally of Siberian origin himself[5]), who in 1822 noted that the yearning to regain the Amur was nearly universal in Siberia and was shared by "many in Russia itself." He expressed his own skepticism, however, by indicating ironically that the actual "experience [of settlement] had destroyed the fascination" that had drawn the settlers to the region in the first place.[6]

Beginning in the 1830s, however, all this began to change. Discussions of the Amur grew more numerous, took on a distinctly new tone, and began to appear in journals, books, and even newspapers directed not to specialists of any sort but to a broader reading public. This transposition of the Amur question into what might be called the popular literature of the day was a development of considerable importance to our subject, for it signaled the transition of the issue from an insignificant historical footnote into a matter of meaning and concern to educated public opinion at large. As part of this, the question of the Amur gradually became imbued with an attraction that was entirely novel, and readers were increasingly drawn to discussions of it out of no practical need or interest to know about Siberia and the Far East. What did interest them, rather, were the problems and challenges of Russia's contemporary social, and political life, and it was within the context of these thoroughly contemporary concerns that the Amur question was now resurrected. In a manner that was often highly subtle, the issue began to be infused with a tangible ideological significance that served – despite the remoteness and obscurity of the region – to make it freshly relevant to the country's ever-expanding and increasingly restless reading public.

An early example of this new interest can be seen in the book*Siberian Fragments* by the Arctic explorer Matvei Gedenshtrom, which appeared in 1830. Inspired by the new nationalist enthusiasm, Gedenshtrom's work was among the first to seek to heighten educated Russians' appreciation of the glory of their homeland by introducing them to exotic and unknown realms of the empire outside of European Russia. As part of this, he was determined to make the issue of the Amur meaningful to the emerging national mood. Gedenshtrom began by going through the various aspects of the traditional perspective on the importance of the Amur which we noted at the beginning of this study. The region was first of all one of fabulous natural wealth. "In its fertile soils and the convenient communication which it affords between the rivers of Siberia and the Pacific," he affirmed, "the Amur contains inexhaust-

[4] Martos, *Pis′ma.* [5] Bykonia, "Vzgliady," p. 65.
[6] [Baten'kov], "Obshchii vzgliad," pp. 120–121.

ible riches."[7] He emphasized the importance of navigation on the Amur for the support of Russia's North Pacific colonies, and beyond this depicted the bright prospects for expanded commerce with the Pacific Far East which this navigation would make possible. "The connection of the Baltic to the Pacific by means of the blessed (blagoslovlennyi) Amur and other Siberian rivers could provide a new and entirely secure route [into Russia] for the abundant products of the East," which would in turn assure that "the South Pacific and the Indian Ocean would cease to be so remote from [European] Russia." This would bring immediate benefit to the homeland, for it would remove the necessity for Russians to pay foreign intermediaries to transmit the goods of these regions.

In addition to these familiar points, however, Gedenshtrom introduced several new themes. The acquisition of the Amur, he asserted, would have the effect of transforming Siberia in its entirety. "The Amur, and only the Amur, which was a part of Russia in earlier days, can transform the entire character of Siberia, and out of a vast desert create a rich land . . . What a wonderful (priiatneishii) dream for a soul who is passionately devoted to the fatherland!" And once the Amur valley was opened for colonization, he continued, its gentle climate and fertile soils would support a flourishing agriculture, which would in turn provide a major stimulus for the cultural and social development of Siberia as a whole. "Then Siberia would [no longer] seem to be a dungeon; willing settlers would hurry to occupy this cornucopia, and the highest level of moral life, arts, and crafts would no longer be such a rarity there."[8] Moreover, the acquisition of the Amur could serve as the basis for restructuring Russia's relations with its neighbors in the Far East. While this theme had figured in earlier discussions of the river, Gedenshtrom presented it with an assertiveness and bellicosity that was quite new. The Amur had been taken from the Russians unfairly at the negotiations in Nerchinsk, he asserted, but reassured his readers that it could be "easily" wrested back from the Manchurians, who after becoming fully assimilated into Chinese society had degenerated along with the rest of the country. Russian possession of the Amur would make it impossible for the Chinese and Japanese to persist in their resistance to entering into full commercial relations, and they would be "compelled" to open their borders to the Russians. There is nothing in either of these two Asiatic countries, he concluded, which could or should obstruct the desires and needs of "all-powerful Russia."[9]

Perhaps the most striking novelty in Gedenshtrom's presentation was the intensity and passion with which he pressed his points. He made it clear that his position on the issue was conceived entirely in the context of his "passionate devotion" to his homeland and that this devotion corresponded fully to his overarching desire to do everything possible to promote Russia's greater

[7] Gedenshtrom, *Otryvki*, p. 148. [8] Gedenshtrom, *Otryvki*, pp. 14, 146–148.
[9] Gedenshtrom, *Otryvki*, pp. 145–149.

welfare. He claimed the right, as a Russian patriot, to entertain such hopes and expectations of the country's glorious future, and he concluded his comments on the Amur with the following declaration:

> It is really the case that all of these sweet dreams about the benefit to my beloved fatherland and about the resurrection of Siberia will be considered only as the warped thoughts of a frivolous mind, only as an impertinent and baseless notion? Do the advantages of trade with China [a reference to the acquiescent position of the government] really outweigh the innumerable benefits of the possession of the Amur in the future? If so, then my ardent love for Russia and for its glory will serve as my excuse.

An excuse, that is, for his boldness and determination in pressing a perspective which was clearly at odds with the directives of official policy at the time. Gedenshtrom's enthusiasm, indeed, was irrepressible, and the prospect of the reacquisition of the Amur spawned in him an intoxicating vision of impending imperial magnificence and conquest. Referring to himself in the third person, he went on:

> In its inexperience and unfamiliarity with careful calculations, an inexperienced mind has become captivated by fantasy. In heartfelt joy it drew for itself a marvelous picture of a revived Siberia, it imagined this region's compatriots – humiliated [for the moment] by the haughty arrogance of China – regaining for themselves the accomplishment of their forebears, and it saw Peking and Paris equally accessible for the great Russian Monarch. His warriors do not have to cross the oceans: with a firm foot [on dry land] he can conquer kingdoms half-a-globe away. Can a dreamer be held guilty for the fact that his most precious dreams seem to him easy to accomplish, useful, and capable of enhancing the glory of the fatherland?[10]

Gedenshtrom's prudent tribute to Russia's ruler as the supreme representative of the nation was in full harmony with the spirit of Official Nationality. His "precious dreams" however – fantasies which envisioned the aggressive expansion of Russian authority in the Far East and Europe in order to embellish its national stature and insure a glorious future – had nothing whatsoever in common with the legitimatist and thoroughly conservative inclinations of Nicholas I.[11] This tension between these two perspectives was one that would grow ever greater.

From the end of the 1830s and throughout the following decade, the production of popular literature in Russia expanded dramatically. During this period, the country experienced what one scholar has termed a "journalistic explosion," which brought well over 100 new periodical publications into exis-

10 Gedenshtrom, *Otryvki*, pp. 149–150 (emphasis in original).

11 Gedenshtrom returned to this theme some years later, in a short article in the newspaper *Moskovksie Vedomosti*. With the same indifference to the actual historical record, he depicted the initial occupation of the Amur basin by the Russians in glowing terms, described the great successes of agriculture there (which in fact had never been), and characterized the Russian loss of the Amur in the Treaty of Nerchinsk as involuntary, and forced by the Chinese through guile and superior military might. Gedensh[trom], "Ob Amure," p. 518.

tence.[12] The role that these publications played in the intellectual and cultural development of the country was important, for they served as the principal fora in which the various views of Russia's national life were articulated and argued.[13] To be sure, these new journals did not escape the stringent surveillance of Nicholas I's army of censors, but imaginative writers in Moscow and St. Petersburg who were interested in circumventing these ideological watchdogs developed stylistic devices that proved highly effective.[14] The result was that the political content of this literature, consumed voraciously by virtually all of the country's educated elite, was quite substantial. In this manner, a highly charged ideological debate was conducted despite heavy censorship, and it was sophisticated enough to provide something of a political education not only for conservative supporters of Official Nationality but for reformers and revolutionaries as well. Many of the ideas set forth at this time would be carried over into the reign of Alexander II.[15]

Quietly but persistently, the Russian Far East figured among the subjects discussed on the pages of these so-called "thick journals," and in the process the nationalist resignification of the issue which Gedenshtrom had initiated was made ever clearer. In Mikhail Pogodin's conservative–nationalist *Moskvitianin*, for example, a series of letters from the Siberian writer Mikhail Zenzinov were published in the early 1840s. Writing from the town of Nerchinsk in the upper Amur basin, Zenzinov gave voice to the characteristic nationalist impatience with those of his compatriots who were more interested in visiting foreign countries than the no-less-foreign reaches of their imperial homeland. "Few Russians travel around their motherland, hurrying off instead to go abroad and marvel at the wonders they find there," he sighed, such that they inevitably "know very little about their own native regions and the life of the people there." Yet it was precisely the latter that ought to be most meaningful to the Russians. "How very much food would these topics supply for the questing mind, for the gifted pen, and for the ardent imagination! . . . When will our journals begin to acquaint us more with the diverse pictures of the great Russian Empire? It is a shame [to have] to say that we still know our Russia so poorly."[16] For his own part, Zenzinov sought to fill this gap with evocative depictions of the regions east of Lake Baikal where he lived. He described them not merely as a natural–geographical wonderland but as a rich repository of geniune native Russian life and folkways, about which he was for the most part gushingly enthusiastic. Indeed, his positive evaluation of life in this distant province grew darker at only one point, in order more effectively to evoke the shimmering prospect of regeneration and revival that the future portended. In a letter about the anticipated expansion of gold mining in

[12] Riasanovsky, *Parting*, pp. 276, 278.
[13] Evgen′ev-Maksimov, "*Sovremennik*", p. 10; Dement′ev, *Ocherki*, *passim*; Kuleshov, "*Otechestvennye Zapiski,*" *passim*. [14] Berlin, "Russia," p. 7. [15] Vyvyan, "Russia," p. 368.
[16] Zenzinov, "Pis′mo" [1844], pp. 234–235.

"Dauria," he speculated on the effects that the increased flow of wealth into the region would have.

> [A] new era of renaissance will begin for Dauria. [Its inhabitants] will throw off the fetters of ancient prejudices, centuries of mouldy traditions, and the life of the young generation will not be eternally soporific, timid, and torpid . . . Instead, life will begin to flow vibrantly, brightly, happily, and richly, the ray of beneficial genuine enlightenment will penetrate to the lower layers of society, and the direct and frank word of a man thirsting for the common good will find its place and with be valued according to its merit.[17]

Even if this flamboyant image of regeneration was conceived with Eastern Siberia in mind, it nevertheless fits precisely into the nationalists' anticipatory vision of popular revival throughout the country as a whole.

In regard to the issue of the Amur river itself, Zenzinov submitted a collection of documents about the Russian occupation and abandonment of the region in the seventeenth century. Although he did not annotate or comment upon these materials, the clear implication emerged from them that the river had belonged rightfully to the Russians at the time.[18] Other writers, however, pressing on in the spirit of Gedenshtrom, were rather more outspoken on this point. In 1840 an unsigned article appearing simultaneously in two journals attempted to present the historical background of the Amur question in the framework of nineteenth-century concerns. This publication was a collection of documents taken from archives in Yakutsk relating to Russian activity on the Amur in the seventeenth century, but in contrast to Zenzinov, the unnamed complier apparently felt that these materials would have little meaning for most readers if left unexplained, and thus additionally supplied an introduction for this purpose.[19] In it, the significance of Russia's presence in the Far East to contemporary nationalist sensibilities was presented in the context of the larger project of recapturing the glorious accomplishments of Russia's cossack past. The great glory accorded to Yermak as the "conquerer" (zavoevatel′) of Siberia was not entirely justified, the author indicated, for Siberia had not really been conquered until the Russians emerged victorious on the shores of the Pacific. This latter accomplishment had however been the work of two other valiant native sons: Yerofei Kharabov, the hero of the Amur, and Semen Dezhnev, who circumnavigated the Chukotsk peninsula in 1648 and demonstrated, three-quarters of a century before the Dane Vitus Bering, that Asia and America were not connected.[20] The fact that the nineteenth century had forgotten these noble warriors of the Russian Far East, making it necessary to explain to the contemporary reader just who they were,

[17] Zenzinov, "Pis′mo" [1843], pp. 555–556.
[18] Zenzinov, "Istoricheskie vospominaniia," pp. 106–118.
[19] "Istoricheskie akty." These documents also appeared verbatim in the journal *Zhurnal dlia chteniia vospitannikam voenno-uchebnykh zavedenii*, no. 105.
[20] Lantzeff and Pierce, *Eastward to Empire*, pp. 190–191.

was sad evidence of how the Russians neglected their national past. To Khabarov and Dezhnev, the anonymous author insisted, belonged nothing less than "the honor of being the first to carry the name of Russia between America and Asia, and to pronounce it formidably on the banks of the Amur, thereby alarming immobile China."[21]

Thus the attention of the reader was directed away from the Urals to the Far East as the true scene of heroic Russian exploits. It was moreover notable that "pronouncing" the name Russia on the Amur had been such a heroic act, and the transparently anachronistic reference to China as "immobile" suggested that it bore a contemporary significance as well. This latter point became clearer as the introduction went on. Why, it demanded, has Khabarov's name been "blotted out" of the memory of his descendants? "Is it not because a timid policy subsequently surrendered all the fruits of his exploits to a cunning and powerful neighbor?" But, it concluded bitterly, should Russia not be aware of its brave son anyway, "even if the misunderstandings and failures of his successors destroyed everything he had begun, and when his exploits moreover were characterized by courageous and genuine *Russian daring*?"[22] Although the author's chagrin remained judiciously focused on the seventeenth century, the point that he wished to make was perfectly clear. The loss of the Amur through the Treaty of Nerchinsk was the result of timid and servile policies, thanks to which the splendid accomplishments of the cossacks on the Amur were forfeited and great damage done to Russia's glory. This critical perspective on the remote historical past had an unmistakable contemporary resonance in terms of the nationalist impatience in the 1830s and 1840s with what they identified as an identically "timid" foreign policy pursued by Nicholas I and his ministers. It was a resonance which his readers would not have failed to detect.

The fullest presentation of the Amur issue in the popular periodical literature of the 1840s was a lengthy article by Nikolai Shchukin appearing in *Syn Otechestva* (Son of the Fatherland) in 1848. Shchukin was one of a group of Siberian writers in the 1830s and 1840s who, fascinated by the works of James Fennimore Cooper about the American frontier (which were quickly translated and voraciously consumed in Russia), were inspired to write novels of their own which similarly glamorized the rugged frontier life of Russian settlers east of the Urals.[23] Like Zenzinov, Shchukin had nothing but contempt for those European Russians who were more interested in points west than points east, and he prefaced one of his stories with the following admonition:

[21] "Istoricheskie akty," p. 55. [22] "Istoricheskie akty," p. 86 (emphasis in original).

[23] On Shchukin see Diment, "Exiled *from* Siberia," pp. 47–50ff. Vissarion Belinskii, who was quite taken with the American original, was less than impressed with this *genre* in Russia and referred to its practioners derisively as "our Siberian Coopers" (nashi sibirskie Kupery). "Literaturnye mechtaniia," p. 10. Also see Kungurov, "V. G. Belinskii," pp. 21–24.

My dear countrymen! . . . If you are so willing to waste money inherited from your ancestors in the cities of Europe from which you mostly bring back harmful innovations alien to our fatherland, then why not come and behold the picturesque and virginal nature of Siberia?[24]

In regard to the Amur, however, Shchukin's attention was fixed on something of more explicitly political moment. His 1848 essay was essentially a narrative history of the Russians on the Amur in the seventeenth century. Despite his claim to have consulted fresh archival material, Shchukin's facts come largely from Müller' histories of the mid-eighteenth century,[25] but in his interpretation of the material he sounded a very different tone. Above all, he was interested in bringing out the nationalist significance of the issue for the present day. This preoccupation was apparent already in his Introduction, where Shchukin took care to describe the geography of the region so that his readers – few of whom, he rightly reckoned, would ever have even heard of the river – would be certain to appreciate just how valuable it was.

This river is navigable for its entire length, and is not plagued by rapids. [I]ts current is gentle, [so much] so that boats can travel upstream by oar, something unknown on other Siberian rivers. The land along its banks is fertile, and in the forests roam sable and other animals.[26]

Most of the article, however, dealt with the seventeenth century, and did so in the same spirit we have witnessed in other examples, namely by depicting Russian activities on the Amur as a glorious chapter in the annals of the nation's history. Shchukin presented Khabarov once again as a great but forgotten figure representing Russia's valiant national heritage, and once again he made a great deal of the comparison with Cortez, inveighing heavily against those of his countrymen who were not prepared to appreciate this. "Our Europeans," he wrote in derisive reference to his westward-looking compatriots,

will laugh at the comparison of Khabarov with Cortez. How is it possible, they will want to know, to compare a hero in rich Spanish attire to a Russian *muzhik* dressed in a velvet caftan, girdled with a silken sash, and in a sable cap? The important point [however] is not in clothing but in what was in their heads . . . All of fashionable Russia knows about Cortez, but who of them has heard of Khabarov?[27]

Shchukin sounded the familiar theme that the ultimate failure of Khabarov's enterprise was not the cossack's fault. "If Khabarov's exploits did not bring favorable results to the state, it was not he who was to blame. He understood the importance of his undertaking, but was not allowed to complete it."[28]

[24] Quoted in Diment, "Exiled *from* Siberia," p. 50. [25] Middendorf, *Puteshvestie*, I, p. 139n.
[26] Shchukin, "Podvigi," p. 4. [27] Shchukin, "Podvigi," pp. 27–28 (quote); 6–7.
[28] Shchukin, "Podvigi," pp. 27–28.

Some day, however, Russian glory will be vindicated and the feats of its dauntless sons acknowledged. "The enthusiast of Russian glory will seek out their names and show Europe that even in the times of the tsars [of old] our forefathers were distinguished by virtues which today seem to be pure invention."[29]

Although all of the writers we have considered presented the exploits of the cossacks on the Amur as evidence of glorious national achievement in the past, Shchukin alone went beyond this to project this same pattern directly into Russia's future. "In all nations there are certain periods of discovery and conquest, and from time to time geniuses arise who carry the masses on with them" to spectacular accomplishments. We are used to thinking, he continued, that Russia's national potential in this regard has been exhausted, and that these times have passed forever. To this discouraging suggestion he responded emphatically.

> No! These potentials exist now and they always will; a genius will find them under his very hands and will [once again] produce miracles, the likes of which we cannot even suspect. The spirit of conquest and enterprise was dominant among the Russians from the middle of the sixteenth to the beginning of the nineteenth centuries. Now it has grown quiet and is resting, awaiting the appearance of new Yermaks, Khabarovs, Dezhnevs, and Shelekhovs.[30]

We have seen that it was precisely such a "spirit of conquest and enterprise" that was felt to be stirring once again in Russia at the moment these lines were being written. Shchukin made the connection between a noble past and portentous future explicit, and in so doing bathed the still indeterminate vision of Russia's approaching messianic destiny in the bright and reassuring luminescence of seventeenth-century Muscovy's charismatic heroes of the Amur river.

At the end of his article, Shchukin again emphasized the contemporary import of Russia's formative experience in the Amur valley. With the Treaty of Nerchinsk, Russia "lost a river which connects all of the Transbaikal region to the Pacific, and in this way forfeited important advantages." At present, he affirmed bluntly, the need to reacquire the Amur has become essential in order to supply Kamchatka and Okhotsk effectively. Shchukin also mentioned trade with Japan, which would benefit from a water route to the Pacific, and reminded his readers of the numerous indigenous peoples living along the river, who – despite Chinese control – nevertheless by historical rights "now belong (prinadlezhat) to us." There was even an oblique reference to the prevailing concern about disturbing relations with China. Shchukin dismissed any fears that the views put forth in his article could disturb the "Chinese in Kiakhta" (by which he meant Chinese in general) with the assurance that they "are not able and do not want to read Russian." In general, he maintained,

[29] Shchukin, "Podvigi," p. 52. [30] Shchukin, "Podvigi," p. 12.

China "should not be measured with a European yardstick: everything there is turned inside out."[31]

In conclusion, a point might be brought out that has been implicit throughout our discussion. It is striking that almost all of the discussion of the Russian Far East in the periodical literature of the 1840s was *historical* in nature. The fact that the authors chose to make a point about the present through discussions of the past was no doubt related to considerations of censorship, for the Amur issue was at that time a delicate question of foreign policy and thus hardly suitable for open debate in the press. However, there was more involved than this. A strong sense of history played a central role in the development of Russian nationalism, as the national past was scoured for virtues and positive examples which could inspire the course for the future. In this way, history assumed a greatly enhanced significance and immediacy. As Sidney Monas has observed, for the Russians "the past has always held a compelling power over the future, exerting a force so constraining that it might foster the illusion, in an extreme instance, that if one changed the *accounts* in which the past is recorded and interpreted, one might well lay a magical hold on the future."[32] This is precisely what we have observed in the accounts of the Russians on the Amur in the seventeenth century. The printing of obscure archival documents was not merely a ploy for the censors, but an act which in itself seemed natural and entirely meaningful. The *Denkweise* of the Russian reading public would of its own volition invest the historical account with the appropriate contemporary significance.

"An irrepressible desire to visit this region"

We have already noted that although the measures undertaken to deal with the problem of supplying Russia's North Pacific coast with provisions – the establishment of an agricultural colony in California or the transport of supplies by sea from European Russian ports – had some limited effect, they in no way solved the problem entirely. For this reason, interest in the overland supply route across Eastern Siberia to the Okhotsk coast remained very much alive. This route was far from satisfactory, but there was hope that if an alternative harbor along the coast could be found to replace the port of Okhotsk, this might well be the best solution. Accordingly, a considerable amount of exploratory activity at the time was devoted to the task of locating such a facility. The southern coast of the Sea of Okhotsk was scoured by number of expeditions, including those of Prokopii Koz′min in 1829–1831, and Vasilii Zavoiko in 1842–1843.[33] While both of these expeditions determined that the the set-

[31] Shchukin, "Podvigi," pp. 51–52. [32] Monas, "Foreword," p. xiii (emphasis in original).

[33] Alekseev, *Russkie geograficheskie issledovaniia*, pp. 20–21, 46; Alekseev, *Amurskaia Ekspeditsiia*, pp. 8–9; Sgibnev, "Vidy," pp. 673–674, 676.

tlement of Ayan to the south of Okhotsk would represent something of an improvement over the latter as a port facility, neither was particularly enthusiatic, and the old notion that the Amur river could offer the ideal solution to the problem remained as strong as ever. This, however, was a solution which could not be pursued, for the government in St. Petersburg was adament in its refusal to allow any exploration of any part of the Amur. Despite repeated entreaties of officials in Siberia and the Far East, Russians were strictly forbidden to venture any further south than the Shantar islands.

Thus when the young naturalist Alexander von Middendorf, at the conclusion of a brilliant scientific expedition to Siberia in the early 1840s, flaunted government policy and the specific instructions of his sponsors in the Academy of Sciences by venturing south across the Stanovoi mountains to the Amur basin, it seemed a natural continuation of the endemic interest of the Russian fur trade in exploring the river. Yet while there indeed was an element of continuity with the search for a new port, no less important was the contrast between Middendorf and explorers such as Koz′min and Zavoiko. The motives which drew Middendorf to the Amur, and his own perceptions of the river's importance, marked a clear departure from those which had characterized the Amur question up to this point, and betrayed the clear influence of the new ideological climate which had developed in Russia since the accession of Nicholas I.

The original idea for the expedition had been conceived over a decade earlier by the naturalist Karl von Baer, one of the outstanding representatives of Russian and European science in the first half of the nineteenth century. Baer's underlying question was essentially biogeographical in nature and focused upon the relationship between organic life and climate, which he was particularly interested in investigating in an Arctic environment. The Taimyr peninsula on the Arctic coast of central Siberia – the northernmost continental protrusion on the globe – was selected as the most suitable location for such a study, insofar as its equidistance from both the Atlantic and Pacific oceans would reduce their moderating influence to a minimum. Baer's proposal for an expedition was supported by the Academy of Sciences, and his young colleague Middendorf was assigned the task of carrying it out.[34] In the formal instructions prepared by the Academy, the scope of the expedition was substantially broadened beyond Baer's original plan. Upon conclusion of his investigations on the Taimyr peninsula, Middendorf was directed to travel to Yakutsk in order to study the extent and nature of permafrost. After this, he was to continue on to the southern coast of the Sea of Okhotsk, where he was to investigate the natural–historical conditions of the Shantar island group. Finally, he was instructed to return to Yakutsk and from there back to

[34] Sochava, "Stranitsy," p. 226; Iurgenson, *Nevedomymi tropami*, pp. 9–10, 22.

European Russia.[35] Middendorf left St. Petersburg for Siberia in the spring of 1842. He spent a year and a half on the Taimyr peninsula, and arrived in Yakutsk in early 1844, where he set about immediately with his investigation of permafrost. These studies were completed by the fall of that year, and he set off once again, this time to the Sea of Okhotsk.

It was at this point that his expedition took an entirely unforeseen turn, and one that guaranteed it a notable place in the history of Russian involvement with the Amur. In contrast to Baer, Middendorf was motivated not only by scientific curiosity but by national–political concerns as well. He wanted to press his explorations further south than his instructions specified, for he was intent on including the Amur river itself. His ostensible goal in visiting the Amur was to clear up confusion about the exact location of the Russo-Chinese border established by the Treaty of Nerchinsk in 1689. According to the treaty, border posts were to have been erected that would delimit the boundary from the Gorbitsa river (a tributary of the Shilka) east to the ocean. It was however unknown in St. Petersburg exactly where these markers stood, or indeed even if they had ever been set up. In the absence of more precise information, the Russian government assumed – and most maps depicted – a border with China that ran along the northern flank of the Stanovoi mountains. Middendorf wanted to locate these boundary markers and determine the actual border, which he suspected lay considerably to the south.[36]

Middendorf was well aware that he was departing from the letter of his instructions, and he sent a communication from Yakutsk to the Academy of Sciences in St. Petersburg explaining his intention to alter his itinerary. He requested permission to travel to the mouth of the Amur and continue on upriver, returning to Russia via Irkutsk rather than Yakutsk. The Academy of Sciences, unsurprisingly, rejected his request out of hand, and a letter was dispatched from St. Petersburg forbidding Middendorf to carry out his plan. He was reminded that such an undertaking would go against his instructions, and was warned darkly of the possibility of his falling prisoner to the Chinese "with little hope for a quick return" – although the Academy was in all likelihood less concerned about the personal well-being of the explorer than with not violating official policy toward the Amur and China. For his part, Middendorf understood the situation perfectly well and, no doubt anticipating rejection, left Yakutsk for the Sea of Okhotsk before the negative reply from St. Petersburg could reach him. He boldly proceeded to the Amur without official approval, paying, as he rather boastfully put it, absolutely "no attention to whether or not this corresponded to the intentions of the authorities in the nation's capital."[37] Instead of sailing south along the Okhotsk coast to the mouth of the Amur, he instead crossed the Stanovoi mountain range in

[35] Middendorf, *Puteshvestie*, I, p. 14; Gnevucheva, *Materialy*, p. 203; Leonov, *Aleksandr Fedorovich Middendorf*, p. 69. [36] Middendorf, *Puteshvestie*, I, pp. 158, 164–165.
[37] Middendorf, *Puteshvestie*, I, pp. 113n, 137.

late 1844 and descended the Zeia river south to its confluence with the Amur. From this point, and without actually having visited the estuary of the river, he travelled upstream, and arrived back in St. Petersburg via Irkutsk in March 1845.[38]

In the published account of his expedition, Middendorf went to considerable effort to explain the inordinate attraction which the Amur river exercised over him at the time, leading him as it did to ignore what he clearly realized to be the wishes not only of his sponsoring organization but of the Russian government as well. His comments reveal the extent to which ideas about the river, and about Russia's overall position and destiny and the Far East, had begun to be transformed by the early 1840s. As if to underscore this point himself, Middendorf began by noting the continuity between his expedition and those maritime explorers in the Sea of Okhotsk who had preceded him. He described the problem of locating a suitable replacement for Okhotsk to support Russian activity in the North Pacific, and the common conviction that the mouth of the Amur would be an ideal location for such a port. In this regard, Middendorf presented his expedition as a direct continuation of those of Koz′min and Zavoiko described above. Indeed, he recognized that Koz′min himself would have done what he had done – that is, continue south to the Amur, and earn thereby the distinction of being the first Russian since the seventeenth century to navigate these waters – had he not been categorically forbidden to do so by the excessive timidity of Russian foreign policy, terrified as it was by the prospect of disturbing its neighbor to the south.[39]

As Middendorf developed the point, however, it became clear that his principal interest in the Amur focused on very different factors. The first of these related to the significance of the river to Siberia as a whole, a region which he saw as suffering from a sort of natural imprisonment. By virtue of its physical–geographical configuration it was sealed off from the outside world, locked in on two sides by icy or frozen seas and on all others by mountains, deserts, and endless snowy expanses. The fact that this region did not contain fertile soil in sufficient quantities to allow for a balanced self-sufficiency completed the gloomy picture of Siberia's lot, for under these circumstances it was condemned to stagnation and decline. Against this background, the Amur acquired great importance, for it was the only route which led out of the region's natural encasement and opened "the interior of Siberia to world trade."[40] In Middendorf's depiction, the river in effect became a sort of vital life artery, upon which hung the fate of the entire region:

> Because a habitable climate begins only along the southern border of Russian possessions in Northern Asia, and at the same time the Amur is the only route which Nature gave to a Siberia [otherwise] sealed off on all sides, I was possessed by an irrepressible

[38] Shul′man, "Puteshestvie," p. 83; Alekseev, *Russkie geograficheskie issledovaniia*, pp. 22–23.
[39] Middendorf, *Puteshvestie*, I, p. 113. [40] Middendorf, *Puteshvestie*, I, p. 188.

desire to visit this region, into which navigation – and together with it civilization – was sooner or later bound to penetrate.[41]

The image Middendorf suggested here of the Amur as the sole line of salvation to Eastern Siberia's geographical entombment was to become a popular one in the years that followed.

On a deeper level, Middendorf perceived a significance in the territories of southeastern Siberia in terms of the centuries of steady Russian–Slavic movement to the east. He understood this movement almost teleologically as something akin to America's Manifest Destiny, and suggested that from the initial penetration of the cossacks eastward across the Urals, it had had an ultimate, if not always conscious goal, namely to traverse and settle the Eurasian continent and secure Russia on the shores of the Pacific.[42] From this historical perspective, the advance into the Amur region in the nineteenth century could be envisioned as the natural culmination of a millenial process of movement and settlement by the Russian nation. This process had an obvious importance for the development of Russia itself, but beyond this represented an expression of a tendency that was characteristic for all of Western civilization, and one that was being mirrored at that very moment on the other side of the Pacific. Middendorf drew an explicit parallel between Russia and the United States:

The great but quiet migration of peoples of our time has now closed into a circle around the earth. From hospitable shores, Americans and Russians are looking at each other across the ocean as neighbors. The Slavs, who for three centuries were fated to press against the sun, to the east, now stand with their immeasurable realm in front of the United States . . ., which at the same time and in a like manner has irrepressibly pressed on with the sun, to the west.[43]

The fraternity of interests which Middendorf suggested here between the United States and Russia as sister countries across the Pacific was quite novel, and it was obviously at odds with the dreams from the 1820s that Russia might extend its own dominion beyond Alaska over the North American continent. It was our explorer's vision, however, which was to become dominant throughout the 1840s and 1850s, and more specifically was to influence evolving perceptions of the significance of the Amur region.

A result of the inevitable Russian penetration onto the Pacific, Middendorf continued, would be the civilizing and enlightenment of the Far East. In developing this idea, Middendorf was giving voice to the new awareness of Russia's special mission to bring cultural, social, and economic development to uncivilized Asia. In specific regard to Russian involvement on the Amur, these civ-

[41] Middendorf, *Puteshvestie*, I, pp. 137 (quote), 169.
[42] Middendorf, *Puteshvestie*, I, pp. 139–140. [43] Middendorf, *Puteshvestie*, I, p. 188.

ilizing activities assumed two different aspects. On the one hand, and most dramatically, he identified them as the penetration of human culture and industry into a wild and primeval region, and the taming of the brute forces of Nature by man. With the beginning of Russia's modern penetration onto the Amur, the obstacles that barbarism and savagery presented to progressive development had been definitively overcome. Writing some years after his expedition, at a time when the first moves toward Russian occupation had already been made, he noted with great satisfaction that the primitive deserts of the Amur region were being brought to life and revived out of what had seemed to be an "enchanted slumber lasting thousands of years." "Man has declared war . . . against Nature," he proclaimed ceremoniously, "he is searching the oceans for whales, bringing down the primeval forests with fire and iron, building houses, and establishing homesteads."[44] The second aspect of Russia's mission in the Far East was its role in the enlightenment of a stagnant Asian society outside of Russia's imperial boundaries. In the conclusion to the first volume of his report, he noted that as Russia carried on with its work in the Amur region, "the Middle Kingdom will complete its revival," and he asserted that future generations in Russia were destined to play a role in the "rebirth of an ossified and obsolescent Asia."[45] Throughout all of this, Middendorf betrayed the same enthrallment with activity, with construction and reformation, and with planting or fostering culture where it did not previously exist, that we have noted in Herzen's thinking and which indeed was characteristic of the entire period.

The information that Middendorf brought back from the Amur established conclusively that there were basic inaccuracies in the reigning view of Russia's Far Eastern border with China. To his "enormous surprise," as he not entirely convincingly insisted, the boundary markers which he had been able to locate ran along the southern, and not the northern slope of the Stanovoi mountains. This circumstance, in turn, indicated that the Chinese possessed legitimate jurisdictional rights over a far smaller, and the Russians consequently over a far larger, amount of territory than assumed up to that point in St. Petersburg. In any event, the entire issue now clearly stood in need of further investigation.[46] Beyond this, the importance of his expedition for focusing general attention on the Amur was immense. His return to St. Petersburg was triumphal in the extreme: he was hailed from all sides as a hero, and in banquet after banquet paeans were sung to his exploits and to Russia's imperial–geographical grandeur, of which they had delivered the freshest demonstration. In a

[44] Middendorf, *Puteshvestie*, I, pp. 27, 173. [45] Middendorf, *Puteshestvie*, I, p. 188.

[46] Middendorf calculated that the boundary indicated by the markers he located would increase Russian territorial holdings in the Far East by at least 50,000 sq. versts. Middendorf, *Puteshvestie*, I, pp. 159, 166–167; "Peschtschuroffs Aufnahme," pp. 474–477; Sukhova, "Sibirskaia ekspeditsiia," p. 146.

speech delivered at one of these occasions, von Baer himself gave colorful expression to the very thoughts on Asian society which we have just noted in Middendorf's text, using metaphors from the animal kingdom. The Russian empire he likened to an eagle, while the Middle Kingdom was by contrast a *Lindwurm* or dragon, whose eyes seemed to be "closed to the present and turned only to the past. A heartbeat is nowhere to be detected, as is the case with amphibians if they are hibernating or have been given opium." "For the warm-blooded Russian eagle," he concluded, "we wish a powerful youthful life, while for its neighbor the dragon: undisturbed slumber!" In fact, as Middendorf himself intimated, the prospect of leaving China undisturbed was not exactly what von Baer's Russian audience had in mind.[47] In any event, in the excitement of the moment even the fact that Middendorf had ignored his official instructions was forgotten, or at least forgiven. His reports about the Amur apparently attracted the attention of the tsar himself, who in a private interview inquired with interest about details of his findings in the Far East.[48] Yet the primary effect of Middendorf's expedition was not on the government, which continued to remain skeptical about projects for Russian activity on the Amur. Above all, it was for Middendorf's fellow scientists, and the Russian educated public in general, that his expedition was portentous, for it supplied the first major impetus toward the large-scale resurrection of the Amur question.[49]

"A blessed location will not remain empty!"

In 1847 Alexander Panteleimonovich Balasoglo, an archivist in the Ministry of Foreign Affairs and a member of the recently formed Russian Geographical Society, was approached at the Society by Nikolai Nikolaevich Murav′ev, the newly appointed governor-general of Eastern Siberia. Murav′ev presented Balasoglo with a request that must at first have seemed rather curious. He asked Balasoglo to compose a memorandum about Eastern Siberia and the Far East and discuss in it anything and everything relating to these regions which he, Balasoglo, deemed to be important. Balasoglo agreed, and after several weeks of concentrated effort was able to deliver a thick manuscript to Murav′ev, which the governor-general then took with him to Irkutsk.[50] This remarkable document is a noteworthy landmark

[47] Karl von Baer, "Feier zu Ehren des Herrn v. Middendorff [sic] (bei Gelegenheit seiner Rückkehr von der Reise . . .)," *St. Petersburger Zeitung*, 82 (1845), quoted in Wittram, "Das russische Imperium," p. 577. [48] Middendorf, *Puteshvestie*, I, p. 187.

[49] Vernadsky, "Expansion," p. 413. For confirmations of the importance of Middendorf's expedition by two leading actors in the subsequent annexation of the Amur, see Semenov, Review of A. Middendorf, p. 11; Veniukov, "Ob uspekhakh," p. 51. Also see Leonov, *Aleksandr Fedorovich Middendorf*, pp. 74–75; Bartol′d, *Istoriia*, pp. 421–422.

[50] Balasoglo, "Ispoved′," pp. 250–251.

in the development of attitudes toward Russia in the Far East during the 1840s.[51]

In fact, Murav′ev's request made a good deal of sense. The young governor-general, as will be seen presently, had no experience in Siberia and knew practically nothing about the region. He did know enough, however, to turn to the right person for some enlightenment. Balasoglo, the son of a petty naval officer from Kherson, had been deeply interested in Asia and the Pacific since early childhood. He followed his father into the navy, but after completing his course at the naval academy in Sevastopol in the 1830s his hopes for an assignment on a voyage to the Far East were disappointed and he was sent instead to St. Petersburg. In the capital he audited courses at the Faculty of Oriental Languages at St. Petersburg university, and after repeated unsuccessful attempts to enroll there as a student he finally took a job in 1841 in the archive of the Ministry of Foreign Affairs. Here he devoted himself to the history of Russia's diplomatic and commercial relations with the East. The access which this gave him to the accounts of diplomatic missions, commercial caravans, and scientific expeditions across all of Asia provided tantalizing material for his imagination, but unfortunately this was the closest he was to come to satisfying his obsessive urge to visit these regions of the globe.[52] Working in the richest collection of information about these subjects in the empire, he quickly gained an expertise of some renown, and was properly regarded as one of the most knowledgeable experts of the time on the Far East. He was personally acquainted with leading specialists on Asia, and no less a luminary than Karl von Baer himself sponsored Balasoglo for membership in the Geographical Society.[53]

Balasoglo was an enthusiastic partisan of the nationalist–oppositionist currents of the 1840s. He was a good friend of Mikhail Petrashevskii and an articulate member of the latter's clandestine circle, which in 1845 began to gather weekly to discuss Fourier's ideas on utopian socialism and to criticize Nicholas I's despotic regime.[54] In his affirmations of nationalist devotion to the fatherland he was emphatic in the extreme, as the following rather giddy pronouncement well indicates.

[51] Balosoglo's essay was published in 1875 in the journal of the Society of Russian History and Antiquities, with the title "Eastern Siberia: A report on a mission to the island Sakhalin by Captain Podushkin" ([Balasoglo], "Vostochnaia Sibir′.") It was one of two manuscripts found in the papers of Nikolai Speshnev, another member of the Petrashevskii circle who was among those exiled to Siberia for their activities in the circle. During his time in Irkutsk in the 1850s, Speshnev became an assistant of Murav′ev (see chapter 5), and presumably received the documents from him. B. P. Polevoi has established that this essay is not the work of Podushkin, but is rather Balasoglo's manuscript of 1847. Polevoi, "Opoznanie," pp. 157ff. Also see Val′skaia, "Petrashevtsy," p. 56; Tkhorzhevskii, *Iskatel′*, p. 174; Balasoglo, "Ispoved′," p. 251.

[52] Balasoglo, "Ispoved′," pp.242–243 (quote); Leikina, *Petrashevtsy*, p. 113. For a sketch of Balasoglo's exceptionally unhappy and frustrated life, see Tkhorzhevskii, *Iskatel′*, *passim*.

[53] Tkhorzhevskii, *Iskatel′*, p. 44; Polevoi, "Opoznanie," p. 155.

[54] Evans, *Petrasevskij Circle*, pp. 54, 76; Egorov, *Petrashevtsy*, pp. 57, 62–63, 90–92.

It is time for Russia to understand its future, its [particular] mission among humanity . . . Russia is another entire Europe unto itself, a Europe that is intermediary between Europe and Asia, between Africa and America: a marvelous, unknown, and new country . . . In Russia, and only in Russia are concentrated all the threads of universal history. Only the Slavic element in Russia represents that rational milieu which is predestined by Nature to absorb everything into itself and transform the thought of the world into the thought of Russia. The Slavic soul is the chosen vessel in which all peoples will combine into humanity.[55]

We have already had occasion to note that Balasoglo fully shared the conviction in Russia's legitimate dominion over Asia and in its messianic mission there that was part of this general mood, and in his own professional activities he sought to do what he could to further this cause. His attention was clearly focused on the Amur region, and at the moment Murav′ev approached him he was laying plans together with a young naval officer Gennadii Nevel′skoi for a naval expedition to the estuary of the Amur. Murav′ev's request offered him a rather different sort of opportunity to address and perhaps to convince a powerful official of his own view of the situation confronting Russia in the region.

Although the greater part of Balasoglo's lengthy essay was devoted to a history of the administration and economic development of Eastern Siberia, the underlying intention was to leave the governor-general with a clear sense of the political measures that would have to be taken to insure that the interests of both the region itself as well as Russia as a whole would be promoted. He left no doubt that the acquisition of the Amur river valley stood at the very top of this agenda. Possession of the Amur, he assured Murav′ev, was "just as crucial for all of Siberia as the salutary Nile was necessary for Egypt,"[56] a fact which he believed to have been amply demonstrated during the half-century of Russian occupation of the region in the seventeenth century. Balasoglo described this early period as one of creative and successful Russian activity along the river, and he argued that it had been interrupted only by the illicit efforts of crafty Manchurians and unscrupulous Jesuit intermediaries. In offering their services as interpreters, the latter effectively tricked the Russians into ceding the river to China during the negotiations at Nerchinsk in 1689. Had the Russian boyar-diplomats been able to read Latin, he insisted, they would never have agreed to surrender the Amur valley and the subsequent history of Siberia and even Russia itself would have evolved very differently.[57] Since this time, he observed, the Russian government had paid attention "neither to the extreme need on the part of [Eastern Siberia] for this river and its valley, nor to the immeasurable advantages that all of Russia would enjoy

[55] Balasoglo, "Proekt," II, pp. 41–43. Also see his poem "Vozvrashchenie."

[56] [Balasoglo], "Vostochnaia Sibir′," p. 181.

[57] [Balasoglo], "Vostochnaia Sibir′," p. 144. Balasoglo's intimations about the bad offices of the Jesuits have been questioned in John Stephan's recent study, which indicates among other things that the Russian negotiator Golovin did read Latin after all. Stephan, *Russian Far East*, p. 359n.

from possessing it,"[58] and he was outspoken in his hope that his influential reader would not follow this pattern.

Toward this end, Balasoglo attempted to convince Murav′ev of the need for the Amur by demonstrating just what these "immeasurable advantages" consisted of. He began in traditional terms, and spoke of the river's potential relevance to the chronic problems of access and supply that beset Russian settlements on the Okhotsk coast, Kamchatka, and North America. He discussed at great length the inadequacies of the port facilities at Okhotsk and the almost insurmountable difficulties associated with the overland route from the coast to Yakutsk, and argued as many others had before him that the single satisfactory answer to these problems was navigation on Amur river.[59] With some bitterness, he referred to the worthy but unrealized projects of past decades to expand Russian dominion across the Pacific to the temperate western shores of the North American continent. Just 10 years earlier, for example, the same Californian territories that were currently being fought over by the United States and Mexico "would on their own have voluntarily surrendered [and agreed to accept] full Russian dominion," and although he recognized that Russia could no longer have any realistic pretensions on these remote lands, he clearly felt that only a short-sighted and misguided governmental policy had prevented the country from from realizing its natural "geographical" potential as a truly trans-Pacific empire.[60]

Along with these familiar points, however, Balasoglo introduced two new considerations into his argument for the significance of the Amur. The first of these, already foreshadowed in Middendorf's comments, related to the role the river could play in the revival of Siberia as a whole. The abysmal condition not only of Siberia's economic life but of its official administration were notorious in European Russia, and Balasoglo noted ironically that the so-called "Siberian plague" referred not only to a medical affliction but to the corrupt and callous bureaucrats with whom the region was infested.[61] Extensive political reform and economic revitalization of the region east of the Urals were consequently imperative, and the acquisition of the Amur river would represent the first and most important step in this direction. To begin with, it would enable provisions to be supplied not only to remote maritime settlements but to vast expanses of continental territory in the Siberian north, where owing to the impracticality of agriculture settlement up to that point had been restricted to a few paltry nomadic tribes. While substantial food supplies from western regions of Russia could never be expected to penetrate overland in significant quantities to Eastern Siberia, they could be shipped from European

[58] [Balasoglo], "Vostochnaia Sibir′," p. 182. [59] [Balasoglo], "Vostochnaia Sibir′," pp. 173–178.

[60] [Balasoglo], "Vostochnaia Sibir′," p. 167. In his view, nothing less than the Rocky Mountains could have served as the "natural boundary" for Russian dominion over the westermost margin of the North American continent. "Vostochnaia Sibir′," p. 165.

[61] [Balasoglo], "Vostochnaia Sibir′," pp. 186–187.

Russian ports to the Pacific and then up the Amur, which would act as a sort of Siberian St. Lawrence to enable maritime penetration deep into the continental interior. Moreover, Russian control of the Amur valley would serve to link Eastern Siberia to the agricultural markets of China, which could in a similar way serve as an important source of food for less-favorably endowed regions to the north. And when the Amur valley was successfully colonized, Balasoglo prognosticated, it would be unnecessary to import any supplies at all from outside the region, for an "abundance of goods . . . will flow (cherpat′sia), as if from a brimming cup, from the luxuriant valley of the Amur river" across all of Siberia.[62]

With reliable sources of sustenance secured in this fashion, the large-scale settlement and economic development of Siberia's boundless expanses could be undertaken. As if anticipating what was certain to be raised as a major obstacle to the prospect, Balasoglo assured Murav′ev that the human resources for this venture would be readily available. Indeed, he suggested that the masses of many countries were waiting in anxious anticipation for just such a opportunity. In regard to China and Japan, for example, the opening of the Amur for free navigation would "instantly produce an unavoidable overflow of entire millions of hard-working, resourceful, and sober individuals from these two empires into all parts of Eastern Siberia and Russian America." Europe would also make its contribution to this movement. The first news about the opening of the Amur and access into Siberia would attract "crowds of Irish," the same unfortunate souls who were at that moment perishing from starvation in their own land. "Nor will Germany or the rest of Europe delay in sending their surplus countrymen who are in need of work and land in order to feed themselves."

> With the opening of the Amur and the first serious navigation upon it, the initial flow of information about the valley . . . will be seized upon by hungry journalists with a greedy interest that is unimaginable. [These reports], publicized in all of Europe's many journals, will call forth entire crowds of all possible sorts of people.

Finally, he affirmed that Russia west of the Urals, once it had been renovated and revived, would be an important source of settlers as well.

> Russia itself, after it has been aroused out of its intellectual stupor . . ., will supply the newly opened region with its surplus entrepreneurs, administrators, and capitalists (kapitalistov) some of whom do not know at present what to do with their hands and talents.[63]

An abundance of potential manpower would be an important precondition for promoting the rapid economic development of this unsettled region, and its utilization depended entirely upon the reacquisition of the Amur river.

[62] [Balasoglo], "Vostochnaia Sibir′," pp. 182–183.
[63] [Balasoglo], "Vostochnaia Sibir′," pp. 184–185.

The second new element that Balasoglo stressed was the importance of the Amur to Russia's international position in the Far East. He drew attention to this point in the opening passage of his essay.

Eastern Siberia, bordering on the Arctic on the north, the Pacific Ocean on the east, [and] the Chinese Empire on the south . . . represents a natural link connecting all of these regions. This fact defines its significance in the world and its importance [to Russia] . . . Across it and to some extent because of it, Russia enters into its multifarious relationships with China, Japan, innumerable Pacific islands, the whole American coast, and all of the Asiatic and European possessions *which are playing any sort of part at all in the general life of the globe*. The activity of any region is determined by its geographical position.[64]

With this, Balasoglo was calling Murav′ev's attention to the Pacific basin as an burgeoning center of global commerce and world civilization, a prospect that as we will see presently had begun to take recognizable shape only a few years earlier. He returned to this point in his conclusion. The "needs and expections" that Russia had for Siberia in the near future were "colossal," and would be most acute precisely in regard to Russia's international standing in the Pacific Far East. Indeed, "this future has already arrived, as England, France, North America, and all the developed nations of the world have laid a road for themselves into China, and are already close to achieving access into Japan." Russia would simply have no choice but to take its part in this activity and project itself fully onto the Pacific arena. The geographical means to accomplish this, of course, lay in Russian control of the Amur valley.

Having set forth this invigorating prospect, Balasoglo chose nonetheless to end his essay on an ominous note. The transitions that were taking place in the Far East had the effect, among other things, of intensifying the competition among the powers involved, in order to enhance their own relative strategic position. Above all, this competition took the form of struggle for territorial advantages. "The whole world was a witness to the struggle over the Marquesas Islands, Hawaii, and the Oregon territories, and everyone is now absorbed with the fate of California and even of Mexico itself."[65] Everyone seemed to recognize the importance of what was taking place on the Pacific – everyone, that is, except the Russian government. Russia had criminally neglected both its duties as well as its opportunities in the Far East, and without doubt would live to regret it, for rather than these contests subsiding, the future promised instead that they would become ever sharper and more intense. The threat that this raised for Russia's fate in the Far East was immediate and critical, and Balasoglo was blunt in sketching out the potentially disastrous consequences.

[64] [Balasoglo], "Vostochnaia Sibir′," p. 103 (emphasis in original).
[65] [Balasoglo], "Vostochnaia Sibir′," p. 87.

> No more than a year will pass, or perhaps only several months, before St. Petersburg will read in its newspaper – but only after the entire world has already learned – that the English or the French, with the voluntary agreement of China, have taken control of the mouth of the Amur river and received permission to sail up and down the river, to Nerchinsk.
>
> This news will be strange, but it is inevitable. If Russia does not wish to recognize its own treasures, then there is no question that someone else will recognize them and take appropriate measures! A blessed location will not remain empty![66]

Balasoglo's urgent message to the individual about to assume supreme control over affairs in the Russian Far East could not have been clearer. There was an imperative for swift and decisive action on the Amur, action which was part of Russia's broader need for national regeneration and self-assertion. The expansion of international activity on the Pacific, which was taking place as he wrote his essay, indicated that the Far East was to be an arena of a great future civilization, with which Russia must be connected in order for it to realize its own national destiny and not lose out to the other great powers. The annexation of the Amur would assure this, and at the same time would supply Russia with a firm link to Asia, where it would find an opportunity to assume its messianic duties.

Balasoglo was frustrated in his own attempts to travel to the Far East, for his involvement with the Petrashevskii circle led to his arrest in 1849, and he was not able to take part in the naval expedition that he and Nevel′skoi were planning.[67] Nevertheless, he at least had the satisfaction of expressing his views to a powerful and sympathetic listener. Murav′ev paid rapt attention to him, read and reread the essay, and stated that he was completely convinced of its accuracy on Russia's position and needs in the Far East. "It is pleasant for me to recall in my heart," Balasoglo later wrote, "that from the first page Murav′ev was so delighted that he even wanted to read it to the Emperor himself, . . . but never found a convenient moment to do so."[68] However that may be, there is no question that from the moment the governor-general assumed his new duties in Eastern Siberia he worked incessantly to translate Balasoglo's recommendations into reality, in which endeavor he was ultimately to enjoy remarkable success. Alexander Balasoglo deserves full credit for helping to plant the seed in the governor-general's mind.

"Something like central Africa"

Beginning around 1845, a loose collection of young men in St. Petersburg began to gather regularly on Friday afternoons in the apartment of one

66 [Balasoglo], "Vostochnaia Sibir′," pp. 187–188.

67 This expedition proved to be of great importance to the Russian advance into the Amur region. See below, pp. 127–129.

68 Balasoglo, "Ispoved′," p. 251. In fact, Balasoglo's contact with Murav′ev left him with considerable bitterness, for despite his efforts on the governor-general's behalf, the latter nevertheless apparently refused to help him advance his own position in St. Petersburg.

Mikhail Vasil′evich Butashevich-Petrashevskii, on the Bol′shaia Sadovaia Street, to drink tea and discuss political ideas. This diverse assemblage, which has come to be known as the Petrashevskii circle, was one of the most interesting manifestations of the intellectual ferment taking place during the 1840s. Among its participants figured individuals who subsequently were to become major public figures, most notably Fedor Dostoevskii and the Pan-Slav ideologue Nikolai Danilevskii. Although the political affinities of the Petrashevskii circle, like those of most dissident groupings in Russia in this early period, were vague and cannot be categorized with any precision, in general its members were attracted by Fourier's ideas about utopian socialism that were dribbling at that time into Russia. Beyond this, however, their thinking was highly eclectic. The geographer Peter Semenov, who attended a few meetings of the group in 1845, emphasized this point in a characterization of its organizer that he wrote some time later. Petrashevskii was "extremely eccentric, if not to say crazy (sumasbrodnyi) . . . an extreme liberal, a radical, an atheist, a republican, and a socialist" – all apparently at the same time![69] What bonded the *petrashevtsy* together were the convictions we have already seen in the case of Balasoglo: a intense sense of nationalism and devotion to the Russian homeland on the one hand, and an uncompromising opposition to the despotic reign of Nicholas I on the other.[70]

In addition to the group's regular members, who gathered faithfully over the course of several years, it was not unusual for more peripheral participants to take part on occasion as well. The appearance in 1848 of one such unfamiliar face, however, caused quite a stir. Rafail Aleksandrovich Chernosvitov, a native of Yaroslavl′, had begun his career as a military officer. In 1830, he was stationed in the western provinces of the empire and participated in the suppression of the Polish revolt, during which he was taken prisoner by the rebels and lost a leg. He was subsequently transferred to the Urals, where in 1841–1842 he once again took part in the suppression of a revolt, this time on the part of the state serfs who worked in the mines around Perm′. In 1842 Chernosvitov left the army, traveled to Eastern Siberia to seek his fortune, and within a short time had become a successful entrepreneur in the gold-mining industry.[71] Despite his impressive resumé of activities in suppressing popular revolts, however, as well as his not inconsiderable success in his business ventures, Chernosvitov was nevertheless attracted to radical ideas, and even enjoyed something of a reputation as a revolutionary.[72] On a business trip to St. Petersburg in October 1848, he attended a meeting of the Petrashevskii circle.

It was not only the fact that this enterprising ex-officer, who walked using an artificial leg of his own design and construction, was somewhat older than the other members of the group that served to focus their interest on him. The

[69] Semenov-Tian-Shanskii, *Memuary*, I, p. 195.
[70] On the political views of the Petrashevtsy, see Leikina, *Petrashesvtsy*; Evans, *Petrasevskij Circle*; Egorov, *Petrashesvtsy*; Seddon, *Petrashesvtsy*.
[71] Chulkov, "Rafail Chernosvitov," p. 82. [72] Semevskii, *Iz istorii*, p. 56.

members of the Petrashevskii circle were for the most part not particularly worldly, and the weighty considerations about political and social reform that they passed back and forth in endless discussions were little more than rarified debates bearing little recognizable relation to the real world around them. Chernosvitov by contrast – a charismatic gold-digger from the rugged Siberian frontier who was rumored to wield great influence east of the Urals, to enjoy not only the confidences of the exiled Decembrists but the personal favor of governor-general Murav′ev himself, and even to be in the capital as an emissary of some sort of secret society – was obviously from a very different mold, and he appeared before them as a credible fount of practical knowledge and worldly experience.[73] By virtue of this, he automatically gained their attention and respect and, with a little effort, was able to command their awe. He was after all one of the very few *petrashevtsy* to have actually witnessed an insurrection, and indeed not once but on two occasions.[74] Moreover, he had traveled throughout the Russian empire and lived in its various corners, and could thus speak with authority concerning regions and peoples about which the others could only fantasize. It is in this latter regard that he is most interesting for our subject, for through Chernosvitov's exhortations the attention of the Petrashevskii circle was directed to the Russian East.

In the course of the three or so months that he attended the Friday gatherings in Petrashevskii's apartment, Chernosvitov joined with another member Nikolai Speshnev to form circle's most radical "wing."[75] Together they looked for indications of potential insurrection throughout the Russian empire. Chernosvitov responded skeptically to Speshnev's enthusiastic descriptions of peasant unrest in the Ukraine, however, and told him instead to pay attention to Siberia. At the iron-ore works in the Perm′ guberniia, he claimed, there were 400,000 potentially insurgent peasants, all deeply discontented and all armed. The situation was ripe for revolt, and awaited only a spark. Chernosvitov laid out an elaborate scenario for the coming Russian revolution, which accorded Siberia the honor of being the point where the conflagration would erupt. First there would be an uprising in Eastern Siberia, he maintained, which would compel the government to send troops from the Russian heartland. No sooner would they reach the Urals, however, than in true Pugachevean fashion the entire region would arise in revolt. The successful insurgents could then stream across the Urals from Siberia into European Russia where, aided by timely insurrections in Moscow and St. Petersburg, they would ultimately be successful in toppling the government of Nicholas I.[76]

Chernosvitov's depictions of Siberia, however, were not limited to this sort

[73] Chulkov, "Rafail Chernosvitov," p. 83. Chernosvitov himself claimed to wield "unlimited authority" over Murav′ev. Seddon, *Petrashesvtsy*, pp. 197, 212; Shchegolev, *Dekabristy*, III, pp. 235, 247; Desnitskii, *Delo*, I, 462.

[74] Evans, *Petrasevskij Circle*, pp. 46, 91.

[75] Seddon, *Petrashesvtsy*, p. 213.

[76] Semevskii, *Iz istorii*, pp. 59–61; Evans, *Petrasevskij Circle*, pp. 83, 92; Shchegolev, *Dekabristy*, III, pp. 27–30.

of whispered conspiratorial fantasizing with Speshnev about armed uprisings. He spoke about the region to the group at large as well, and at the inquisition held after the arrest of the circle he gave some idea of the tantalizing image of the lands beyond the Urals that he had depicted for them.

> Speaking about Siberia in general, I often referred to it as America, California, El Dorado, a Russian Mexico, and so on. By virtue of its geographical position, Eastern Siberia really is a country separate from Russia. Becoming carried away in my speculations about Siberia's future I sometimes referred to it as a great empire, but not at all separate. However, maybe I did use this expression as a possibility, for when Siberia becomes a great empire, who knows what will happen in Europe?[77]

It is not difficult to appreciate the appeal that descriptions of this sort were certain to have exercised on the young men listening to him, disaffected as they were from all aspects of the contemporary status quo in Russia and searching for alternative solutions and directions. There is some evidence that they had already taken note of Siberia in this regard. Petrashevskii in particular appears to have shared Herzen's conviction about Siberia's rejuvenating potential for Russia, having noted in a youthful diary in 1840 that "Siberia will replace the real Russia. A pure Russian nationality will develop there, and it will be under a republican administration."[78] Still, these were no more than the musing of an untraveled schoolboy. What Chernosvitov offered was something entirely different, namely a realistic and credible account from someone who could speak with the ultimate authority of experience. Siberia, he told them, was a country quite foreign to the rest of Russia: a massive and exotic empire unto itself, destined for its own great and glorious future independent of Europe and the Russia of Nicholas I. He was well aware of the complete ignorance of his audience about the realities of the Russian East – after his arrest he insisted to his interrogators that in the imaginations of the circle's members, Eastern Siberia was "something like central Africa" – and he used their lack of knowledge to his own good advantage.[79] It is indicative that he should have described Siberia as an America, California, Mexico, or El Dorado, apparently unconcerned by the fact that the final image was not a real place at all, but an imaginary paradise.

Finally, Chernosvitov spoke specifically about the Far East and the Amur valley. He told the members of the group that the eventual acquisition of the river by the Russians was inevitable, and like Balasoglo he was convinced that this would connect Russia with the dynamic arena of world activity taking shape on the Pacific.

> If the Amur were Russian, which no doubt one day it will be, then steamships would sail up it bringing . . . all the goods of the south and the east; California, India, and

[77] Quoted in Desnitskii, *Delo*, I, p. 462; Semevskii, *Iz istorii*, p. 59n; Shchegolev, *Dekabristy*, III, pp. 26, 235, 244.

[78] Quoted in Semevskii, *Iz istorii*, p. 15; Semevskii, "M. V. Butashevich-Petrashevskii," no. 3, pp. 43–44.

[79] Shchegolev, *Dekabristy*, III, p. 251; Desnitskii, *Delo*, I, p. 481.

Canton would all be close at hand. [T]his immeasurable region, which lies empty today, would revive with a miraculous life. Beyond Lake Baikal, Russia is separated [from the Pacific] by an enormous stretch of land, a region which by its geographical position is agricultural, and through which Russia should carry on trade with the [Pacific] East, America, and even India.[80]

Chernosvitov was certainly held in awe by the Petrashevskii circle, which listened intently to his descriptions and exhortations. He offered his listeners a vision of progress and development in the Russian East that they were not only willing but indeed anxious to believe, and with his self-assured exuberance he was able to fire their imaginations about what the future might hold for these remote and ill-understood regions. Although his entrance into the circle had been fortuitous, and it was not many months before he lost interest and returned home to his gold-mining activities,[81] his ideas about Siberia and the Amur fit perfectly into the general framework of extreme nationalist–oppositionist sentiment which the circle shared. The enthusiastic reception by his fellow *petrashevtsy* provides an indication of how the specific issue of Russia in the Far East could be absorbed into a larger framework of political conviction and in this way judged to be critically important, even by those who knew nothing of the region and had no practical sense of the problems actually involved there. The *petrashevtsy* continued to meet for some time after Chernosvitov departed, until the heightened vigilance of the police in the wake of the European revolutions of 1848 led to the arrest of its members. Chernosvitov, arrested the following year while traveling in the Urals, was interrogated and released "under strict observation."[82] Other members of the group, however, were not so fortunate. After a lengthy inquisition, several – including Speshnev and Petrashevskii himself – were sent in exile to Eastern Siberia. Here they were finally able to experience Siberia first-hand, and as we will see they were both to take active, if very different, roles in the events surrounding the annexation of the Amur river.

"Science . . . is *samopoznanie*"

The Russian Geographical Society was founded in 1845. Almost immediately thereafter, discussions were initiated concerning the possibility of sending a major geographical expedition to the Far East. Planning continued over the following decade and enthusiasm mounted steadily, such that by the time an expedition was finally dispatched in 1855, it was to represent the single biggest

[80] Desnitskii, *Delo*, I, p. 457.
[81] Evans, *Petrasevskij Circle*, p. 46. In the course of the following decade, Chernosvitov became absorbed with technological innovations, writing articles extolling the virtues of a new sand-carrying machine for gold-mining and of "airborne locomotives," by which he apparently meant dirigibles. Chernosvitov, "O dorogovizne," pp. 3–5; *idem*, "Korrespondentsiia," pp. 7–9; Egorov, *Petrashesvtsy*, p. 218n.
[82] Leikina, *Petrashesvtsy*, p. 128; Chulkov, "Rafail Chernosvitov," pp. 87–90.

scientific endeavor of the Society's early history. The actual exploratory activities of the expedition, and in particular the views of certain of its participants, will have a definite interest at a later point in this study. For the moment, it is the peculiar evolution of the plans for the expedition that is significant, for through it we can trace the same pattern of conceptual reorientation in regard to the Amur question that we have already noted in the thinking of Middendorf and Balasoglo. The Great Siberian Expedition offers a further indication of the extent to which this issue was acquiring a significance within the framework of the nationalist perspective of the day, and at the same time represents the most conclusive demonstration of the perceptual disengagement of the Amur river valley from the problems of Russia's fur colonies in the North Pacific.

In order to understand the background to the expedition itself,[83] it is necessary to appreciate the character of the organization that conceived of and sponsored it. The establishment of the Russian Geographical Society was the product of a variety of concerns, civic as well as scientific, but one of the major impulses that inspired the entire enterprise was the surge of nationalist sentiment under Nicholas I. Within its organizational framework, a significant section of the membership hoped that their urge to engage in constructive activism could find a useful outlet and be directed toward the goal they held most dearly, namely the reform and regeneration of the fatherland. Their intention was to create a center for research as independent as possible from the existing government and academic bureaucracies, which would be devoted not to the cosmopolitan goal of the enrichment of Western science in general, but rather exclusively to the study of the Russian homeland for the good of Russia alone. The fanaticism with which they came to insist upon this latter objective was striking. A year after its founding, for example, in a debate over revisions in the Society's basic charter regarding the scientific focus for its activities, Vasilii Grigor′ev exclaimed in disgust that "had it been said [at the outset] that the main goal of its establishment was the fostering of geography as a [universal] science, for the benefit of mankind as a whole, and Western Europe in particular," it would never have occured to him to "seek the honor of becoming its member." "I have never understood," he continued,

> the use for us Russians of worrying about . . . the enlightenment of western Europe: why does it concern us whether or not they know us? . . . It is time for us to stop judging ourselves on the basis of other peoples' evaluations, and to find in ourselves the standards for our own accomplishments and shortcomings.[84]

These sentiments were widely shared, and within a few years they were to result in a major organizational upheaval. Germans of Baltic origin made up an important element of the intellectual and administrative elite of Nicolas I's

[83] For a fuller discussion, see Bassin, "Russian Geographical Society," *passim*.
[84] Quoted in Veselovskii, *Vasilii Vasil′evich Grigor′ev*, pp. 96, 245–247.

Russia,[85] and they were particularly important in the development of the sciences. Outstanding scholars such as Baer, Georg Wilhelm von Struve, G.P. Helmerson, Peter Köppen, Ferdinand Lütke, and others had been a major force behind the founding of the Russian Geographical Society, and they played a highly visible role during its early days, occupying almost all of the positions of leadership within it. This circumstance was the source of considerable irritation for the nationalists in the Society, who felt it entirely inappropriate that their Russian organization should be so dominated by individuals who were not themselves ethnically Russian. As one participant later explained it, they longed "to follow a unique and independent path, a precious nationalist path," and regarded the founders of the Society as a "collection of German teachers, who kept the Society in its outgrown, and in spirit foreign swaddling clothes."[86] There was broad support for this view, with the result that in the Society's first general elections in 1850, virtually the entire leadership was replaced by individuals deemed more satisfactory. The scientific qualifications of the latter might not have matched those of the old guard, but their names were in all events impeccably Russian.[87] With this, the energies of the Russian Geographical Society could be concentrated exclusively on the priorities of the nationalist–oppositionist movement, and it quickly became the main gathering point in Russia's capital for nationalist–oppositionist sentiment, attracting critics of Nicholas I's regime from across the political spectrum.[88] Throughout the 1850s the Society provided the main organizational forum for discussion and the early planning of the far-reaching reforms social and political that were to be promulgated in the following decade by Alexander II.[89]

The opposition to the old guard was stimulated by more than just the latter's non-Russian ethnic origin, however, for the nationalists believed that a fundamental principle in the practice of science itself was at stake. This principle was expressed through a distinction they drew between two contrasting types of research, one termed "cosmopolitan" and the other "national." Cosmopolitanism referred to an understanding of science as detached and "objective" inquiry, preoccupied with nothing more than a scholastic concern in determining the true nature of the universe. Nationality in science, on the other hand, meant that the scientific endeavor had blended successfully into the vital life of the nation, ceasing in the process to be a dead symbol and becoming rather a living and breathing principle. Cosmopolitan science of course represented the legacy of the Enlightenment, and although no one denied that it had served an important function in the eighteenth century, its

[85] Presniakov, *Emperor Nicholas I*, pp. 4–6; Laserson, *American Impact*, p. 8; Riasanovsky, *Nicholas I*, pp. 144–145.

[86] Semenov-Tian-Shanskii, *Memuary*, I, pp. 2–4, 191; Veselovskii, *Vasilii Vasil'evich Grigor'ev*, pp. 94–96.

[87] Lincoln, *In the Vanguard*, pp. 96–99.

[88] Val'skaia, "Petrashevtsy," p. 65.

[89] Lincoln, *In the Vanguard*, pp. 91, 98–101.

had nonetheless outlived its relevance. Indeed, it was now distinctly at odds with the progressive spirit of the day, which sought to infuse scientific inquiry precisely with the dynamism of national awareness and give it thereby a new direction and meaning. All of these points were outlined by Peter Semenov, who became secretary of the Physical–Geographical Section of the Society in 1850. In his introduction to the translation prepared by the Society of *Erdkunde von Asien* by the German geographer Carl Ritter, he wrote that science in the present day

> is no longer a foggy distraction of scholastic minds; it is rather *samopoznanie*, the recognition of the objects and forces of Nature and the ability to subject them to our own power, to use them for our needs and demands . . . The striving of every scholar, if he does not wish to remain a cold cosmopolite but rather wants to live a single life with his countrymen, has to be . . . the desire to introduce the treasures [of human knowledge] into the life of the nation.[90]

With this, Semenov sought to infuse the pursuit of scientific knowledge, and geographical knowledge in particular, with the invigorating spirit of nationalist activism. The extent to which this perspective was endorsed at all levels within the reorganized Society was apparent in its annual report for 1852. In it, the Society declared its resolution to "work always and in every way *for the benefit of Russia*" and to dedicate all of its efforts "primarily to the study of the *Russian* land and the *Russian* people in all of their varied relationships, striving to place even those tasks which do not relate immediately to Russia into the closest correspondence with the practical interests and needs of the fatherland." Members were exhorted not to waste their energies studying "irrelevant questions of science" but rather to direct their attention to problems that were "vital and *practical*" and which could help satisfy the "current needs of Russian society."[91] This meant that members, by directing their efforts primarily toward the eminently practical concerns of rebuilding Russia, could now envisage making their own full contribution to the national cause in their professional capacity as geographers. Indeed, the founding of a new society afforded them a truly splendid array of opportunities to realize their potential and help the country deepen and perfect its *samopoznanie*. These opportunities came in various forms. The ethnographic section of the Society, for example, elaborated an extensive program for the study of the

[90] Ritter, *Zemlevedenie*, I, p. 3. The pointedness with which Semenov made these comments is partly to be explained by the fact that Ritter's work itself was about as quintessential an example of "cosmopolitan" science as one could imagine. Semenov defended the expenditure of the Russian Geographical Society's energies and resources on this monument of European science with the argument that it could be "nationalized" by the Russians in that they translated only those volumes dealing with regions geographically contiguous to the Russian empire and thus of immediate political relevance. Moreover, Ritter's information was corrected and expanded with their own geographical research. On this ambitous translation project, which lasted several decades, see Sukhova, *Karl Ritter*, pp. 98–105; Vucinich, *Science*, p. 304.

[91] "Otchet . . . za 1851," pp. 16–18 (emphasis in original).

Russian people and their culture, while the statistical section set about collecting and assembling the data on the economic, demographic, and social conditions in Russia that would be necessary for the impending social reforms.[92]

A rather less obvious, but no less significant contribution to the nationalist project that members of the Society could make was through the dissemination of geographical information concerning remote, little-known regions of the Russian realm or immediately contiguous areas. One way to accomplish this was through the translation of useful reference sources such as Ritter's work, the publication of which in fact represented one of the major projects of the Society in its early years. Far more interesting and satisfying, however, was the exploratory activity that the Society itself carried out. It is significant to note in this regard that it was the great excitement marking the return of Middendorf from Siberia that provided the immediate stimulus for the organization of the Russian Geographical Society in the first place, and the exploration of the Russian East to which he had so brilliantly contributed remained one of its most enduring preoccupations.[93] The annual report from 1851 just cited was outspoken about the patriotic dimension of this enterprise.

> Siberia forms one of the most important parts of the Asian continent, the study of which is primarily the calling of *Russian* science, and by virtue of our close ties with Asia it is a matter of great interest and importance for us Russians.[94]

It went without saying that such investigation could not be directed at solving the sterile and abstract questions of "cold and cosmopolite" scientists, but would rather have to concern itself with practical problems of direct relevance to the reformation of Russia. These included the search for exploitable natural resources, the evaluation of possibilities for agricultural development – and, importantly, the detailed topographical survey and mapping of foreign regions that might be of potential political significance to Russia in the future. In the 1850s, the Society had sponsored a number of expeditionary forays into Russian Asia, the most important of which were Semenov's own explorations in the Tian-Shan mountains in what was soon to become Russian Turkestan,[95] and – on a far grander scale – the Great Siberian Expedition to the Russian Far East. In all cases, the explorers were intensely concerned that their activ-

[92] Lincoln, *Petr Petrovich Semenov-Tian-Shanskii*, pp. 16–17.

[93] Indeed, attempts have been made to present Middendorf's expedition itself as an illustration of the two types of science Semenov referred to. The first part, to Taimyr and Yakutsk, reflected the direct inspiration of Baer, and was concerned with questions of a general scientific nature – in other words, "cold cosmopolite" science. With his penetration to the Sea of Okhotsk and south to the Amur, however, the expedition became animated by practical political concerns, and it thus was an example of "nationality in science." For this intriguing (if dubious) analysis, see Savitskii, "Iz proshlogo."

[94] "Otchet . . . za 1851," pp. 18–19 (emphasis in original).

[95] An authoritative study of Semenov notes his resolution to devote himself as he put it "completely to scientific work and . . . to search for some sort of civic activity connected with science," but relates his exploration to the Tian-Shan only to this first part of this committment. Lincoln, *Petr Petrovich Semenov-Tian-Shanskii*, pp. 54–55 (quote), 16–17. In fact, the

ities should have a civic and national as well as a purely scientific significance.[96]

The evolution of the Great Siberian Expedition provides an excellent illustration of the principled distinction between different types of science described above. In 1846 Admiral Lütke, vice-president of the Society, made the first suggestions for sending an expedition to the Russian Far East. He conceived of this expedition as part of the legacy of Russian maritime exploration in the North Pacific, and focused his attention accordingly on the Bering Sea, the coasts of Kamchatka, and the waters surrounding the Aleutian and Kurile island chains.[97] The goal of the expedition, moreover, was directed entirely toward problems of what Semenov would have called "cosmopolite" science. Lütke was struck by the fact that, although these bodies of water lay well to the south of the Arctic, they nevertheless displayed close similarities to polar seas and correspondingly exerted an Arctic influence on the weather over all of Russia's eastern coast.[98] There was immediate interest in this project, but because of the considerable expense involved it remained in the planning stages for several years. By 1851 instructions were finally prepared and printed for this "Kamchatka–America" expedition.[99]

The following two years saw fundamental changes in the thinking about this expedition. At first it was decided to limit its geographical scope and concentrate exclusively on the peninsula of Kamchatka, but this became increasingly problematic as the approach of the Crimean War threatened to include Russia's main naval port of Petropavlovsk on Kamchatka in the theater of military action.[100] At the same time, other factors as well were working to modify the original plans. At this very moment, the governor-general of Eastern Siberia, Murav′ev, was agitating in unsympathetic government circles for a reexamination of the Amur question. In order to focus attention on the region and bolster his cause, he was anxious to attract geographical exploration to Eastern Siberia and the Far East. Such activity would fill a significant need for cartographic, physiographic, and geodesic surveys of this unknown territory, and at the same time the attention would help promote his cause in the capital.[101] Toward this political end the governor-general, who was a

geographer understood all of the work of the Geographical Society in Asia, very much including his own expedition, to be an important part of its civic activity. Bassin, "Russian Geographical Society," pp. 243–244, 253–254n. For an excellent example of how the contrast between these two perspectives was expressed in regard to the study of ethnography, compare the "Old Guard" (Ber, "Ob etnograficheskikh issledovaniiakh") with a nationalist (Nadezhdin, "Ob etnograficheskom izuchenii").

[96] Russian geographers were of course not alone in the desire that their scientific–professional activities should serve the greater national interest in this way. On the contribution of geographers and geographical societies in other countries to the project of political expansion in the nineteenth century, see Brockway, *Science*; Lejeune, *Les sociétés*; McKay, "Colonialism."

[97] Alekseev, *Russkie geograficheskie issledovaniia*, pp. 19, 54; Aleskeev, "Amurskaia Ekspeditsiia," pp. 95–96. [98] Semenov-Tian-Shanskii, *Istoriia*, I, p. 26. [99] *Svod instruktsii, passim.*

[100] Radde, *Bericht*, p. x; Semenov-Tian-Shanskii, *Memuary*, I, p. 193.

[101] "Otchet . . . za 1852," p. 37; Mancall, "Major-General Ignatiev's Mission," p. 57.

member of the Russian Geographical Society in St. Petersburg, had sponsored the establishment of its first filial, the "Siberian Branch," in Irkutsk in 1851.[102] Through his adjutant M.S. Korsakov, who was an old schoolmate and good friend of Semenov, Murav′ev was able directly to influence the deliberations in the Russian Geographical Society about the fate of their far-eastern expedition. He insisted that a geographical expedition could "conveniently be sent into the Amur basin," and promised every assistance of the Russian Geographical Society in Irkutsk.[103]

The Russian Geographical Society in St. Petersburg was entirely receptive to Murav′ev's suggestions. The vision of the Pacific as a future arena of world civilization was becoming an inspiring one for the membership,[104] and the need to reacquire the Amur in order to enable Russia to participate in this activity seemed uncontestable. For the Russian Geographical Society, Murav′ev in Siberia represented a dynamic nationalist force in the struggle against Nicholas I's Official Nationality, actively seeking to redress the injustices of the present and to lead Russia on to its new and positive destiny. In a thoroughly appealing contrast to the cosmopolitanism of Lütke's proposed expedition, with its exclusive focus on scientific questions of a universal nature, the prospect which Murav′ev offered of exploring a region so politically vital to Russia's future prosperity corresponded ideally to their vision of nationality in science, and specifically in geography.

By the middle of 1853, the original idea of a maritime expedition to the waters of the North Pacific had been entirely discarded. Instead, the expedition was to be a large-scale exploration of the continental interior of southeastern Siberia. The incipient occupation of the Amur valley by Murav′ev provided the Russian Geographical Society with a "seductive opportunity," as Semenov put it, by opening a new geographical arena for the activities of the expeditionary group which possessed a "national" significance of a very unique sort.[105] "At the time when the instructions for the . . . expedition were completed, it was impossible to foresee with any certainty the degree to which [changes in] the political situation would allow the expedition to broaden its work to the east."[106] The practical concerns that motivated this shift of venue were underscored in the expedition's official report. The intention was now to study "exclusively that southern part of this vast region which, by virtue of its more favorable climatic conditions and political situation, could hope for a

[102] Semenov-Tian-Shanskii, *Istoriia*, I, pp. 65–66; *Ocherk dvadtsatipiatiletnei deiatel′nosti*, p. 4. In the eyes of one founding member, the establishment of a branch in Eastern Siberia, which could greatly facilitate the geographical study of Russian Asia and contiguous regions was proof of the fact that the Society was "not only a geographical, but precisely a *Russian* society, and that it was not merely a scholarly society but rather a scholarly–patriotic society." Struve, *Vospominanii*, p. 115 (emphasis in original).

[103] Semenov-Tian-Shanskii, *Memuary*, I, pp. 194 (quote), 164.

[104] E.g. Semenov, "Opisanie Novoi Kalifornii," II, pp. 81–82; III, p. 56.

[105] Semenov-Tian-Shanskii, *Istoriia*, I, p. 81.

[106] *Trudy. . . . Matematicheskii otdel*, p. 2.

more rapid development of its industry and commerce" – more rapid, that is to say, than Kamchatka or Russian settlements in North America. The problems it would investigate were correspondingly reformulated along practical lines. The territories indicated were to be surveyed topographically, with the determination of astronomical points, on the basis of which the first comprehensive map of the region was to be prepared. Beyond this, the expedition had a section devoted to natural history, involving a descriptive study of the flora and fauna of the area. Finally, the geological conditions and soils of the region were to be examined.[107] In the form of the *Bol'shaia Sibirskaia Ekspeditsiia* or Great Siberian Expedition, which set out in 1855 and carried on with its work over the following eight years, Russian geographers took the best advantage possible of the opportunity to render their civic duty to the fatherland and make a unique professional contribution to the process of national rejuvenation.

[107] *Trudy. . . . Matematicheskii otdel*, p. 1; Semenov-Tian-Shanskii, *Istoriia*, I, pp. 73–76, 79. The only remaining trace of the spirit of Lütke's original plans can be seen the work of the topographer Karl von Ditmar, who from 1851 to 1855 carried out geological surveys on Kamchatka. Ditmar's remarkable account of his observations, published only a half-century after the fact and relatively unknown still today, has been compared with justice to Krasheninnikov's classic *Opisanie Zemli Kamchatskoi*. Von Ditmar, *Reisen*. Also see Semenov-Tian-Shanskii, *Istoriia*, I, p. 80; Alekseev, *Russkie geograficheskie issledovaniia*, pp. 46–48.

4

The push to the Pacific

"A useless river"

In the late summer of 1842, a treaty was concluded by the British and the Chinese in the city of Nanking. This agreement brought to an end the hostilities of the so-called Opium Wars, and at the same time initiated a new era in East Asia's relations with the Western world. In their negotiations at Nanking the British had demanded, and were granted, the full cession of the port of Hong Kong and the opening of a further four ports for their commercial activities, to be supported by consular representation.[1] The Russians followed these events carefully, for they understood that the consequences of the British victory would ultimately affect their own position and interests in China. Exactly how these interests would be affected, however, was far from clear. Seen from one standpoint, the course of developments in the Far East offered definite cause for encouragement. In striking contrast to the image of confident power with which the Chinese armed forces had impressed and overwhelmed the Russians in the seventeenth century and kept them at bay ever since, the relatively easy British victory now exposed the Middle Kingdom for the "paper dragon" or helpless giant it had become, and observers in Moscow and St. Petersburg began to appreciate that the strategic balance of forces on the Russia's southeastern frontier had shifted decisively. Their confidence in their military parity with the West, reconfirmed most recently by the victory over Napoleon and Alexander I's triumphal entry into Paris in 1814, had yet to be been shattered by the Crimean débâcle, and thus they were inclined to see China's defenselessness against one of the European powers as a defenselessness against them as well. For the first time in the history of relations between the two countries, it appeared not at all unrealistic for the Russia to entertain notions about its own domination – cultural as well as military and political – over China. Given the spread of messianic sentiment at the time, this new awareness was portentous indeed.

[1] Fairbank, *Trade*, pp. 57–132; Hudson, "Far East," p. 692.

Viewed from a different perspective, however, the new situation in the Far East stirred new concerns and fears which acted to offset and even undermine the positive aspects just described. China may have been exposed in all its weakness, but it seemed obvious that the Western European powers, chief among them Great Britain, were going to be the main benefactors. In the first place, there was apprehension that Britain's success was just the beginning, and would quickly lead to a scramble on the part of other West European countries to secure their own access in order to establish commercial dominance over as much of the "China market" as possible. In the event, they had not long to wait, for agreements extending similar commercial concessions were signed by China with the United States and France in 1844, Belgium and Denmark in 1845, and Sweden and Norway in 1847.[2] As a result the Russians, whose still largely feudal economy was no match at all for the rapidly industrializing West, felt the need to scramble not necessarily to extend their influence but merely to protect their own limited interests there from the inevitable Western attempts at encroachment. Beyond the potential erosion of their position in China itself, moreover, the Russians were troubled by the disturbing thought that their own territories in the Far East, especially the Okhotsk coast and Kamchatka, would be exposed to the expansionist designs of the Western powers. Up to this point, the safety and security of these regions had seemed insured by their relative remoteness and isolation, both of which would be significantly reduced by the expansion of European political and commercial activity in the Pacific basin. In view of these considerations, the government was compelled to take careful note of developments in the Far East and to give some thought to redefining a policy for its own activities there.[3]

The most immediate and tangible consequence for the Russians of the European incursion into China was its negative impact on the trade at Kiakhta. The relative advantage of oceanic transport from ports in south China, which could supply Europe with imports from China more quickly and cheaply than the long overland haul from Kiakhta, had already been apparent in the 1820s. Russian customers west of the Urals were occasionally able to obtain Chinese goods more cheaply from West European suppliers than from their own Siberian merchants, putting an early end to Russian hopes for developing an overland transit trade of Chinese tea across Siberia and Russia for reexport into Europe.[4] The effects of the Treaty of Nanking in this regard were particularly dramatic, however, as the British took hasty advantage of their newly won privileges and poured their goods onto the Chinese market. The following year, 1843, was catastrophic for Russian trade: the total volume of fabric sold by the Russians to the Chinese at Kiakhta fell by nearly 50

[2] Clubb, *China*, p. 72; Cameron, *China*, pp. 173–174

[3] Lin, "Amur Frontier," pp. 5–6; Clubb, *China*, p. 72.

[4] Sladkovskii, *History*, pp. 64, 68; Sladkovskii, *Istoriia*, p. 199.

percent, and Russian goods in general sold in China for even less then they obtained in Moscow.[5] To be sure, this was a low point, and the Kiakhta trade subsequently regained some of its vigor, but the ominous implications for the future were plain enough. By the late 1840s, for example, Chinese tea – Russia's most important import – was selling in London for one-fifteenth of the price of tea brought to Moscow via Kiakhta.[6] In blunt terms, the issue had become the very survival of Russia's China trade and the revenues it generated.

The decline of the Kiakhta trade in 1843 was sufficiently precipitous to attract attention at the highest governmental levels in St. Petersburg, and a special committee was convened to consider means of rectifiying the situation and protecting Russian commercial interests.[7] A member of this committee, Admiral Efimii Putiatin, put forward an elaborate project for a Russian naval mission to the Pacific. Russian suspicions of British designs in the Far East were so intense that the mission was to be kept strictly secret, to the extent indeed that it was to be dispatched not from Russia's principal – and presumably well monitored – naval bastion at Kronstadt, but from the Black Sea instead. Putiatin outlined three principal tasks for this expedition – namely, to evaluate the situation in Canton and other Chinese ports now open for European trade, to make an attempt to conclude a trade agreement with Japan, and finally to conduct further reconnaissance along the southern coast of the Sea of Okhotsk and northern Sakhalin, with the goal of surveying the Amur estuary and locating a port which could replace the unsatisfactory facility at Okhotsk. The Russo-Chinese border was to be surveyed as well.[8] Nicholas I approved this project, which was then sent to his ministers for consideration.

Despite the elaborate precautions for insuring secrecy, Putiatin's proposal foundered nonetheless on the opposition of Nicholas I's ministers toward undertakings of any sort which might upset the ever-more fragile status quo in the Far East. As had been the case with the Ladyzhenskii expedition, both Nesselrode and Kankrin strongly opposed Putiatin's project. The foreign minister feared that it might precipitate a decisive break with China or England, or both,[9] while Kankrin abhorred the thought of devoting scarce state funds on an endeavor he felt to be so pointless. As he wrote in his formal objection to the expedition,

> [i]n view of the underdevelopment, or better yet the non-existence of our trade on the Pacific, and also of the fact that we do not anticipate that this trade can ever exist without Russia establishing itself in the Amur region, the only useful function [of the

[5] Korsak, *Istoriko-statisticheskoe obozrenie*, p. 219; Clubb, *China*, p. 72; Kabanov, *Amurskii vopros*, p. 64, 107; Quested, *Expansion*, p. 23–24.

[6] Sladkovskii, *Istoriia*, p. 205–206.

[7] Alekseev, *Amurskaia ekspeditsiia*, pp. 10–11.

[8] Romanov, "Prisoedinenie," pp. 374–378; Sgibnev, "Vidy," pp. 693–694.

[9] Lensen, *Russian Push*, p. 263.

proposed expedition] would be to confirm the existing belief that the mouth of the Amur is inaccessible. This is the factor which determines the degree of value of this river and its basin for Russia.

Kankrin allowed that the resolution of this question about the Amur estuary was not without significance, but it hardly merited the expense of the proposed full-scale naval expedition, and he recommended instead that an expedition in the Sea of Okhotsk be conducted locally under the auspices of the Russian–American Company.[10] Nicholas I relented in his initial enthusiasm for the Putiatin project and deferred to his ministers. The expedition was put off, and instructions were sent to Eastern Siberia for a smaller local expedition to take its place. On these instructions, Nicholas I noted in his own hand: "Take all measures above all to determine whether or not ships can enter [and navigate] the Amur river, for this is the entire question which is of importance for Russia."[11]

In Petropavlovsk the director of the Russian–American colonies M. Teben′kov entrusted this mission in 1846 to Alexander Gavrilov. Secrecy was at such a premium that Gavrilov himself was not told the actual nature of his mission, and was instructed simply to look for Russian renegade camps at the mouth of the Amur. Gavrilov sailed on the brig *Konstantin* to the west coast of Sakhalin, reached the mouth of the Amur, and continued up it in a rowboat. In his report on his expedition he concluded that the estuary of the Amur was so shallow that it could not be practically navigated by even small boats.[12] This information was relayed back to the capital, where it was taken as the final word on the murky question of the navigability of the river. The skeptics seemed to have been justified, and in the light of Gavrilov's negative conclusions the Amur really did forfeit any potential significance as a communications and transport line which could help strengthen Russia's position in the Far East. In December 1846, Nicholas I wrote into the margin of the report from the Russian–American Company: "I regret this very much (ves′ma zhaleiu). Drop the question about the Amur, which is a useless river."[13]

Nicholas I's resolution of 1846 had two immediate effects. It was decided officially to acknowledge once and for all Chinese suzerainty over the Amur region, and to relocate the unfortunate port at Okhotsk, not – as originally considered – to the mouth of the Amur, but rather to Ayan, a harbor south of Okhotsk already identified as a possible alternative in the 1820s. Both of these decisions were set forth in a resolution made by a special committee meeting in 1848 under the chairmanship of Nesselrode. The committee determined

[10] Quoted in Nevel′skoi, *Podvigi*, p. 43; Sgibnev, "Vidy," pp. 694–695; Lensen, *Russian Push*, p. 264. [11] Nevel′skoi, *Podvigi*, p. 44.

[12] Sgibnev, "Vidy," p. 706–707; Romanov, "Prisoedinenie," pp. 378–379; Alekseev, *Amurskaia ekspeditsiia*, pp. 12–13.

[13] Quoted in Nevel′skoi, *Podvigi*, p. 51; Romanov, "Prisoedinenie," pp. 387–388.

to establish our border with China along the southern slope of the Khingan and Stanovoi ranges to the Sea of Okhotsk . . ., and in this way to give up permanently (otdat′ . . . navsegda) all of the Amur basin to China. [The Amur is] useless to Russia due to its inaccessibility for ocean-going vessels as well as the lack of a harbor at its mouth. All attention [should be] directed to Ayan as the best port on the Sea of Okhotsk, and to the port of Petropavlovsk [on Kamchatka], which should become Russia's main naval bastion on the Pacific.[14]

In accordance with this resolution, an expedition was dispatched to Eastern Siberia in 1849 headed by Nikolai Akhte to carry out the necessary topographical surveys as a preliminary step toward a formal border conference with China. This was the state of affairs which Murav′ev encountered soon after his appointment as governor-general of Eastern Siberia in 1847, and with which he had to do battle in his remarkable struggle to acquire the Amur basin for Russia. Thanks in no small measure to his efforts, not only was Akhte's expedition not fated to carry out its work in the intended spirit, but within a few short years, St. Petersburg's resolute far-eastern policy was to be turned inside out.

"This is one courageous, enterprising Yankee!"

Nikolai Nikolaevich Murav′ev (1809–1881) was appointed by Nicholas I as governor-general of Eastern Siberia in 1847, and served in that capacity for fourteen years. Although it is clear by this point that he was not the first to raise the question of the reacquisition of the Amur, this issue became his overriding obsession during his tenure as governor-general. It was due to his unrelenting efforts that attention in St. Petersburg came to be focused upon it, preparing the way for the general acceptance in the mid-1850s of his policy which he urged. He may thus be seen, if not as the instigator of this issue, then at least as its most important consummator, and it was entirely fitting that he be honored for this with the title Graf Amurskii, or Count of the Amur. Before examining the arguments he used to motivate the Amur issue, it would be useful to establish, to the extent possible, the nature of his broader social and political convictions. This will help better to understand the general relationship of the Amur question to the political climate in Russia during this period.

Like the Amur epoch itself, its principal actor was something of an enigma. Indeed, in a sense the controversy that raged around this individual may be seen as a microcosm of the larger debates that his most important accomplishment was to engender. For many of those who knew and worked with him, he was a noble and heroic figure, a reformer and a democrat who represented the very apogee of selfless service to the fatherland. Such unreservedly positive

[14] Quoted in Nevel′skoi, *Podvigi*, p. 56; Lin, "Amur Frontier," pp. 9–10.

evaluations owed not a little to the adulatory aura that developed around him, a product of his considerable magnetism on the one hand and his undeniably irrepressible devotion to his chosen cause on the other. The power of Murav′ev's charisma was remarkable, indeed "legendary" in the evaluation of one contemporary,[15] and it gave rise to something resembling a cult among many of those who were close to him. The writer Ivan Goncharov, for example, who spent time with Murav′ev during a sojourn to Eastern Siberia in the mid-1850s, was entirely captivated by the governor-general, and left the following impression:

What energy! What breadth of horizon, what quickness of wit! [He burns with an] inextinguishable fire throughout all of his body . . . struggling with the obstacles – the *batons dans les roues* as he calls them – with which [his opponents in the St. Petersburg] try to dampen his zealous ardor! This is one courageous, enterprising Yankee! Short in stature, he is nervous and active. Not once did I ever see him cast a tired glance or make a limp movement. This is a fighting, zealous champion, filled with an inner fire, and ebullient in speech and movement.[16]

The anarchist Mikhail Bakunin, a distant relative of Murav′ev who spent some two years in exile in Eastern Siberia while the latter was governor-general there, was even more enthusiastic. To his friend Alexander Herzen in London he wrote

I have met many people, but have never known one who concentrated in himself so many mutually complementary gifts and capabilities: a mind which is bold, broad, fervent, decisive; an innate eloquence which is compelling and firey, and an ability to express himself and be understood with an amazing simplicity.[17]

To others who knew him, however – fewer in number but no less trenchant – he was a self-serving charlatan, a despot and a megalomanic, whose motives were dictated at all times by his "two principal passions: ambition and vanity."[18] The exiled Decembrist Dmitrii Zavalishin was probably the most famous of Murav′ev's critics (he was certainly the most vociferous) and in his estimation Murav′ev represented the "embodiment of the very worst of base egoistical strivings."[19] A few people on the scene recognized Murav′ev's contradictory qualities, and offered rather more balanced appraisals. Peter Kropotkin, who served in the Far East during Murav′ev's tenure there, observed that although the governor-general "was very intelligent, very active, extremely amiable, and desirous to work for the good of the country," he was at the same time "a despot at the bottom of his heart."[20] Herzen himself,

15 Miliutin, "General-gubernatorstvo," p. 595.
16 Goncharov, "Po Vostochnoi Sibiri," p. 9.
17 [Bakunin], *Pis′ma*, p. 114. For another enthusiastic appraisal, see the memoirs of Peter Struve's father Berngard Struve, who served for several years as an assistant to Murav′ev in Irkutsk. *Vospominanii, passim.*
18 Quoted in Butsinskii, *Graf . . . Amurskii*, p. 5.
19 Quoted in Svatikov, *Rossiia*, p. 39; Zavalishin, *Zapiski*, II, p. 314.
20 Kropotkin, *Memoirs*, p. 169. Also see Veniukov, *Iz Vospominanii*, I, pp. 223, 234, 273; Koz′min, "M. B. Zagoskin," p. 193.

whose exile in London located him half-a-globe away from Murav′ev in Irkutsk and afforded him a uniquely detached perspective on events and personalities in Eastern Siberia, concurred. In a true Faustian manner, he concluded, Murav′ev was at once both a "democrat and a tatar, a liberal and a despot."[21]

There seems to be little question that, to a significant extent at least, Murav′ev shared the nationalist–reformist sentiments which were developing in the country in the 1840s. After a distinguished military career that included service in the war with Turkey in the late 1820s, the suppression of the Polish insurgents in 1830–1831, and in the campaigns in the Caucasus, he was appointed as military governor of Tula in 1846. In this capacity, Murav′ev was bold enough to voice his convictions about the need for the reform of Russian society, and he presented an address to the tsar concerning the emancipation of the serfs.[22] He expressed moreover the characteristic nationalist sensitivities regarding Russia's relation to Europe. During trips to the West, he wrote letters back to his family complaining about the stifling world he encountered there, where everything was "good, smooth, calculated, computed, conceived, and fitted in measure and weight" with such extraordinary precision that nothing remained to the imagination or spirit. The West may dominate in the present, he maintained, but there could be no doubt that "the future belongs entirely to Russia."[23] The realization of this future, and the resurrection of Russia's world stature, provided an important inspiration for his subsequent activities in Eastern Siberia.

In his position as governor-general of Eastern Siberia, which placed him in one of the more powerful offices in the country and at the same time removed him far from the watchful eye of St. Petersburg, Murav′ev seemed to confirm the reputation as a liberal and a democrat which he had developed in his early career. Especially in the first years of his administration, he stirred a considerable amount of excitement with the prospect of liberal reform in the region.[24] A certain Mr. S. Hill, an English traveller who happened to be in Irkutsk when Murav′ev assumed his post in 1847, testified to the optimistic anticipation that his arrival occasioned among the local population. In the new governor-general they sensed an unanticipated and quite unprecedented attitude of sympathy and genuine committment to the region – a perception which Murav′ev enhanced by such gestures as serving local Siberian cherry wine when he entertained Irkutsk society, instead of the champagne or burgundy which an illustrious administrator from St. Petersburg might have been

[21] Quoted in Gollwitzer, *Europe*, p. 128; Shtein, *N. N. Murav′ev-Amurskii*, p. 3. The controversy around Murav′ev was to continue long after his departure from the scene: see below, p. 260n.

[22] Barsukov, *Graf . . . Amurskii*, II, pp. 17–27; Veniukov, "Graf . . . Amurskii," p. 524; Shtein, *N. N. Murav′ev-Amurskii*, pp. 11–12; Sullivan, "Count . . . Amurskii," pp. 105–106.

[23] Barsukov, *Graf . . . Amurskii*, I, pp. 153, 466.

[24] Miliutin, "General-gubernatorstvo," pp. 619–620.

expected to prefer. "There was at least the conviction among them," Hill observed, "that a vigorous government was about to relieve the oppressed and ameliorate the condition of the people of Eastern Siberia."[25]

Among the encouraging signs was his interest in the numerous political exiles in the region, individuals who had been banished to Russia's remote provinces often enough for the very same liberal and reformist covictions that Murav′ev himself now claimed to espouse. Immediately upon assuming his position, he made it a point to establish close and friendly ties to these exiles, and he drew upon them rather heavily for advice and even for service in his administration. The first such group were the Decembrists, who had been living in the region since the late 1820s, and Hill reported that Murav′ev's very first act as governor-general was to pay a visit to one of them.[26] In the 1850s, he also established relations with the members of the Petrashevskii circle who had been exiled to Eastern Siberia, including Petrashevskii himself, Fedor L′vov, and Nikolai Speshnev.[27] For a time, these contacts were quite close, and although they ultimately soured for the first two, Speshnev remained his loyal assistant and even accompanied him on a trip to Japan in 1859. The most celebrated of Murav′ev's connections with political exiles, however, was his friendship with Bakunin. With all of these individuals the governor-general was friendly and open: he received them in his home and shared his library and ideas with them.

The contact between the governor-general and Bakunin, who was transferred from Tomsk to Irkutsk in 1859 and remained until 1861, produced the most extensive description of Murav′ev's political and social views. In 1860, Bakunin wrote a number of letters to Herzen in London stridently defending Murav′ev against criticism of his administration that Herzen had published in his *Kolokol*.[28] Bakunin had a definite personal stake in defending Murav′ev, who acted as something of a patron for him. It was Murav′ev who arranged for the transfer to Irkutsk, and once there the governor-general secured a comfortable sinecure for Bakunin in a newly formed company for commerce on the Amur and petitioned repeatedly on his behalf in St. Petersburg for a pardon.[29] Nevertheless, taking into account these circumstances as well as Bakunin's characteristically hyperbolic manner of expression, a picture of Murav′ev emerges which is not at odds with other available evidence and which identifies him clearly as an opponent of Nicholas I's Official

[25] Hill, *Travels*, I, pp. 441, 444.

[26] Hill, *Travels*, I, p. 438; Zavalishin, *Zapiski*, II, p. 313; Sullivan, "Count . . . Amurskii," pp. 275–277.

[27] Leikina, *Petrashevtsy*, pp. 59–60; Semevskii, "M. V. Butashevich-Petrashevskii," no. 2, pp. 18–21; Aref′ev, V. "M. V. Butashevich-Petrashevskii," p. 178.

[28] It was apparently at Murav′ev's request that Bakunin undertook to respond. In all, four letters were written, only the first of which was published in *Kolokol*. Dulov, "K. Marks," p. 19.

[29] Steklov, *Mikhail Aleksandrovich Bakunin*, I, pp. 500–501; Carr, *Michael Bakunin*, p. 228; Adrianov, "Tomskaia starina," pp. 122–125.

Nationality. In all of Russia, Bakunin wrote to Herzen, there is but one official who has made his name with a genuinely patriotic act, by which Bakunin meant the annexation of the Amur river. This official is dedicated with all his soul to the good of Russia, not – he hastened to explain – as a "bearded reactionary Slavophile," but rather as an advocate of social justice. "He is a decisive democrat, in the same way that we are democrats."[30]

Murav′ev's political program, Bakunin went on, included the following main points: unconditional and full freedom for the serfs with land allotments, a public judiciary based the jury system, full freedom of the press, and universal public education.[31] Bakunin took particular care to emphasize Murav′ev's implacable opposition to the bureaucracy of Nicholas I, and maintained that the governor-general had a radical, even revolutionary alternative. Murav′ev desired nothing less than

> the destruction of the ministries [in St. Petersburg] . . ., and for the initial period thereafter not a constitution, not a long-winded parliament of the nobility, but a provisional iron dictatorship, under any sort of name. For the achievement of his goal [he envisions] the complete destruction of Nicholas I's . . . servile St. Petersburg bureaucracy.

Murav′ev had no trust in the nobility or the privileged classes, and placed all of his hopes in the people. He "believes only in the humble people, loves them, and sees in them alone Russia's future." He did not believe that Nicholas I's government could solve the problem of emancipation on its own, and therefore hoped that the "the peasant axe will sober up (vrazumit) St. Petersburg and make possible that rational dictatorship (razumnaia diktatura) which he is convinced is the only way to save Russia."[32]

In addition to the far-reaching internal reformation of Russia, Murav′ev believed it necessary for the country to recreate for itself the strong international position it had lost through its recent defeat in the Crimean War. This could be accomplished only with the adoption of a resolute and unintimidated, indeed even aggressive, stance in its dealings with foreign powers. This goal also required a "rational" dictatorship, which was needed

> in order to reestablish the power of Russia in Europe. This power should be directed first of all against Austria and Turkey in order to liberate the Slavs, and for the establishment, not of a centralized Pan-Slav monarchy, but of a firmly unified but voluntary Slavic federation. Murav′ev is a friend of the Hungarians and a friend of the Poles, and is convinced that the first step of an intelligent Russian foreign policy should be the resurrection and liberation of Poland."[33]

(Bakunin conveniently neglected to mention that this "friend of the Poles" had participated in the suppression of their uprising in 1831.) The anarchist of course entertained his own independent vision of a pan-Slavic amalgam.

[30] [Bakunin], *Pis′ma*, p. 112. [31] [Bakunin], *Pis′ma*, pp. 113, 177–178.
[32] [Bakunin], *Pis′ma*, pp. 113–114. [33] [Bakunin], *Pis′ma*, p. 114.

He bragged to Herzen that it was he who had convinced Murav′ev of the necessity of such a Slavic federation, and in his own mind apparently assigned the governor-general the role of its leader. In this way, Carr notes, Murav′ev became for Bakunin the principal carrier of Russia's messianic mission as "the predestined saviour not only of Russia but of Europe."[34] In conclusion, Bakunin wanted to leave no doubt as to where Murav′ev stood, and stated enthusiastically: "He is entirely one of us (krepko nash) and is the best and strongest of us, in him is the future of Russia . . . Murav′ev is the only person in Russia with power and authority whom we can and should *without the slightest exaggeration* and in the full sense of the word definitely call *ours*."[35]

It is not possible to determine the depth or sincerity of Murav′ev's devotion to all of these various ideas and projects. To be sure, the bombastic claims of Bakunin were exaggerated and at points fundamentally incorrect.[36] The governor-general left ample evidence to make clear his overwhelming personal ambition, his tyrannical proclivities, and his capacity for cynical opportunism. What is important, however, is not Murav′ev's character, but rather the principles and policies he represented in the eyes of the country as a whole, and the sentiments with which he was able – with admirable success – to imbue his activities on the Amur. From this standpoint, Murav′ev emerges as a brilliant example, indeed as something of a prototype, of the liberal or reformist expansionism discussed earlier in this study.[37] He clearly stood for the resurrection of Russia's national integrity not only through fundamental internal transformations of its social and political structure, but by means of an energetic policy of external territorial and political expansion as well. Moreover, with his tangible accomplishments in the Russian Far East, he gave this

[34] Carr, *Michael Bakunin*, p. 230; Bakunin, *Pis′ma*, p. 188. On the political program Bakunin sketched in Siberia, see Vyvyan, "Russia," p. 367 ("proto-fascist dictatorship"), and Kohn, *Pan-Slavism*, pp. 78–79.

[35] [Bakunin], *Pis′ma*, p. 177 (emphasis in original). In 1861, after it became clear that Murav′ev was going to be replaced as governor-general, Bakunin made a dramatic escape from Eastern Siberia, down the Amur river to Japan and further to San Francisco. His questionable behaviour in Irkutsk, his close relations with Murav′ev and other high administrators there, and his acceptance of their favoritism (it was even claimed that they assisted him in his escape) earned him a good deal of scorn from many of those who knew him there. See Zavalishin, *Zapiski*, II, pp. 411–412; Svatikov, *Rossiia i Sibir′*, p. 36; Miliutin, "General-gubernatorstvo," p. 630; Kun, "Gertsen," pp. 105–106ff; Steklov, *Mikhail Aleksandrovich Bakunin*, I, pp. 509ff; Carr, *Michael Bakunin*, p. 232. A decade later, as part of a lengthly exposé aimed at discrediting Bakunin in the First International, Karl Marx publicized unflattering details of the Russian anarchist's personal affairs in Siberia, and referred mockingly to his escape down the Amur as a "hegira." Marx, "Die Hegira Bakunins," pp. 442–444; Dulov, "K. Marks," p. 18.

[36] It would nevertheless appear that the anarchist was convinced of the picture of Murav′ev he conveyed to Herzen. In an unpublished letter, Bakunin described with sorrow how he lost faith in the governor-general only in 1863, when during a meeting in Paris (where Murav′ev emigrated after leaving Eastern Siberia) his former mentor voiced his convinced support for the repression of the insurrection in Poland. Bakunin's biographer Steklov emphasizes the genuineness of Bakunin's devotion to Murav′ev in Irkutsk, which he characterizes as "perhaps the darkest page of Bakunin's biography." *Mikhail Aleksandrovich Bakunin*, I, pp. 505, 521, 527n.

[37] Gollwitzer, *Europe*, pp. 127–128.

program substance and brought it to life in a special way, as effectively as anyone else in the Russia of his time.

"To rule the entire Asiatic coast"

Immediately upon assuming office as governor-general of Eastern Siberia in 1847, Murav′ev undertook a reevaluation of the new situation on the Pacific and in China and its significance for Russia. With the assistance of Balasoglo's memorandum, he arrived quickly at the conclusions which set the tone for his entire tenure as governor-general. In view of Russia's pitifully weak position in the Far East, the incursion of European powers into China represented a basic threat to Russian commercial interests – and, indeed, to its security and territorial integrity in these regions. Russia's position on the Pacific must be substantially enhanced, and the sole means for accomplishing this so that it would be permanent was to gain Russian jurisdiction over the Amur river. This position was expressed in a steady stream of reports, letters, and memoranda to the government in St. Petersburg, where as we have seen it had been firmly decided to grant formal recognition of Chinese authority over the river. It was Murav′ev's persistence and vehemence in the face of official resistance that prepared the government for the change of policy which the onset of the Crimean War then forced upon it.

The central theme around which Murav′ev developed his case for the Amur was the specter of the expansion of British influence on the Pacific. The example of Britain's occupation and colonial subjugation of India was still relatively fresh, and it seemed that now the *ostrovitiane* or "islanders," as he derisively referred to them, were once again seeking to augment their influence, moving northward along the East Asian coast in the search for new markets to conquer and new territories to absorb into their already vast empire. With their victory in the Opium Wars, they had made a decisive inroad into China, and all that remained now was for them to fortify their position there. Toward this end, Murav′ev claimed, Britain would inevitably be constrained to turn its attention to the Amur, for this river was a necessary logistic route which it would require as a part of its overall consolidation in the Far East. "For their secure and firm control of trade with China," Murav′ev wrote to Nicholas I in 1849, "there is no question that the English need the mouth of the Amur and navigation on this river."[38] As evidence of these designs, Murav′ev could point to the increasing number of British travelers who crossed Siberia and, passing themselves off in Irkutsk as botanists or geologists, requested his assistance in continuing east to the Amur.[39]

[38] Barsukov, *Graf . . . Amurskii*, I, p. 211.

[39] Barsukov, *Graf . . . Amurskii*, I, 191, II, 35; Struve, *Vospominanii*, pp. 32–33; Kabanov, *Amurskii vopros*, p. 105; Stanton, "Foundations," pp. 9–10. As early as 1841, the foreign minister

Murav′ev had no doubt that in fact they were reconnaissance agents – he was convinced that the admiring Mr. Hill was one of them![40] – and he assured St. Petersburg that this permission was invariably refused. Nonetheless, he pointed out that if the British were to discover that the Amur region was *de facto* under the firm control of no one, they would certainly lose no time in occupying it.[41]

It might have been possible for Nicholas I's ministers in St. Petersburg to react with indifference to these pretensions, insofar as they had no designs of their own on the river and were quite prepared to relinquish it to China. However, having established this point, Murav′ev pursued the logic of the situation yet further and tried to demonstrate that British control of the Amur would lead to consequences intolerable for St. Petersburg. Through their domination in the Amur region, he argued, they would gain a fateful influence over all of Siberia, which could lead ultimately to its breaking away from Russia west of the Urals. In another report to the tsar, Murav′ev developed this geopolitical logic in the following manner:

> If the Amur were not the only river flowing from Siberia into the Pacific Ocean, we could afford to be more condescending to the British undertakings, but navigation on the Amur, the single convenient route to the East, is an age-old dream of the local inhabitants . . ., and I . . . dare to say that whoever shall control the mouth of the Amur shall control Siberia as well, at least to Baikal, and control it firmly, for it is enough to have the mouth of this river and navigation on it under lock and key in order that Siberia, which is more populous and richer in agriculture and industry [than the Amur region itself], will become an irretrievable tributary and subject of that power which holds the key.[42]

Given its preoccupation with European affairs, the government in St. Petersburg may not have identified any of its most vital interests in the Russian East, but Murav′ev nevertheless understood the weight of this sort of argu-

Nesselrode was reporting on an unusual increase of in the number of requests received by the Russian embassy in London for transit privileges across Siberia. Kuznetsov, "Sibirskaia programma," pp. 9–10. On British interest, or lack of it, in the Amur, see Costin, *Great Britain*, p. 195; Gillard, *Struggle*, p. 104; Quested, *Expansion*, pp. 43–44.

40 Struve, *Vospominanii*, pp. 32–33.

41 Barsukov, *Graf . . . Amurskii*, II, p. 35.

42 Barsukov, *Graf . . . Amurskii*, I, pp. 211 (quote), 288; Kabanov, *Amurskii vopros*, p. 103. It is interesting to note that nearly a half-century earlier, the American president Thomas Jefferson made an identical geopolitical argument in regard to French intentions on the mouth of the Mississippi. In 1802, he wrote to the American ambassador in France:

> There is on the globe one single spot, the possessor of which is our natural and habitual enemy. It is New Orleans, through which the produce of three-eighths of our territory must pass to market, and from its fertility it will ere long yield more than half of our whole produce, and contain more than half of our inhabitants. France, placing herself in that door, assumes to us the attitude of defiance.

Writings of Thomas Jefferson, X, p. 312; also see Turner, "Significance," p. 188.

ment. The prospect of a Siberia broken away from Russia was in fact a terrifying one for Nicholas I's ministers and for the tsar himself, and one which they could not possibly ignore.

To bolster this point, Murav′ev suggested ominously that in such an eventuality British intentions would not necessarily be greeted unsympathetically by the local Siberian population. In February, 1849, he submitted a special note to the tsar with a description of separatist sentiments in Siberia. "I had more than once encountered the fear in St. Petersburg," he wrote, "that Siberia will sooner or later break away from Russia, but before my arrival here I considered this fear unfounded." Now, after two years in the region, he had come to appreciate the danger. "Sire, I became convinced that these fears are entirely natural." The entire population of Eastern Siberia, he reported, was under the influence of the rich merchant and industrial classes: powerful and self-confident groups whose loyalty to the empire was not to be trusted. He explained carefully that

> because of their origin and their distance from the center of the empire, [they] do not share at all the same feelings of devotion to their tsar and fatherland which internal regions of the empire take in with their mother's milk. [These Siberians] are indifferent to everything except their own advantage, and . . . there is almost no hope of awakening in them those noble feelings which are the pride and glory of every Russian.

The Siberians' subordination of patriotic fealty to their own entreprenurial advantage was in itself reprehensible, but it did lead them to a practical and realistic appraisal of evolving local circumstances. They saw clearly that the future well-being of Siberia rested entirely on the need to secure a reliable and convenient connection with the Pacific, and that without navigation on the Amur any efforts at the economic development of the region would be doomed from the outset. "In recent years, therefore, . . . the not-unfounded notion has taken hold that the English are going to occupy the mouth of the Amur. On the surface, many of them express fear and regret for this, while in their hearts they are completely indifferent to whoever opens the Amur for navigation [as long as it is opened]." Murav′ev left these thoughts hanging to play on Nicholas I's endemic suspicion of sedition lurking throughout the empire, a fear which by the time this note was written had been excited to near-paranoiac dimensions by the European upheavals of 1848 and the discovery of clandestine dissident groups in St. Petersburg itself. Murav′ev concluded the note by pointing out that timely Russian occupation of the mouth of the Amur could easily and cheaply spare Russia the misfortune which he depicted, and thereby secure firm Russian control over Siberia "for eternity."[43]

According to Murav′ev, British territorial designs were not limited to the Amur region. It was clear that they were additionally interested in the penin-

[43] Barsukov, *Graf . . . Amurskii*, I, pp. 205–206.

sula of Kamchatka and the entire Okhotsk coast as well, which together could offer bases of support for their activities on the Pacific and virtually complete their hegemony on this ocean. Moreover, in view of the logistical problems experienced by the Russians in the Far East and the lack of an adequate naval base of their own there, communications between their outposts were tenuous at best, and it could by no means be guaranteed that an aggressive maneuver on the part of the British could be successfully defeated. Indeed, Murav′ev asserted that under the present circumstances it would be quite impossible for the Russians to resist an English advance. Once again, he pointed to the Russian occupation of the mouth of the Amur, enabling Russia to secure its connection with Kamchatka, as the only way out of this perilous situation. "Kamchatka and the Sea of Okhotsk can be defended by us and preserved under Russian dominion only if we are in possession of the Amur river, or, at the very least, of the right of navigation upon it."[44]

The frightening prospect of the loss of Russian territory as a consequence of European incursion into China appeared to be the most immediately menacing aspect of the situation, and he stressed it above all else. There was however another consequence of the activities of the Europeans in China which Murav′ev included as part of his argument for a reformulation of Russian policy in the Far East. This was the fate of Russian commercial relations with China, which as we have seen formed the central concern in St. Petersburg for Russian policy in the Far East at the time. In harping on this theme, Murav′ev was once again touching on what he understood very well to be a sensitive nerve in the Russian government and, as with the vision of a Siberia broken off from the Russian empire, he did not fail to make good use of it. Murav′ev's basic analysis of Russia's China trade followed the same lines as that concerning Russia's allegedly endangered territories. With the forcible opening of China to Western commerce, as a result of which no less than five ports were not only made accessible to the British but practically "converted into English cities," it was inevitable that Britain would come to dominate the entire Chinese market.[45] This in turn would enable them to achieve political dominance throughout the country as well, which spelled disaster for Russian commercial interests there.

The negative effects Murav′ev was referring to were in fact already apparent in the decline of the volume of trade at Kiakhta, and he predicted grimly that Russia's overland trade with China "is fated to lose all significance in the face of English activities in internal Chinese markets."[46] Moreover, a China which under foreign influence abandoned its traditionally favorable disposition toward Russia could represent yet another territorial threat for the Russian East. "Our neighbor, populous China, which is at the moment pow-

[44] Barsukov, *Graf . . . Amurskii*, I, p. 258; also see II, pp. 47, 105.
[45] Barsukov, *Graf . . . Amurskii*, II, p. 48 (quote), I, p. 258.
[46] Barsukov, *Graf . . . Amurskii*, I, p. 206 (quote), 252, 263.

erless in its ignorance, can easily become dangerous for us under the influence and direction of the English and the French, and then Siberia would cease to be Russian."[47] If Russia was not only to maintain a foothold for trade in China – which may well have already been lost – but simply to preserve its own empire intact, then it had to take immediate and decisive action. This would take the form of the occupation of the mouth of the Amur, the assertion of Russian suzerainty over its entire left bank, and the acquisition of rights of free navigation along it. Thus, for Murav′ev the securing of the Amur was not only a defensive entrenchment against anticipated aggrandizement, giving Russia a border which was clearly delimited and defensible,[48] but at the same time something of an offensive measure, aimed at securing for Russia its share of the potentially lucrative China trade. For with the Amur, or more correctly the Amur basin, Russia would have direct river access to the heavily populated regions of northern China. This was an area relatively far removed from the scene of the most intense British commercial activity, and it thus presented the most auspicious opportunities for the free development of Russian commerce.[49]

In this way, Murav′ev's call for an active policy in the Far East was founded on two very different sets of considerations, and was consequently characterized by an essential duality. On the one hand, he felt that Russia's far-eastern interests were threatened by the European move into China. On the other, however, he saw an opportunity for Russia to take its stand with the other European powers and join with them in carving up the rich Chinese market into spheres of influence. From this latter perspective, the Amur was important not so much in itself but by virtue of the fact that it served as a sort of gateway to northern China and offered, through the network of Manchurian rivers which drained into it, convenient access to markets which could conceivably form a "Russian" sphere. This in turn implied an aggressive policy on the part of Russia toward China, one that was a clear departure from the traditional pattern of relations between the two countries but entirely in line with the new European imperialist stance after 1840. Russia thus saw its interests threatened by this stance, but – at least in the hopes of Murav′ev and others – felt that it could benefit from it as well. The product of this duality was a fundamentally ambiguous attitude on the part of Russia toward China, which was to characterize Russian views throughout this period and indeed the rest of the nineteenth century. Russia was never slow to offer its sympathies to its fraternal neighbor suffering under European attack and occupation. At the same time, however, it was highly conscious of the benefits which it could obtain from China's weakened position, and without any question desirous of realizing these. Murav′ev expressed this ambiguity with admirable candor on

[47] Barsukov, *Graf . . . Amurskii*, II, p. 105.
[48] Barsukov, *Graf . . . Amurskii*, II, p. 48.
[49] Barsukov, *Graf . . . Amurskii*, II, p. 48.

the occasion of the formation of the cossack regiment in the Transbaikal region in 1851, at which time he remarked on the need for a show of Russian might in China, "both to help them, and to frighten them (i dlia pomoschi im, i dlia strakha)."[50]

In a special note to Nicholas I in 1853, Murav′ev presented a sort of capstone to his arguments for the reacquisition of the Amur by offering a novel picture of the geopolitical situation confronting Russia in the Far East. Developing the points we have already examined in Balasoglo's memorandum, Murav′ev depicted an immanent transition from what might be called an "American" to an "Asian" perspective on Russia's position in the Far East and on the Pacific. In his conclusions, however, the governor-general went significantly beyond Balasoglo. The first and most fundamental insight for understanding the new state of affairs confronting Russia on the Pacific, he explained to the tsar, was the complete pointlessness of continued Russian territorial pretensions in the New World. Most obviously, this referred to the grandiose notion of a trans-Pacific Russian empire, and Murav′ev showed little patience with the dreamy visionaries who had spoken evocatively of such a possibility only a few decades earlier.

> It was impossible [in the 1820s] not to foresee the rapid spread of the dominion of the North American states across North America. It was also impossible not to foresee that these states, having established themselves on the Pacific, would quickly surpass all other sea powers in this region and would require the entirety of the northwest coast of America. The dominion of the United States across all of North America is so natural that we should not regret that twenty five years ago we did not establish ourselves in California – we would sooner or later have been compelled to give it up.

It was easy enough for Murav′ev to make such an assertion about the inevitable forfeiture of imperial realms Russia never in fact possessed, but having made it he went on to lay out a corollary prognostication which was not so painless. Because with its "rapid spread" the United States would indeed ultimately cover entirety of the North American continent, it would necessarily involve territories which Russia did in fact control.

> [N]ow, with the possession and development of the railroad, we may be even more convinced than ever that the United States will of necessity spread throughout all of North America. *We must not lose sight of the fact that sooner or later we will be compelled to give up all of our North American possessions.*[51]

With his careful emphasis on the "naturalness" and inevitability of American continental dominion, Murav′ev was advancing a position that was in total contrast to that of Balasoglo. We will see presently just why he took such delib-

[50] Barsukov, *Graf . . . Amurskii*, I, p. 288; Hudson, "Far East," p. 702.
[51] Struve, *Vospominanii*, pp. 153–154 (emphasis added).

erate pains to insure that the expansion of the United States not be perceived as aggressive or in any way harmful to Russia's true interests on the Pacific.

With this, Murav′ev was clearly predicting the imminent eclipse of Russia's imperial presence in North America, an eventuality which would obviously undermine Russia's presence on Kamchatka and the Okhotsk coast as well. This picture was not intended to be a gloomy one, however, for he depicted it merely in order to set the stage for an entirely new prospect for Russia in the Far East. "We must nevertheless not lose sight of yet another point," he continued, namely "that it is entirely natural for Russia, if not to control all of East Asia, then [at least] to rule (gospodstvovat′) the entire Asiatic coast of the Pacific Ocean." Thus, in what amounted to a direct exchange, Murav′ev suggested that Russia should turn its attention to a new arena of expansion and activity in Asia in order to replace the one fated to be lost in North America to the expansion of the United States. And although Murav′ev chose not to emphasize the point at that moment, it was nevertheless clear that this "Asian perspective" was potentially far grander than anything the icy waters of the Bering Sea or the snows of Alaska could offer. Above all, such a shift in perspective involved a dramatic metamorphosis in the function of the Amur river, from a logistical supply route for the North Pacific fur trade to the critical artery facilitating Russian emergence as a major imperial power in the Pacific basin.

In concluding his memorandum, Murav′ev returned to the problem presented by the penetration of British influence in East Asia, which he described as Russia's most serious and substantial obstacle. Motivated by no interest other than profit, the "islanders" have forced their presence on unwilling societies throughout the world, and will continue to do so until checked by equal force. Their latest moves in China demonstrated without a doubt their intention to capture for themselves the dominant role on the Pacific, which in the natural course of things should fall to Russia. As part of this grand scheme, they planned to "conquer Kamchatka, or at least leave it as a desert, to rule the coasts of China and Japan, and in this way, so to speak, *to tear Russia away from the Pacific*." We may well predict the means which Murav′ev proposed as the only way to prevent this eventuality and to win for Russia its rightful and natural position in the Far East: navigation on the Amur and the establishment of the river as the boundary with China. To further offset the British challenge, he added the necessity for maintaining "*a tight connection between ourselves and the United States*," a sort of trans-Pacific alliance which Britain would certainly spare no effort to disrupt.[52]

It might be noted that the exchange Murav′ev was offering here between American and Asian biases in Russia's Pacific policy was not entirely as straightforward as it might appear, for the proposed new perspective rested on

[52] Struve, *Vospominanii*, pp. 154–156 (emphasis in original).

very different foundations than the one it was to replace. Down to the 1840s, the Russian presence on the Pacific and in North America had been animated by the same mercantile–colonial interest that had initially drawn the Russians into Siberia centuries earlier and then lured them ever further to the east, an interest which as we have seen focused on the quest for furs. The perceived value of Russia's territories beyond the Urals was identified almost exclusively in terms of this resource, and their value was sustained only as long as the region could be profitably exploited. The decline of the Russian fur trade in the second quarter of the nineteenth century, therefore, was an important precondition for the general acceptance of Murav′ev's arguments. The significance of his proposed alternative, however – that of Russian dominion "over the entire Asiatic coast" – was different, and considerably more complex. Rather than belonging to an essentially outdated tradition of mercantile colonialism, it represented instead an early, but nonetheless thoroughly characteristic example of the more "modern" imperialism of the latter half of the nineteenth century, what Dietrich Geyer called the "new imperialist conditions of international politics."[53] The political–territorial expansion that this new movement involved would lead not to the incorporation of yet more vast barren expanses of taiga, tundra, and ice, but rather ultimately to the establishment of Russian political and commercial hegemony over densely populated societies in the Far East and central Asia. It was in this spirit that Murav′ev initiated, with the formal annexation of the Amur region in 1858 and 1860, imperial Russia's final and decisive move into Asia. The spirit of the enterprise was appropriately celebrated in the latter year with the name given to the newly-founded naval bastion on the Pacific: *Vladivostok*, or Ruler of the East. His memorandum from 1853 was significant in that it was the first deliberate and more-or-less coherent articulation of this perspective, which within the brief space of a few years was to serve as the basis for formulating Russian far-eastern policy. Once again, at the very center of it all stood the need for the Russians to establish their authority over the Amur.

"You will not hold back Russia's universal destiny"

Throughout the late 1840s and early 1850s, the Amur question became ever more of a major issue for the Russian government. In part, this was a natural reaction to events in the Far East, but it was due also to the special efforts of Murav′ev in pressing the question. In the course of the ensuing conflicts and debates around this issue in St. Petersburg, the existence of two distinct groups within the government became clear, one of which opposed the cause of

[53] "Es war wohl kein Zufall, dass die Besitzungen der Russisch-Amerikanischen Kompagnie in Alaska aufgelassen wurden (1867), als sich das Zarenreich den neuen imperialistischen Bedingungen internationaler Politik anzupassen begann." Geyer, "Russland als Problem," p. 341.

annexation while the other vigorously supported it. An examination of these conflicting tendencies will help reveal the extent to which the differences around the specifics of Russian policy in the Far East had come to be a reflection of the more fundamental tensions and controversies that characterized Russian educated society in general during the period preceding the Great Reforms of the 1860s.

Neither of the two groups enjoyed the wholesale sympathy and support of the Emperor himself, for in regard to Russian policy in Asia Nicholas I was ultimately ambivalent.[54] That he was aware of the Amur question and interested in it there can be no question. Middendorf had had a special audience with the tsar upon his return from his Siberian expedition in 1844, during which the explorer was questioned with what he described as "great interest" about his findings on the Amur.[55] Two years earlier, the tsar had approved the project for Putiatin's expedition, which included among its tasks a survey of the mouth of the Amur. These points have led most chroniclers of the Amur annexation to conclude that Nicholas I was convinced of the need for Russia to annex the river,[56] a conclusion supported by the comment the emperor is alleged to have made to Murav′ev upon the latter's appointment as governor-general in 1847: "And in regard to the Russian [sic] river Amur, we will talk about this in the future (ob etom rech′ vperedi)." Such evidence, however, is unconfirmed and in fact highly dubious. Gavrilov's expedition of the previous year was accepted in St. Petersburg as conclusive evidence that the mouth of the Amur was inaccessible to ocean-going vessels, thus eliminating its potential value for Russia. Nicholas I recognized this and, as he himself noted, reluctantly gave up the entire question as hopeless. The memoirs of an admiral intimately involved with the Russian advance on the Amur offered a very different (and rather more plausible) account of Murav′ev's interview with the tsar in 1847, claiming that Nicholas cut short Murav′ev's inquiries about the Amur with the retort " 'what use is this river to us, now that it has been positively proven that only small boats are able to enter its mouth?' "[57] The sort of self-consciously forward policy in the Far East which was necessarily indicated by Russian pretensions on the Amur violated the conservative, legitimatist spirit of Nicholas I's foreign policy. These pretensions effectively constituted a challenge to the international status quo, aggression against a neighboring friendly monarch, and an abrogation of the territorial agreement recognized in good faith by both countries in the Treaty of Nerchinsk. All these considerations prevented him from supporting Murav′ev's endeavor, and it was to

[54] Gillard, *Struggle*, pp. 64–65, 104–105. For a dissenting interpretation, see Presniakov, *Emperor Nicholas I*, p. 62. [55] Middendorf, *Puteshestvie*, I, p. 187

[56] Romanov, "Prisoedinenie," no. 7, p. 93; Barsukov, *Graf . . . Amurskii*, I, p. 171. Stanton is particularly emphatic on this point, and makes the highly improbable argument that the entire reformulation of Russian far-eastern policy was ultimately the inspiration of Nicholas I. "Foundations," pp. 103–104. Also see Sullivan, "Count . . . Amurskii," p. 124.

[57] Nevel′skoi, *Podvigi*, p. 59.

take the shock of the Crimean War to bring him finally to approve the Russian occupation of the Amur.

Nicholas I's reluctance, however, was not the only obstacle confronting Murav′ev. All of the governor-general's impassioned pleas for a change in Russian policy on the Pacific met with the unrelenting resistance of the tsar's ministers and many of his closest advisers. We have already noted, in the period preceding Murav′ev's appointment, the strident objections of the Ministers of Finance and Foreign Affairs, Yegor Kankrin and Karl Nesselrode[58] to any suggestion of Russian activity on the Amur. Nesselrode and Kankrin, along with the Minister of Justice Viktor Panin,[59] the head of the Asiatic Department Lev Seniavin,[60] the Minister of War Alexander Chernyshev, and Kankrin's replacement as Minister of Finance (from 1844) Fedor Vronchenko,[61] dominated the upper echelons of Nicholas I's government. As a group, they were firmly united in their unquestioning dedication to their tsar, and for decades worked dutifully to translate the conservative principles of Official Nationality into practice. Accordingly, they were all convinced opponents of internal social change and reform. In their direction of Russia's international dealings, as one contemporary unsympathetically observed, they "gravitated to the West, . . . [and] covered up their occasionally shameful pliability by pointing to the necessity of following a Machiavellian–Metternichian system, devoid of any moral law, for the maintenance of the existing order."[62] The all-important "moral law" which they strove to exclude was in fact the spirit of Russian nationalism and pan-Slavism, which for them was nothing other than dangerous revolutionary agitation.[63] They could hardly approve of suggestions for innovative action which would represent a departure from traditional Russian policy, especially coming from such a youthful and brash upstart as Murav′ev. They were bitter opponents of his projects, and of the governor-general himself.[64]

The specific objections raised by these conservatives to Murav′ev's proposals related to points which have been noted above, namely the concern about disrupting relations with China and, additionally, the desire not to give the impression of Russian expansionist intentions to the European powers.[65] Beyond this, they argued that Russian acquisition of the Amur would actually

[58] Grimsted, *Foreign Ministers*, p. 270; Riasanovsky, *Nicholas I*, pp. 44–45, 237; Presniakov, *Emperor Nicholas I*, pp. 44–45; Lincoln, *Nicholas I*, p. 110; Kabanov, *Amurskii vopros*, pp. 102, 107. [59] Lincoln, *In the Vanguard*, pp. 37–38; "Panin, Viktor Nikitich," p. 786.

[60] "Seniavin, Lev Grigor′evich," pp. 335–336.

[61] On Chernyshev and Vronchenko, see Lincoln, *In the Vanguard*, p. 99; Riasanovsky, *Nicholas I*, pp. 43, 46. [62] Struve, *Vospominanii*, p. 8; Kabanov, *Amurskii vopros*, pp. 102–107.

[63] Pushkarev, *Emergence*, p. 123; Riasanovsky, *Nicholas I*, pp. 237–238.

[64] Kabanov, *Amurskii vopros*, pp. 85, 109; Nevel′skoi, *Podvigi*, p. 102; Lin, "Amur Frontier," p. 7. Accounts of Murav′ev's bitter conflicts with the bureaucracy in St. Petersburg made their way to the West, where they earned the governor-general some admiration from Herzen in London. "Imperatorskii kabinet," *passim*. Also see Bakunin, *Pis′ma*, pp. 119–120.

[65] Barsukov, *Graf . . . Amurskii*, I, p. 192.

have highly detrimental effects for Siberia, and thus for the empire at large. They feared the possible consequences of precisely what Murav′ev wanted to accomplish, namely the linking of Eastern Siberia to the Pacific with a reliable connection. Nesselrode explained that up to this time distant Siberia had represented a "deep net" into which Russia could discard its social sins and scum (podonki) in the form of convicts and exiles. With the annexation of the Amur, however, "the bottom of this net will be untied, and our convicts would be presented with a broad field for escape down the Amur to the Pacific."[66] There was a yet more substantial reason, however, to fear of the effects of direct communication between Siberia and the Pacific. The arrest of the Petrashevskii circle in St. Petersburg in 1849 had brought to light Chernosvitov's pronouncements about an imminent uprising in Siberia. An imperial order was dispatched immediately to the governor-general of Western Siberia P.D. Gorchakov in Omsk to investigate these rumors and report on the political situation in the region. Gorchakov responded in September of that year with a confidential note to Chernyshev. The Minister of War was assured that there was no danger of an immediate uprising, but his attention was called to the ominous traces of incipient separationist sentiment. "Undoubtedly the Siberians are proud of their homeland," Gorchakov wrote, "to which they are strongly tied . . ., and whose abundance and advantages (udobstva) they value to extremes." A class was now developing which looked unfavorably upon the fact that European Russian benefited from Siberian commerce, industry, and raw materials, for it regarded all of these treasures "as if they were the private property of Siberia." These sentiments were shared by many Siberians, the governor-general continued, "and, possibly, [they may] awaken the desire for the establishment of an independent country."[67] Such inclinations were widespread only among the youth, however, and would not pose any essential danger as long as Siberian society was kept isolated from external contacts. On its own, the region lacked the munitions and other resources necessary to undertake an insurrection.

In regard to this final point, Gorchakov directly confronted the question of the Amur. Contact with foreigners, he insisted, was absolutely inadmissable. Happily, under the existing conditions Siberia's "geographical position works against such a project," for the "immeasurable wastes" stretching from Irkutsk to Kamchatka and along the Sea of Okhotsk formed what he called a Wall of China, effectively shielding Siberia from foreign influence without requiring

[66] Quoted in Barsukov, *Graf . . . Amurskii*, I, p. 670–671; Kabanov, *Amurskii vopros*, pp. 78–79, 102. Such a possibility for escape had indeed been discussed by the exiled Decembrists as early as the late 1820s, and no sooner was the Amur annexed than Mikhail Bakunin utilized it in fleeing Eastern Siberia, as we have seen. He did so, however, with the collusion of the local authorities, a malefaction not foreseen by Nesselrode. Basargin, *Zapiski*, p. 111; Azadovskii, "Neosushchestvlennyi zamysel," *passim*. [67] Quoted in Semevskii, *Iz istorii*, p. 65.

special surveillance. Navigation on the Amur, however, would move Siberia closer to the United States and England, and their support – in the form of arms and goods in exchange for Siberian gold, for example – might encourage Siberia to undertake to break away completely from Russia. "It is important above all," he wrote, "to keep the inhabitants of Siberia away from immediate contact with foreigners, contact which could *easily* turn into fatal propaganda."[68] Thus, citing precisely the same danger of Siberian separatism as had Murav′ev, Gorchakov came to precisely the opposite conclusion. He insisted that any thought about the annexation of the Amur and the opening of Russian traffic upon it was inadmissable, indeed self-destructive. With this, he gave an impetus toward even greater resolution on the part of the conservatives in the government to oppose the pleas of the headstrong governor-general in Eastern Siberia.[69]

For his part, however, Murav′ev was not entirely alone in his struggle to resurrect the Amur question, and this circumstance proved to be critical to his ultimate success. The early years of his tenure in Eastern Siberia witnessed the coalescence in government and court circles of a loose group sharing some of the popular sentiments for the reform and rejuvenation of Russia, and they were convinced accordingly of the fundamental need for a reordering of the country's priorities and policies. Although this group was comprised of members who for the most part were not as politically powerful as the conservatives discussed above, the support of important members of the royal family added significantly to their influence. They were far from radical, and their very presence in the government and close proximity to the tsar indicated an ability to work within Nicholas I's system. Nevertheless, in their convictions about Russia's needs and future, they were opposed to Official Nationality and to the conservatives who ruled the country in its name.[70] The most important of the governmental officials who figured in this grouping was Lev Perovskii, who served as Minister of the Interior from 1841 to 1852. He had demonstrated his liberal inclinations in the early 1820s through his brief participation in various Decembrist organizations, and like Murav′ev, he submitted a petition to the tsar in the mid-1840s concerning the need for the abolition of serfdom.[71] His

[68] Semevskii, *Iz istorii*, p. 65 (emphasis in original); Svatikov, *Rossiia i Sibir′*, pp. 29–30; Iadrintsev, *Sibir′*, pp. 708fn.

[69] Kuznetsov, "Sibirskaia programma," pp. 10–11. Although he did not stress the point in this report, Gorchakov had no affection for Murav′ev. Like many other officials, he had been scandalized by the appointment of someone so young – in 1847 Murav′ev was just 38 – to such a major position. Upon first hearing this news, Gorchakov is reported to have flown into a rage, danced around the table at which he had been sitting, grabbed a youthful adjutant and screamed: "You! You've been made a minister! The proof? Murav′ev, a kid just like you, has been made a governor-general!" Veniukov, "Graf . . . Amurskii," p. 524.

[70] Riasanovsky, *Nicholas I*, p. 138.

[71] Lincoln, *In the Vanguard*, pp. 30ff; Riasanovsky, *Nicholas I*, p. 229; "Perovskii, Lev Alekseevich," p. 51.

brother Vasilii, an officer and the military governor of Orenburg, shared his brother's sympathies.[72] Pavel Kiselev, the Minister for State Properties from 1837 to 1856, had like Perovskii been close to the Decembrists, and was even more outspoken about the desirability of the abolition of serfdom, having submitted a project as early as 1816 concerning emancipation. In the late 1830s he experimented with gradual reform of the state serfs, and some years later played a key role in the acutal emancipation.[73] Exercising less authority in the government, but central to the developments around the Amur question, was Yegor Kovalevskii, who from 1856 to 1861 headed the Asiatic Department of the Ministry of Foreign Affairs. He had long served on diplomatic assignments to central Asia and the Near East, including a mission to China in the late 1840s.[74]

Among the members of the imperial family, Grand Prince Konstantin Nikolaevich and Grand Princess Elena Pavlovna may be counted as part of this nationalist–reformist group. The daughter of the Prince of Württemberg, Elena Pavlovna had come to Russia at the age of 17, and had fully adapted herself to her new environment in St. Petersburg. She became one of the great inspirations for enlightenment and emancipation during the reign of Nicholas I, and at her famous "salon" gathered together many of the luminaries of Russian and European science and letters. Her guests included Karl Baer, the poets Fedor Tiutchev and Vladimir Odoevskii, and even Alexander von Humboldt, as well as liberal-minded figures from the government, such as Kiselev and the future governor-general of Eastern Siberia himself. During the 1840s, these gatherings became a center for discussion of ideas of reform in Russia, specifically of the emancipation of the serfs. Elena Pavlovna enjoyed a particularly close relationship with Nicholas I, who held her in high esteem and greatly respected her opinion.[75] An equally strong and consistent voice for liberal reform was that of Konstantin Nikolaevich, who had received a naval education under the tutelege of Admiral Lütke and was appointed as the head of the Naval Ministry in 1853. Under his direction, a series of progressive reforms were carried out in it, and its journal *Morskoi Sbornik* became one of the most important printed organs for reform in the country.[76] At the same time, the Grand Prince served as the vice-president of the Russian Geographical Society. Both he and Elena Pavlovna provided much support for

[72] "Perovskii, Lev Alekseevich," p. 51; "Perovskii, Vasilii Alekseevich," pp. 50–51.
[73] "Kiselev, Pav. Dmitr," pp. 762–763; "Kiselev, Pavl Dmitrievich," pp. 293–294.
[74] Kovalevskii, *Puteshestvie*; "Kovalevskii, Egor Petr.," p. 120.
[75] Lincoln, "Circle," pp. 375, 381–387; Kipp, "Grand Duke Konstantin Nikolaevich," pp. 107–108; "Elena Pavlovna," p. 27.
[76] Indeed, as Kipp notes in the early years of Alexander II's reign "naval affairs became something of a microcosm of the politics of reform in Russia." "Grand Duke Konstantin Nikolaevich," p. 177 (quote); Clay, "Ethos," *passim*; Dneprov, "*Morskoi Sbornik*," *passim*. This point will be of considerable importance later in our study.

reformist sentiments in court circles,[77] and Konstantin Nikolaevich's involvement with naval affairs gave him a particularly important role in the evolution of the Amur issue in the 1850s. As a final member of this group, the tsarevich Alexander might be mentioned. His role in the Amur question is discussed below.

By the late 1840s, this reformist tendency was becoming recognizably distinct within the government, and along with it a general policy orientation which contrasted sharply to that of the conservatives. In regard to internal affairs the points of contention are already apparent, as one group favored social reform – most importantly, the abolition of serfdom – while the other resisted this with all available means. In regard to foreign policy, the contrast was no less evident, for the perspective of the reformists betrayed some influence at least from the various elements of Russian nationalism discussed above. They called for a foreign policy which, if not aggressive would at least be resolute and not reluctant to undertake decisive action in order to defend and promote Russia's interests internationally. They were critical of the flaccid and accommodating policies of Nicholas I, whom they felt had been intimidated into inactivity by a fear of arousing the suspicions and antipathy of the other European powers.[78] They were disgusted in particular with Nesselrode and Kankrin, who were both of non-Russian origin and thus tangible enbodiments of precisely that non-native, "German spirit"[79] that was felt to dominate in the government and to be doing so much damage to Russian interests. The poet Tiutchev, an ardent Pan-Slav who had himself served for some time in the foreign ministry, expressed these sentiments scornfully in 1850 in a poem dedicated to his former director:

> No, my little dwarf, my coward without precedent.
> . . .
> You will not mislead Holy Russia
> With your faithless soul.
> . . .
> No, you will not hold back
> Russia's universal destiny.[80]

It was however not simply the manner in which foreign policy was to be carried out that needed to be altered. Some of the reformers we are discussing were caught up in the new ideology of a Russian mission, and felt that this element – this "moral law" Struve spoke of – should be recognized as an

77 "Konstantin Nikolaevich," pp. 67–69; Lincoln, "Circle," p. 381; Kabanov, *Amurskii vopros*, pp. 109–110. 78 Struve, *Vospominanii*, p. 9.

79 This was Murav'ev's evaluation. Barsukov, *Graf . . . Amurskii*, I, p. 671; Presniakov, *Emperor Nicholas I*, p. 91.

80 "Na grafa Nesselrode," in Tiutchev, *Stikhotvoreniia*, pp. 131–132. On Tiutchev and Nesselrode, see Tiutchev, "Pis'ma." pp. 524–534.

inspirational basis in the formation of Russia's international policy. Pan-Slav sympathies among them were pronounced, but together with these the reformers entertained an explicit interest in expanding the scope of Russian activities in Asia. A number of the individuals we have named were directly involved with Asian affairs. These included Vasilii Perovskii, Kovalevskii, who concluded the secret Treaty of Kul′dzha in 1851 and opened Western China to Russian trade, and not least of all Konstantin Nikolaevich, who as Minister of the Navy bolstered Russian naval communications with far-eastern waters. These individuals were appreciative of the developing international competition in Asia, and convinced of the importance of Russia's participation in it.[81] These sentiments insured that their response to Murav′ev's appeals would be quite different from that of Nesselrode and his colleagues.

From an early age, Murav′ev had been associated with the members of this group, and indeed to a considerable degree he owed his career to them. He had been a chamber page in the service of Elena Pavlovna, with whom his relations were particularly close. In St. Petersburg in the 1840s, he was a frequent visitor at her salon before his appointment to Eastern Siberia.[82] He was also close to Lev Perovskii. His entry into the civil service in 1845, after his retirement from the military, had been under Perovskii's wing, and it was the latter who arranged for Murav′ev's appointment in 1846 as governor of Tula. It was apparently also Perovskii who first conceived of sending Murav′ev to Eastern Siberia, once it became clear that the former governor-general V. Ia. Rupert would have to be removed becaue of corruption. Perovskii then convinced Elena Pavlovna to use her influence with Nicholas I to arrange for Murav′ev's appointment.[83] Thus even before assuming his new position as governor-general, the attention of the reformist tendency in the government was fixed upon him, and their hopes only intensified thereafter. The degree of Konstantin Nikolaevich's support, for example, was suggested in a message he sent Murav′ev in 1851, requesting to be kept fully informed of the governor-general's activities: "every item of news about what is being done for the good of Russia is comforting (uteshitel′no) for Russians."[84]

Once installed in Eastern Siberia, Murav′ev not only did not disappoint his patrons, but with his unbounded enthusiasm and irresistable energy in raising and pressing the Amur question must have surpassed their every expectation. For at this point in our discussion it is clear that the issue of the Amur was

[81] Veniukov, *Iz Vospominanii*, I, p. 246; Kabanov, *Amurskii vopros*, pp. 109–110; Stanton, "Foundations," p. 138; Gillard, *Struggle*, p. 105.

[82] Stanton, "Foundations," pp. 107–108.

[83] Stanton, "Foundations," pp. 116, 120; Butsinskii, *Graf . . . Amurskii*, p. 13; Struve, *Vospominanii*, p. 7. Zavalishin, however, offered a different version of Murav′ev's appointment, claiming it had been entirely unintended and mistaken: the tsar had actually wanted to name a different N. N. Murav′ev (Karskii) to the post, and the confusion of identities was sorted out only after it was too late. Zavalishin, *Zapiski*, II, p. 318. There was a confusingly large number of Murav′evs, to be sure, but this sort of mix-up was unlikely.

[84] Quoted in Barsukov, *Graf . . . Amurskii*, I, pp. 291–292.

adopted by this group as a whole, in the sense that they supported Murav′ev and looked upon his cause as their own. In precisely the same way that the Russian Geographical Society had seen the developments in southeastern Siberia as an example of Russia setting out on a new and positive course, so the nationalist tendency in the government believed that here was a means to effect a definitive break with Nesselrode's stifling system of appeasement and pandering to Europe. The policies pursued by Murav′ev would secure Russia's position in Asia and on the Pacific, and insure that neither its own territorial integrity nor any imperialist advantage in China and East Asia would be sacrificed to the other European powers. From this standpoint, the acquisition of the Amur represented an excellent – and, indeed, a necessary – basis for a reformed and renovated Russia.

"The Amur is going to drive you insane"

In our discussion of Alexander Balasoglo it was noted that at the same time he was preparing his essay for Murav′ev, he was helping to plan a naval expedition to the Far East. On this latter project he collaborated with his close friend, the naval officer Gennadii Nevel′skoi.[85] In view of the prominent role that Nevel′skoi was to play in the events in the Far East, it is unfortunate that there is not more information on his background. What we do know, however, confirms his strong sympathies with the nationalist–reformist movement. Nevel′skoi was friendly with other *petrashevtsy* in addition to Balasoglo, and it is possible that he had some marginal involvement in the *kruzhok* himself.[86] He was in any event very active in the Russian Geographical Society, where as one of its "young Russian forces" he was determined to make a patriotic contribution in the form of a naval mission to the Far East.[87] The intention of the expedition was to clear up the confusion about the estuary of the Amur: specifically, Nevel′skoi and Balasoglo wanted to disprove Gavrilov's findings and demonstrate both that Sakhalin was an island and the Amur a fully navigable river. Balasoglo's arrest did not daunt Nevel′skoi in his determination to see this project realized. In 1847 he had been given command of the ship *Baikal*, commissioned to be built in the following year in Helsingfors, with orders to deliver provisions and naval supplies to Russian settlements on the Sea of Okhotsk and Kamchatka. Nevel′skoi resolved to try and include the exploration of the mouth of the Amur as part of his itinerary. He described

[85] Evans, *Petrasevskij Circle*, p. 54. [86] Alekseev, *Amurskaia ekspeditsiia*, p. 17.

[87] Boris Polevoi maintains that it was Balasoglo who first interested Nevel′skoi in the prospect of a voyage to the Far East, while Alekseev suggests that Nevel′skoi first heard about the Amur from Kruzenshtern during his studies in the Naval Academy in the 1830s. Polevoi, "Opoznanie," p. 156; Alekseev, *Delo*, pp. 43, 51–52; Val′skaia, "Petrashevtsy," p. 60. Whichever is true, these points indicate the important role which naval officers played in keeping the Amur question alive. Murav′ev's association with it, by contrast, was far more fortuitous. Also see Alekseev, "Gennadii Ivanovich Nevel′skoi," pp. 8–9.

his intention to the head of the Navy A. S. Men′shikov, who expressed reservations about the advisability of the undertaking but did not expressly forbid it.[88] It was at this time that Nevel′skoi first made the acquaintance of Murav′ev at the Geographical Society, and the new governor-general expressed his full support for his endeavor.[89] With considerable effort, Nevel′skoi managed to arrive at Kamchatka early enough in 1849 to allow time for his expedition to the mouth of the Amur, which he carried out during the summer. In his haste to get underway, however, he did not wait in Kamchatka to receive the official sanction for the expedition, which was on its way from St. Petersburg.[90]

The results of Nevel′skoi's 1849 expedition were exactly what he had anticipated. There was no isthmus connecting Sakhalin to the mainland, and consequently passage from the Sea of Okhotsk into the Tatar straits was not blocked. Moreover, he determined that the mouth of the Amur was in fact navigable by ships of any size.[91] When this news reached St. Petersburg, however, the government was incensed. In a special meeting of the so-called Giliak committee in February, 1850, Nevel′skoi was confronted by the conservatives Nesselrode (who chaired the meeting), Chernyshev, Seniavin, and Vronchenko, all of whom denounced his insubordination in not awaiting the arrival of official permission and instructions for his expedition, and demanded that he be broken in rank. They also challenged his findings, and invoking the authority of European voyagers and the expeditions of Krusenstern and Gavrilov declared his conclusions in regard to the mouth of the Amur to be invalid. His only support at this meeting came from Men′shikov and Perovskii. Their defense of Nevel′skoi and his actions was strident, however, and Nicholas I chose not to follow the extreme course called for by his ministers. Nevel′skoi was sent back to the Far East to establish a wintering post for the Russian–American Company on the southern coast of the Sea of Okhotsk, but explicitly forbidden from venturing south to the mouth of the Amur.[92]

[88] Nevel′skoi, *Podvigi*, pp. 58, 60. Although Men′shikov, a former governor of Finland whose inept command led to military débâcles early in the Crimean War, cannot be counted among the nationalist tendency in the government, he was an important early supporter of Nevel′skoi's and Murav′ev's efforts. Riasanovsky, *Nicholas I*, p. 43; "Men′shikov, Aleksandr Sergeevich," pp. 355–356. Kabanov mistakenly identifies him in the group of Murav′ev's opponents. *Amurskii vopros*, p. 109.

[89] In Nevel′skoi's account it was *he* who described the project to Murav′ev and convinced him of its importance. According to Barsukov, however, "the inception and resolution of the question of the location and occupation of the mouth of the Amur . . . was entirely the work of N. N. Murav′ev alone." Most subsequent accounts concur with this. *Podvigi*, p. 59; Barsukov, *Graf . . . Amurskii*, I, p. 172. In the light of our earlier discussion of Murav′ev and Balasoglo, however, it seems entirely likely that Nevel′skoi was justified in claiming the honor for himself. There was to be a yet more significant disagreement between Nevel′skoi and Murav′ev: see below, pp. 214–217.

[90] Nevel′skoi, *Podvigi*, p. 76.

[91] Nevel′skoi, *Podvigi*, p. 90; Romanov, "Prisoedinenie," p. 113.

[92] Nevel′skoi, *Podvigi*, pp. 95, 101–104; Struve, *Vospominanii*, pp. 94; Stanton, "Foundations," pp. 153–154.

Not to be deterred in his resolve to enhance Russia's position on the Pacific, Nevel'skoi ignored these directives, and in the summer of 1850 founded the Russian post of Nikolaevsk at the mouth of the Amur. He also carried out surveys in the De Castries straits south of the Amur. In December of that year a special committee was convened once again in St. Petersburg for the purposes of reviewing these activities, once again chaired by Nesselrode and dominated by the same conservative forces as before. The scenario of the preceding February was repeated to the letter. Nevel'skoi was roundly condemned for his boldness and audacity in deliberately violating his orders, and his opponents once again insisted that he be broken in rank. Once again Men'shikov and Perovskii spoke in his defense, but to no avail. The committee, fearful of injuring Russia's friendly relations with China and thereby the fragile trade at Kiakhta, resolved to remove the Russian post from the mouth of the Amur.[93]

This time, however, Murav'ev was not to be defeated, and in a special audience brought the situation to the attention of the tsar himself. Nevel'skoi later reported that Nicholas I considered his actions to be "noble" and "patriotic," and ordered that he be decorated. Moreover, in expressing his satisfaction during this interview, the emperor is said to have uttered the words which have become famous in the annals of the Amur epoch: "Where the Russian flag has once been hoisted, it should not be taken down."[94] He did not approve the resolution of Nesselrode's committee, and declared that it would have to convene again, this time under the chairmanship of the heir to the throne, the future Alexander II. This assured that the affair would have a different outcome, for Alexander was fundamentally more sympathetic to the reformist perspective – it was he after all who as the "Tsar-Liberator" was finally to carry out the long-awaited emancipation of the serfs in 1861, along with a series of other civic reforms. Alexander had also been touched in the early 1850s by the enthusiasm Murav'ev generated with his activities in Eastern Siberia and the Far East (if to a lesser extent than his brother Konstantin), and he expressed his support for the governor-general with what one biographer characterized as a "steadfast resolution."[95] The reconstituted committee then convened for a second time in February 1851. Nevel'skoi's actions were endorsed, and it was

[93] Nevel'skoi, *Podvigi*, pp. 108–111; Romanov, "Prisoedinenie," pp. 117–119; Barsukov, *Graf . . . Amurskii*, I, p. 281.

[94] Nevel'skoi, *Podvigi*, p. 112.

[95] Tatishchev, *Imperator Aleksandr II*, I, pp. 273 (quote), 274–277, 127–128; Holborn, "Russia," p. 390. "[I]t was the Tsarevich's personal intervention which turned the scales in [Murav'ev's] favor." Mosse, *Alexander II*, p. 121; Hoetzsch, *Russland*, p. 31. Murav'ev's adjutant Struve recounted that, as Murav'ev was leaving the Winter Palace after his conference with Nicholas I in late 1850, he suddenly heard someone call his name. Turning around, he saw that it was Alexander, and hurried back to him: "It has been ordered that I will be present at the consideration of the Amur question," the tsarevich is said to have breathlessly exclaimed: "We will work and labor (trudit'sia) together!" Whereupon he embraced Murav'ev in a burst of satisfaction and kissed him. *Vospominanii*, p. 102. As we will see, this support was not to last.

decided to maintain the post at Nikolaevsk, but not to expand it.[96] All Russian activity in the Far East was still officially presented as part of the operations of the Russian–American Company, in which way the government avoided assuming direct responsibility and hoped thereby not to stir the suspicions of the other powers.

In September, 1852, Murav′ev's efforts received a serious setback when he learned that the most prominent of his supporters, Lev Perovskii, was to retire from his position.[97] The effect of this loss was offset, however, by the support of Konstantin Nikolaevich. In the same year, the Grand Prince was appointed to the so-called "Siberian committee" overseeing affairs east of the Urals, and on this occasion he reaffirmed to Murav′ev his approval for "the adoption of measures which would give Russia a greater importance in the Far East."[98] Also working to Murav′ev's advantage was the return to St. Petersburg of the Akhte expedition in early 1853. This expedition, it will be remembered, had been dispatched in 1849 after the Russian government had decided to relinquish formally all claims to the Amur and needed more detailed topographical surveys and some sort of precise determination of the extent of Chinese control in the region. Murav′ev in Eastern Siberia was furious when he learned of the nature of Akhte's tasks, and detained him as long as he could in Irkutsk.[99] Akhte was finally permitted to carry out his assignment along the southern slope of the Yablonovy mountains north of the Amur, and returned to Irkutsk in December 1852. He had found that the topography of the region differed significantly from that assumed by the Treaty of Nerchinsk. The existing boundary markers, he reported, were further to the south than even Middendorf had asserted, and his observations among the indigenous peoples of the region indicated that the Chinese exerted no authority whatsoever in this region, and indeed never had.[100] In this way, Akhte's expedition ended up actually supplying Murav′ev with yet more material for his struggle in St. Petersburg.

Thus the course of events down to 1853 is clear. In St. Petersburg the two opposing tendencies in the government were grouped on opposing sides of the Amur question, lending this issue a recognizable political relevance going beyond the specifics of the situation in the Far East. Yet while even the emperor himself was obviously sympathetic to some degree to the exploits of Murav′ev and Nevel′skoi in the Far East, the conservative ministers around him were nonetheless able to prevent the government from directly sanctioning their advances. This was demonstrated on the one hand by the fact that all of these activities continued to be ascribed to the Russian–American

[96] Nevel′skoi, *Podvigi*, pp. 112–113; Barsukov, *Graf . . . Amurskii*, I, pp. 281–282.
[97] Barsukov, *Graf . . . Amurskii*, I, p. 316. [98] Quoted in Stanton, "Foundations," p. 194.
[99] Barsukov, *Graf . . . Amurskii*, II, pp. 45–46.
[100] Alekseev, *Russkie geograficheskie issledovaniia*, p. 37; Romanov, "Prisoedinenie," p. 106; Stanton, "Foundations," pp. 194–195.

Company and presented as the result of their local commercial needs – that is, trade with the natives along the coast of the Sea of Okhotsk – rather than as the determined policy of the Russian government. On the other hand, and more importantly, the government refused to grant approval for Murav′ev's much-promoted proposal for a naval squadron to descend the Amur river from its headwaters to the ocean.[101] For the purposes of establishing Russian claims in the region, such a descent would be far more significant that the occupation of the river's mouth by Nevel′skoi. At the same time, it would be a far more visible and unambiguous demonstration for China and the Western powers. Russia's policies in the Far East, however, were still dominated by the cardinal concern of preserving the Kiakhta trade and of not antagonizing the European powers with seemingly expansionist intentions, and Murav′ev's proposal was successfully resisted.

This state of affairs might have persisted for some time to come had the onset of the Crimean War not sent tremors down to Russia's very foundations. This war, from the nationalist standpoint, represented Russia's great Armageddon, a natural and long-awaited conclusion to the tensions which had been developing between Russia and the West in the course of the preceding decade. There was even a relief of sorts that these tensions had finally come into the open, and that Russia's feeble masquerade as a fraternal European power was exposed for the sham which it was generally felt to be. At the outset, the nationalists saw the war as the first major opportunity for Russia to exercise its liberating mission as savior of their Slavic brethren imprisoned within the Ottoman empire, and aimed at nothing less than the establishment of a grand Slavic federation with its capital at Constantinople.[102] These fanciful dreams were dispelled quickly enough by Russia's dramatic military failures, but these failures in turn took on a constructive meaning of their own as tragic proof of the country's stagnation under Nicholas I's Official Nationality. More than anything else, the possibility of defeat pointed to the fundamental need for the reconstruction of Russian society. The conservative tendency in the government represented by Nesselrode and his colleagues was soon to be thoroughly disgraced, and the way left open for the ascent of the nationalist–reformist forces.

In April 1853, Akhte reported on his findings at a meeting of a special committee in St. Petersburg. Murav′ev, who attended this meeting, had arrived in the capital the previous month, and at that time had presented the tsar with his memorandum on changing Russian perspectives on the Pacific discussed above. After listening carefully to Akhte's report and examining the charts and maps which had been prepared, Nicholas I accepted the governor-general's conclusions that the Amur region should be Russian and ordered that border

[101] Struve, *Vospominaniia*, pp. 126–127.
[102] Riasanovsky, *Nicholas I*, p. 165; Hunczak, "Pan-Slavism," pp. 88–89.

negotiations with China should proceed on this new basis. Moreover, the tsar now realized that his foreign minister had been misleading him in claiming that this region was heavily fortified by the Chinese. His confidence in Nesselrode was shaken, and the latter was removed from all further deliberations on Russian policy in the Far East.[103] Nevel′skoi's ongoing activities in establishing Russian posts at the mouth of the Amur from 1850 to 1853 were now placed directly under the aegis of the Russian government, and became thereby expressions of official Russian state policy.[104] Nevertheless, the Amur question was still not entirely resolved. Nicholas I, dismissing Murav′ev's pleas with the somewhat irritated retort that "the Amur is going to drive you insane some day," persisted in his refusal to permit a naval expedition down the course of the river.[105]

Permission for this was not long in coming, however. The clouds of war, which had begun to gather early in 1853, grew darker as the year progressed. It appeared increasingly likely that the Far East would be a theater of the war, and thus specific attention had to be given the question of military fortification of Russia's far-flung settlements on the Pacific. Murav′ev's repeated warnings about European territorial designs on the Russian Far East, and the likelihood of their success if Russia could not navigate on the Amur, seemed more and more compelling. Accordingly, in January 1854 Murav′ev received orders from the capital which fulfilled his two long-awaited dreams. Negotiations with China on the issue of the Amur frontier were now to take place under his direction as the chief negotiator, and permission was finally given for a naval mission down the Amur, for the purpose of delivering supplies to Kamchatka.[106] This famous first descent took place in May and June 1854, and was followed by a similar voyage in each of the following three years.[107]

The Russians were not disappointed in their anticipation of conflict in the Far East. In August 1854, the British conducted a land assault on the port of Petropavlovsk on Kamchatka. In refreshing contrast to the military débâcles experienced on other fronts, Murav′ev led a heroic defense of the port, and his success in repulsing the attack attracted a good deal of positive attention throughout the country.[108] A second assault was attempted in May of the following year. In the meantime, however, Murav′ev had lost no time in taking

[103] Barsukov, *Graf . . . Amurskii*, I, pp. 324–325; Struve, *Vospominanii*, p. 157; Stanton, "Foundations," p. 198–200.

[104] Nevel′skoi, *Podvigi*, pp. 221–222; Romanov, "Prisoedinenie," p. 129.

[105] Barsukov, *Graf . . . Amurskii*, I, p. 325; Gillard, *Struggle*, pp. 104–105.

[106] Barsukov, *Graf . . . Amurskii*, I, pp. 345–346; Romanov, "Prisoedinenie," pp. 109–113; Lin, "Amur Frontier," pp. 9–10.

[107] The second voyage, in 1855, was accompanied by the naturalists who comprised the first contingent of the Great Siberian Expedition. Maak, *Puteshestvie*, p. 1; Barsukov, *Graf . . . Amurskii*, I, pp. 417–418; Romanov, "Prisoedinenie," p. 129; Bassin, "Russian Geographical Society," pp. 250–251.

[108] Lin, "Amur Frontier," p. 12.

advantage of Russia's recent entrenchment on the Amur. Fearing that Petropavlovsk could not withstand another assault, he ordered the relocation of its entire population to Nikolaevsk,[109] and thus the British attacked a totally deserted town. The superior geographical knowledge provided by Nevel'skoi's expeditions, namely that Sakhalin was not a peninsula and that the Tatar straits were accessible from the north, allowed the company of Russian ships from Petropavlovsk to elude the British and arrive undisturbed at the mouth of the Amur, where they could anchor in safety.[110] The experience in the Crimean War thus appeared to confirm that Murav'ev had been correct in his insistence on the value of the Amur as a key part in the defense strategy of the Russian Far East.

It was appropriate that the death of Nicholas I in 1855 should follow shortly after the collapse of the system with which he had ruled the country for over a quarter of a century. The new tsar was confronted not only with the internal reconstruction of Russian society, but with the resurrection of the commanding international stature that had declined so unhappily from that glorious moment four decades earlier when his uncle Alexander I led a triumphant Russian army down the streets of Paris. Alexander II, moreover, could no longer fail to recognize, as had his father's ministers, the fundamental changes that had occurred in the international situation as a result of the expanding European presence in East Asia. He understood that the geographical arena for Russia's most important international challenge – namely, the contest with Europe to regain prestige and dominance after the Crimean débâcle – had been dramatically expanded across the globe, through a series of actions in which the Russians had failed to participate adequately. He appreciated, moreover, that in the future Russia's international position would have to be secured and maintained in terms of a complex geostrategic calculus which had no precedent in modern Russian history – namely, that from this point on, the struggle with the West would be conducted not only, and often not primarily, in Europe itself, but in Central and East Asia as well.[111] "[J]e prévois," he wrote to his friend Prince Alexander Bariatinskii in 1858, "que c'est là [in Asia] que les destinées futures seront décidées."[112] This premonition insured among other things his preparedness to support a program of extending Russia's imperial dominions in Asia, and in assembling his government he relied not inconsiderably upon individuals who were well inclined to bring such a program to life. These included Bariatinskii himself, whom Alexander appointed as viceroy of the Caucasus, Egor Kovalevskii, promoted at this time

[109] Stankov, "Pochemu," pp. 23–24; Nevel'skoi, *Podvigi*, p. 359.
[110] Stephan, "Crimean War," pp. 263–272; Sladkovskii, *Istoriia*, p. 221; Romanov, "Prisoedinenie," p. 131.
[111] On Alexander's support for an expansionist policy in Asia, see Rieber, *Politics*, pp. 59–93; Gillard, *Struggle*, p. 101–103; Mosse, *Alexander II*, pp. 121, 123; Pereira, *Tsar-Liberator*, p. 88.
[112] Cited in Rieber, *Politics*, p. 122.

to direct the Asiatic Department of the Foreign Ministry, and Nikolai Ignatev, the brilliant diplomat responsible for negotiating the Treaty of Peking in 1860, who was to succeed Kovalevskii as the head of the Asiatic Department and later served as ambassador to Constantinople. All of these individuals were concerned with advancing Russia's position in Asia,[113] and Kovalevskii – together with Konstantin Nikolaevich, who retained his important position as head of the Naval Ministry – became the chief advocate of expanding Russian influence and territories in the Far East.[114] Alexander Gorchakov, who was chosen to replace Nesselrode at the head of the Ministry of Foreign Affairs, did not share the unrestrained enthusiasm of this group for a forward policy in Asia, but he was prepared to lend his strategic support, and certainly did so in the case of Russian pretensions on the Amur.[115]

A few skirmishes remained to be fought out, but from this point on Murav′ev had won his war, and he knew it.[116] An indication of the government's new position regarding its far-eastern territories came in early 1857, when Gorchakov sent the following message to the Russian ambassador in the United States Eduard de Stoeckl: "We have historical rights to the estuary of this river . . . Free navigation on the Amur is an extreme necessity for us, and is not open to dispute."[117] Indeed, not only had Murav′ev's position been accepted without reservation in St. Petersburg and become state policy, but there was a beginning of a shift of interest in the country at large to the Amur. As noted, the unfortunate outcome of the Crimean War in the West cast an unusually bright light on the Russian successes on the Pacific, quite out of proportion with their actual significance. In the immediate aftermath of the war, the Far East seemed to offer the opportunity for Russia to recoup at least some of the glory it had lost elsewhere, and through territorial acquisition could give some compensatory relief for Russia's frustrated designs on Constantinople.[118] In 1856, Murav′ev could note with some satisfaction that for the first time everyone in the government – the imperial family and the various ministers – supported him. In December of that year, the Primorskaia oblast′, comprising Kamchatka, the coast of the Sea of Okhotsk, and the Amur, was established as an official administrative district, and the Kamchatka flotilla was solemnly rechristened as the "Siberian–Pacific Fleet."[119] All that remained was formal international ratification and recogni-

[113] Gillard, *Struggle*, pp. 102–105; Quested, *Expansion*, p. 65.

[114] Veniukov, *Iz Vospominanii*, I, pp. 246–247.

[115] "Gorchakov, Aleksandr Mikhailovich," pp. 599–600; Rieber, *Politics*, pp. 73–76ff; Popov, "Tsarskaia diplomatiia," p. 190; Stanton, "Foundations," p.244n.

[116] Barsukov, *Graf . . . Amurskii*, I, p. 479.

[117] Quoted in Popov, "Tsarskaia diplomatiia," p. 190.

[118] Popov, "Tsarskaia diplomatiia," p. 188; Pushkarev, *Emergence*, p. 377; Sarkisyanz, "Russian Imperialism," pp. 50–51, 67.

[119] Romanov, "Prisoedinenie," p. 138; Stanton, "Foundations," p. 246; "Das neue Armeekorps," p. 387.

tion of the new status quo.[120] In 1858, two separate agreements between Russia and China were negotiated, one by Murav′ev at the border settlement of Aigun (in recognition for which he received the title Count Amurskii[121]), and the other by Putiatin as part of the Treaty of Tientsin. The provisions of these agreements were conflicting, however, and the Treaty of Aigun at least was never ratified by the Chinese.[122] It was only in 1860, when Nikolai Ignat'ev masterfully exploited his position as intermediary between the Chinese and the French and British, that the question of jurisdiction over the Amur was definitively resolved in Russia's favor. Russia received jurisdiction not only over the river and all of its left bank, but over the territory extending from the Ussuri river (a southern tributary) eastward to the sea as well.[123] With this, Murav′ev's dream had at long last been fully realized. The stage was finally set for Russia to test its visions of the Amur.

[120] For a detailed account of the international diplomacy around the Amur issue in this period, see Quested, *Expansion*, *passim*.
[121] Ravenstein, *Russians*, p. 144.
[122] Mancall, "Major-General Ignatiev's Mission," p. 58.
[123] Evans, *Russo-Chinese Crisis*, pp. 1–13; Hudson, "Far East," pp. 701–703.

PART II

Introduction

The death of Nicholas I in 1855 and the accession of his son Alexander, framed as they were against the background of defeat in the Crimean War, heralded the dawn of a very new era for Russia and at the same time marked one of the decisive turning-points in its modern history. The disastrous outcome of the war appeared to confirm beyond any question the damage which 30 years of Official Nationality had done, not only in regard to Russia's military capabilities *vis-à-vis* the West, but more fundamentally to its underlying economic, political, and social structure. Disillusionment with the *ancien régime* was indeed so strong that many thoroughly patriotic Russians actually cultivated a kind of resigned defeatism during the course of the war and viewed its unfortunate outcome as a perversely appropriate and necessary climax to the entire sad period. The historian Sergei Solov′ev, for example, recalled his deep ambivalence at hearing the news that Sevastapol had fallen to the enemy. While his patriotic feelings were "terribly injured by this humiliation of Russia," he explained, at the same time he believed that "only disaster, and precisely military defeat, would be able to bring about the saving transformation and put an end to further decay. We were convinced that success on the battlefield would only draw our fetters yet tighter." "I am glad, glad that we have been beaten," the Decembrist N.R. Tsebrikov tearfully confessed on the same occasion: "Now we shall wake up. You shall see great steps forward . . . It's too bad about Sevastapol, it's too bad about all the spilled blood. But it's all for the best, [because] now our eyes will open up."[1]

Given this degree of anguished desperation, which apparently extended even into the royal family itself,[2] it is hardly surprising that Nicholas I's death should have brought immense relief and even rejoicing. Herzen in London described how champagne was poured and congratulations exchanged with a euphoria that not only brought "tears of sincere joy" to the eyes of the Slavic emigré

[1] Levin, "Krymskaia voina," p. 317; Berlin, "Russia," p. 19.
[2] In the person of Konstantin Nikolaevich. Kipp, J. W., "Grand Duke," p. 138.

community but was shared by the general population as well. "I saw not a single person on the street," he wrote, "who would not have breathed easier, having learned that this sore had been removed from the eye of humankind, and who would not have rejoiced at the fact that this oppressive tyrant in jack-boots was finally returned to the elements."[3] In Moscow the historian K.N.Bestuzhev-Riumin, hardly the radical opponent of tsarism that Herzen was, embraced a colleague upon hearing of Nicholas I's demise, and spoke of the "general delight" among his countrymen that the news occasioned.[4] "It somehow became easier to breathe," recalled Ivan and Konstantin Aksakov's sister Vera: "fantastic hopes were suddenly resurrected; a hopeless situation . . . seemed at once to be open to change."[5]

This final point was significant. It was generally appreciated that the death of the tsar marked the beginning of a new era, in which Russia would be able to develop in new and positive directions and try to call into life the program of reform and renovation about which for so many years it had been possible only to dream and speculate. An air of intense, almost giddy optimism animated the nation, a veritable festival of hope for what the future could and would bring to Russia. As one observer in St. Petersburg reported in 1858 to a friend travelling abroad:

> When you arrive [back] in Russia you will run the risk of not recognizing it. Externally everything seems the same, but you feel an inner renovation in everything, you feel that a new era is beginning . . . Read our newspapers and journals, listen to the conversation in the brilliant *salons* and in modest homes, and you will be amazed at the work which these heads are accomplishing. From all sides, ideas and new perspectives are little by little displacing the old routine . . . We are entering a new era, an era of political and social life.[6]

The common image, articulated time and again as people described their feelings, was that of a country stirring to life or waking up after a protracted period of somnolence and lethargy. The geographer Mikhail Veniukov, who was to play an important role in the events that unfolded in the Russian Far East during this period, compared his homeland at the time of the death of Nicholas I to a Rip Van Winkel-like figure from Russian folklore. "After an involuntary slumber of 30 years, Russia – just like the fairy-tale hero Ilia Muromets – awoke [to carry out] numerous glorious feats, the equal of which I won't see for the rest of my days."[7] It was in this atmosphere of hopeful anticipation and heartfelt desire to work for the rejuvenation of Russia that the final planning and realization of the so-called Great Reforms of the early 1860s took place, most notably the emancipation of the serfs in 1861.

At the same time, this mood of euphoric activism served to focus the vital

[3] Levin, "Krymskaia voina," pp. 331–332. [4] Levin, "Krymskaia voina," p. 333.
[5] Levin, "Krymskaia voina," p. 355.
[6] Letter of 6 May 1858 to from F. (?) Balabin to P. D. Kiselev, quoted in Zablotsskii-Desiatovskii, *Graf P. D. Kiselev*, II, pp. 341–342. [7] Veniukov, *Iz vospominanii*, I, pp. 198–199.

attention of the country upon Russia's remote and little-known frontier in southeastern Siberia. That this was so, it must be recognized, was more than anything else due to the unstinting efforts of governor-general Murav′ev throughout the 1850s: his widely rumored commitment to social reform, his undisputable military victories during the Crimean War, and of course most spectacularly his political and diplomatic success in securing the annexation of an enormous and to all initial appearances fabulously valuable river region to the Russian empire. "The Amur epic (epopeia) was on everyone's lips," wrote a youthful contemporary. "The figure of Murav′ev had become a kind of legend."[8] By the end of the decade, the governor-general's undertaking in the Far East appeared to an ever-increasing number of Russians to be perhaps the most illustrious practical realization of the determination to revitalize and reinvigorate national life and the national spirit – of precisely that program, in other words, that everybody was talking about and yearning to accomplish throughout Russia. The author of a pamphlet written to introduce the unfamiliar territories on the Pacific to a popular audience in European Russia betrayed this attitude in his choice of the following verse by M.P. Rozengeim as an epigraph at the beginning of his account:

> Russia! Wake up – it's time, it's time!
> Arise, my native land
> Cast off the shameful burden of sloth
> Arise – God is with you.
> . . .
> The hour of awakening has struck
> You can hear the call ahead.
> It is the Tsar who is summoning you
> To the cause of [Russia's] renovation.[9]

Here, in a few lines, the direct association of Russia's advance in the Far East to the larger project of national reform was conveyed in a way that no one could fail to appreciate, however little they may have known about the region itself.

Not a few souls indeed were so convinced of the national importance of what was happening on the Amur that they actually decided – often at a definite cost to their career – to venture out to the region and contribute as best they could to this glorious enterprise. Perhaps the most outstanding of these personages was the youthful Prince Peter Kropotkin, who upon graduation from the Cadet Corps in 1862 passed up an easy opportunity for an officer's commission in any of the country's most elite military regiments. Instead, he opted to join the recently formed "Mounted Cossacks of the Amur," a rag-tag and scruffy division hastily assembed by Murav′ev a few

[8] Miliutin, "General-gubernatorstvo," p. 595. [9] *Puteshestvie po Amuru*, 1860, p. 7.

years earlier. He offered the following explanation for this highly unorthodox decision, which scandalized his family and friends.

> My thoughts turned more and more toward Siberia. The Amur region had recently been annexed by Russia; I had read all about that Mississippi of the East, the mountains it pierces, the sub-tropical vegetation of its tributary, the Ussuri, and my thoughts went further – to the tropical regions which Humboldt had described, and to the great generalizations of Ritter, which I delighted to read. Besides, I reasoned, there is in Siberia an immense field for the application of the great reforms which have been made or are coming: the workers must be few there, and I shall find a field of action to my tastes.[10]

If at the end of the day Kropotkin was not exactly to discover that "field of action" that he so earnestly sought, his decision nevertheless insured that he would be a key observer and participant in the development of the Amur epoch, as we will have ample opportunity to witness.

The writer N.A. Mel′gunov was in all likelihood speaking only rhetorically when he observed in 1855 that "before everything else" it was necessary to give "space (prostor) to the Russian mind and Russian strengths, so that there would be some place [for us] to spread out and square our shoulders after a long stagnation."[11] His point, however, could well be taken in a strict geographical sense as well. There was a clear connection between reform, national rejuvenation, and political–territorial expansion in mid-nineteenth-century Russia, a connection that yielded the images of the Amur region to be explored in the following chapters.

[10] Kropotkin, *Memoirs*, p. 155. The "astonishment and pity" which Kropotkin described as being etched on the faces of his superiors in the Cadet Corps upon hearing of his decision to go to Siberia give a good indication of residual "official" views toward the region.

[11] Levin, "Krymskaia voina," p. 354.

5

Dreams of a Siberian Mississippi

"The merchant princes of the earth"

The 1840s and early 1850s witnessed a series of major developments on both coasts of the Pacific. The most momentous of these was the victory of Britain over China in the Opium Wars, which dramatically widened the scope for the penetration of European commerce and influence into the Middle Kingdom. Although it was to take yet more military pressure before the Europeans obtained concessions which fully satisfied their commercial appetite, the lucrative China market now lay more invitingly accessible than it had ever been, and the Western powers wasted no time in moving to secure their respective positions on it. This movement, in turn, gave rise to a contest among these powers themselves for diplomatic and political advantages and spheres of influence. The opening of China was followed in the 1850s by the opening of Japan to Western commerce, exposing a market less populous than that of China but powerfully attractive nonetheless. At the same time, on the opposite side of the ocean this period witnessed the definitive occupation by the United States of its own Pacific margin, an event which as we will see had a very particular significance for the Russians. Each of these factors taken alone would have naturally seemed portentous enough, but they combined to lend the Far East and the Pacific a distinct and quite unprecedented aura of burgeoning development. For those with imagination – and such individuals were never in short supply in Russia – it seemed not at all unreasonable to speculate that this was destined to be the arena of a great world civilization of the future.

These developments, as we have seen, were followed with interest and not a little concern by the Russians. Their initial response had been one of apprehension, in view of the detrimental effects of the Western incursion into China on the Russian caravan trade, but with each ensuing year the potential benefits became increasingly apparent. By the early 1850s, the Russians were making attempts of their own to conclude a trade agreement with Japan,[1] and under

[1] Lensen, *Russian Push*, pp. 308–354.

Alexander II an assertive and forward policy toward China was finally adopted as well. Reaction to the American occupation of Oregon and California was by and large favorable. With his pronouncement in 1853 that the Pacific was destined was destined to become the "Mediterranean of the future," Alexander Herzen gave perhaps the first clear indication of the extent to which the Russians were prepared to share in the international optimism for the future of commerce and civilization on the Pacific.[2] In the euphoria that followed the demise of Official Russia in the Crimean War, the death of Nicholas I, and the accession of Alexander II, these jubilant expectations were much intensified. With the occupation and annexation of the Amur and Ussuri regions, taking place at precisely this moment, it was widely felt that Russia was securing its access to this grand arena of the future.

These feelings reached a high pitch in 1858–1860, when the accords were signed by which China formally relinquished these regions to Russia. The acquisition of these territories and the firm link which they promised to provide to the Pacific arena seemed to insure Russia of yet another means for the process of renovation upon which it was embarking. A good indication of the high expectations accompanying these developments can be seen in the following observations, which appeared in an Irkutsk newspaper in 1858.

> [W]e cannot hide from ourselves [the fact] that over the last 30 years the historical significance of events on the Pacific has expanded more broadly with each decade, and that the time is not at all far off when questions of historical significance for all of humanity will be decided on this arena. The most powerful actors of the contemporary epoch, powers with world-wide influence, have all entered this new historical scene of activity and each is selecting a place for itself. From the northwest, Russia has moved to the shores of the Pacific Ocean, from the northeast – the United States; and from the southwest England has advanced its chain of naval bases, and established colonies in Australia and New Zealand which have already become significant. From the southeast, across the entire Pacific, France has occupied focal points for its future domination. China and Japan, which kept themselves sealed-off and isolated from the rest of the world, have now been forcefully drawn into the circulation of world history.[3]

As one regarded this exhilarating prospect, the author pointed out, it was possible to envision an entirely new era in world history dawning on the shores of

[2] Gertsen, "Kreshchenaia sobstvennost′," p. 110; Svatikov, *Rossiia*, p. 40. Although Herzen later expressed pride in having coined this expression, Karl Marx had in fact said very nearly the same thing several years earlier in a column for the *Neue Rheinische Zeitung*.

> Dank dem kalifornischen Golde und der unermüdlichen Energie der Yankees werden beide Küsten des Stillen Meers bald ebenso bevölkert, ebenso offen für den Handel, ebenso industriell sein, wie es jetzt die Küste von Boston bis New Orleans ist. Dann wird der Stille Ozean dieselbe Rolle spielen wie jetzt das Atlantische und im Altertum und Mittelalter das Mittelländische Meer – die Rolle der grossen Wasserstrasse des Weltverkehrs.

Marx, "Revue," p. 221; Gertsen, *Byloe*, VIII, p. 256n.

[3] S. Untitled communication, pp. 1–2 (quote); Volkonskii, "Pis′ma," p. 99; Azadovskii, "Putevye Pis′ma," p. 211.

the Pacific, a era which on a grander scale corresponded to the new period which Russia itself was entering. The alluring prospect of a great commercial civilization dawning in the East, moreover, offered a neat and effective – if only temporary – escape from the West, with which the Crimean War had left the Russians bitterly disenchanted.

This perspective served to augment the significance of the Amur enormously, for it was this river, as the vital conduit linking the motherland to the sea, that would enable Russia to play a central role in this new Pacific epoch. In the early years of the decade, Murav′ev had spoken specifically of the importance of the Amur to Russia's participation in the new networks of world commerce that were taking shape in the Far East, and after the river's annexation its vital quality as the "single route for the development of Russian military, commmercial, and industrial forces" on the Pacific was stressed again and again.[4] This point was articulated forcefully by Dmitrii Romanov, one of Murav′ev's chief lieutenants in Irkutsk and a prolific propagandist for the Amur, in a speech celebrating the signing of the Treaty of Peking in 1860. The Far East had only recently become the object of general attention of all developed nations, Romanov explained, but the main center of world commercial activity was already shifting very quickly to the Pacific.

> The events of our day indicate that it is here, on this vast ocean, that the tasks of the intercourse of peoples will be solved and new, broader interests will be drawn together . . . California and Panama from one side, newly-opened China and Japan from the other, the Australian continent now being settled in the south, a multitude of various countries touching on the ocean from all sides: all this must necessarily call forth new relationships and situations.

The speaker drew numerous historical parallels, including the discoveries of America and of the route around the Cape of Good Hope, and indicated that Russia's recent territorial advances in the Far East followed in this glorious tradition.

> If we remember that the true center of events has always been the sea – across which the most significant intercourse [among peoples] took place – then the shift of the world historical field to the Pacific is entirely in accord with the laws of history and the examples which it provides. Russia's new acquisitions move it to the shores of the Pacific, and draw it into the sphere of world interests and the general movement of nations. This, in our opinion, is the principal importance of the annexation by *Russia* of the Amur and Ussuri regions.[5]

Romanov's final point was echoed by the geographer Semenov, who asserted that despite its abundant natural resources and good agricultural lands, the Amur region would nonetheless be of only limited value to Russia if it did not

[4] Barsukov, *Graf . . . Amurskii*, I, p. 288; Sgibnev, "Vidy," p. 693.
[5] Romanov, *Poslednie sobytiia*, pp. 33–34 (emphasis in original); *idem*, "Prisoedinenie," p. 357.

"possess a living artery in the form of a titanic river, which . . . leads from the depths of Siberia [first] to a sea whose coasts are crowded on one side with tens of millions of Japanese, on the other with incomparably more Chinese, and further onto the most expansive ocean on the face of the earth."[6]

The particular sorts of activities and international exchanges which these accounts envisioned for the Pacific of the future indicated that it was to be pre-eminently a commercial civilization, and the Russian Far East, accordingly, was seen as an future center of world trade, with the Amur itself as the key artery. Expectations and optimism about the realization of this vision were very strong indeed in the period we are discussing, especially after the signing of the Aigun treaty in 1858. The press in the imperial capital, and not only in Eastern Siberia, was filled with encouraging news of the latest developments in international diplomacy, the growth of commerce in the Far East, and the prospect for trade on the Pacific. "Just look at the blossoming foreign trade in Siberia," declared one article loudly in the St. Petersburg journal *Russkii Vestnik* (The Russian Herald) in 1858: "it gladdens the heart!" Seven foreign ships, it reported, had already appeared on the Amur, and the Russian merchant flag now waves on the Pacific. Thanks to all this activity, "the population of Irkutsk smokes cigars imported from Manila and Havana through Nikolaevsk, and orders for wine come in from Yakutsk to be placed with local American traders."[7] This auspicious picture was confirmed by scattered accounts of foreign traders in the region, of which a great deal was made in the Russian press. Otto Esche, for example, a German–American merchant from San Francisco, visited Nikolaevsk in 1857, and reported enthusiastically that international trade along the river was growing by leaps and bounds. Another German merchant A.F. Lühdorf, who had opened a trading house in Nikolaevsk, echoed Esche's assurances that "everything pointed to . . . a great future" for the Amur region.[8] All of these highly encouraging comments were translated and reprinted in Russian newspapers and journals, among which the *Vestnik* of the Russian Geographical Society carried the following glowing report.

> Nikolaevsk is now supplied with all the items of comfort. Its stores carry the best selection of Japanese and Chinese furniture, expensive cigars from Manila and Havana, sugar, confectionery, pâté, fruits, oysters, sea crabs, pineapples, grapes, rum, porter, and a thousand other items brought by sea and sold cheaply . . . With the exception of a few items, the import and export of all goods is allowed without tariffs. Nikolaevsk is a free port in the full sense of the word. The number of merchants there is growing daily.[9]

[6] Semenov, "Amur," pp. 188–189. [7] D. R. "Pis′ma," pp. 368–369.

[8] "A. F. Lühdorfs Schilderung," p. 336; "Otto Esche's Expedition," pp.161–162; Griffin, *Clippers*, p. 340; Ravenstein, *Russians*, p. vii.

[9] "Ekspeditsiia g. Ottona Eshe," pp. 42–43n (quote); "Iz Pis′ma," *passim*; Collins, *Siberian Journey*, p. 304.

It is not difficult to imagine the enticing effect that accounts such as these about the blossoming of commerce in the Far East must have had on a home-bound reader in Russia's European capital, who may well have never tasted such an exotic fruit as a pineapple.[10]

The enthusiasm stirred by the Russian advance in the Far East was not limited to the prospect of developing the Amur region itself as an international trading emporium. Beyond this, there was a desire for Russia to project its commercial activities beyond its borders by taking part in trans-oceanic trade across the Pacific. The acquisition of the Amur region seemed to bring the entire Pacific rim within easy reach of Russian ships, and the Russians saw no reason why they should take any less advantage of this vast and beckoning basin than the others powers of Europe and North America. The captain of the ship which had brought Putiatin to Japan in 1855 to conclude a treaty with Russia spoke of his desire to sail further into the Pacific, to Macao, Manila, New Guinea, and the New Hebrides. "I am attracted there not by simple curiosity, not by an ambitious hunger for discovery," he explained, "but rather by the desire to acquaint myself and our naval public with those places which, it seems to me, can with time become very suitable for our future ships on the Pacific." The markets of China and the gold of Australia are beckoning, he continued, and great riches will "practically fall by themselves into the mouth" (chut′ ne sami v rot mogut valet′sia) of the enterprising captain who knows how to take advantage of them.[11]

The tempo of activity in the Russian Far East at the time did indeed seem to support the high expectations presented in these depictions. Even before the region was formally annexed by Russia, foreign ships had begun to call at Nikolaevsk, a number of German and American firms opened trading houses there,[12] and in 1857 the Amur Company, a private venture for trade in the Far East, was organized in St. Petersburg. The tasks the company set for itself were broad-ranging, and included maintaining steamship communication along the river, supplying Russian and indigenous settlements with provisions, overseeing the exploitation of the natural resources of the region, and conducting trade in foreign ports. As stipulated in the company's charter, 17 boats were to be ordered from American manufacturers, and permanent agents of the company were to be sent to Shanghai and New York with the intention of establishing regular and direct commercial links through yearly voyages.

[10] In regard to pineapples at least, there was apparently some truth in these depictions. The writer Ivan Goncharov, returning to European Russia via Irkutsk after a mission through south Asia in the mid-1850s, described how Murav′ev, having wined and dined him, insisted on serving "pineapples marinated in their own juice" for dessert. Ironically enough, Goncharov had had his fill of tropical fruits in India, and was less than impressed with this novel indication of Eastern Siberia's new international status. In what must have been a unexpected disappointment for the governor-general, he told Murav′ev he would actually prefer Russian kvas and pickles instead. Goncharov, "Po Vostochnoi Sibiri," pp. 24–25.

[11] V. R. K., "Otryvki," p. 6.

[12] Griffin, *Clippers*, pp. 343–344; Ravenstein, *Russians*, pp. 420ff.

Initial interest in the company in St. Petersburg was extremely lively, and its shares were sold out on the first day.[13] Some orders for American ships were in fact placed, and boat traffic up and down the Amur expanded appreciably, Lühdorf estimating in 1858 that no less than 29 steamships had plied at least some length of the river's waters.[14] While no commercial representatives were apparently ever sent to the United States, some efforts were indeed made toward China. In 1859, several merchants from Kiakhta made their way, with Murav′ev 's support, down the Amur and south across the Pacific to Shanghai, where under the patronage of an American trading firm they explored the feasibility of maritime trade with the Russian Far East.[15] The Russian government, for its part, demonstrated a committment to facilitating the activities of foreign merchants in the region by building a lighthouse in De Castries bay south of the Amur mouth and then marking the navigation route to Nikolaevsk for them with buoys and beacons. No less importantly, a series of edicts were issued in St. Petersburg beginning in 1856 pronouncing the Amur valley to be a "free-trade zone" for foreign as well as Russian merchants.[16] Along with all of this, the settlement of Nikolaevsk grew at a comely rate, and by the early 1860s boasted over 250 buildings and a population that numbered in the thousands.[17] According to Esche, life there was "extremely pleasant": Nikolaevsk was already a center of trade and culture, with a commercial district comparable to that of San Francisco and a library of over four thousand volumes "aus allen Fächern des Wissens."[18]

As positive as the Germans may have been regarding Russia's role in the great commercial transformation of the Pacific, however, their enthusiasm paled when compared to the activities and pronouncements of an American, one Perry McDonough Collins, who visited the Amur during this period. Collins, a New Yorker who went to California during the gold rush and became a lawyer and promoter there, followed the events unfolding on the opposite shore of the Pacific through the accounts of American whalers in San Francisco. His imagination was captivated by what he saw as the fabulous potential of the Russian Far East and its significance for American commerce. "I . . . fixed in my own mind," he later wrote "upon the river Amoor as the destined channel by which American commercial enterprise was to penetrate the obscure depths of Northern Asia, and open a new world to trade and civiliza-

[13] "Russkie na Amure," p. 272; N. N., "Proekt ustava," pp. 5–7; Sychevskii, "Russko-Kitaiskaia torgovliia," p. 160. Not surprisingly, some of the individuals closely associated with the acquisition of the Amur, such as Middendorf and Nevel′skoi, invested heavily in the venture. Kubalov, Letters to M. K. Azadovskii, p. 276; Alekseev, *Delo*, p. 284.

[14] "A. F. Lühdorfs Schilderung," p. 334; Collins, *Siberian Journey*, p. 276.

[15] Noskov, "Izvestiia," p. 10; P, "Svedeniia," p. 109.

[16] Ravenstein, *Russians*, p. 415; Barsukov, *Graf . . . Amurskii*, II, p. 209.

[17] *Geograficheskо-Statisticheskii Slovar′*, III (1867), p. 464; Griffin, *Clippers*, pp. 336n, 340.

[18] "Otto Esche's Expedition," p.161.

[19] Collins, *Siberian Journey*, p. 45.

tion."[19] In March 1856 he brought this vision with him to Washington DC, and after audiences with Secretary of State William Marcy, the Russian ambassador de Stoeckl, and finally President Franklin Pierce himself, he succeeded in being appointed as "American Commercial Agent to the Amoor River." Armed with this commission and wasting not an instant, he arrived in Russia in May of the same year, and after months of delay in the capital finally reached Eastern Siberia the following January. Accompanied by a helpful escort of five cossacks provided by Murav′ev , Collins spent some nine months traversing and inspecting conditions in the Russian Far East before securing passage on a Russian government steamer – once again courtesy of Murav′ev – and returning home across the Pacific via Honolulu.[20]

To say that Collins' overall evaluation of the prospects for the development of the Amur region was positive would hardly do justice to his grandiloquent, awe-inspiring, and even apocalyptic vision. Raising a glass to Mur'avev at a banquet given to celebrate his arrival in Irkutsk, Collins toasted the "true and lasting honor" that Russia and world will eternally owe to the "genius" of his host, thanks to whose "enlightened, sagacious, and far-seeing policy" this hitherto unknown region was destined to become "one of the most important [territories] on the map of the world."[21] Elaborating upon the significance of the region a few years later to an American audience, Collins was no less effusive. In its mighty course, he wrote, the Amur gathered

> the accumulated streams of a hitherto unknown little world, and onward rolling its unbroken and majestic course for 2,500 miles, towards the rising sun, and the mild waters of the Pacific Ocean. Throw yourself with confidence upon its flowing tide, for upon this generous river shall float navies, richer and more powerful than those of Tarshish; mines shall be found upon its shore, richer than those of Ophir, and the timbers of its forests, more precious than the *Almugim* of Scripture; a mighty nation shall rise on its banks and within its valleys, and at its mouth shall arise a vast city, wherein shall congregate the merchant princes of the earth, seeking the trade of millions of people.[22]

Though the millennial language and imagery with which Collins presented this sweeping vision for the future development of the Amur region drew upon a characteristically American idiom of Manifest Destiny, to Russian ears in the late 1850s it did not sound foreign at all. Quite to the contrary, it resonated powerfully with them, for it was in terms of the very same sort of jubilant vision of a glorious future evoked by the expansion of the United States to the Pacific coast of North America that they now wanted to understand their own recent movement to the sea.

The Russian reaction to Collins' ideas was an indication of the high degree

[20] "Donesenie Kollinsa," p. 7. On Collins, see Vevier, "Introduction," *passim*; Griffin, *Clippers*, pp. 335–346. [21] Collins, *Siberian Journey*, p. 90. [22] Collins, *Siberian Journey*, pp. 94–95.

of affinity felt at the time for the United States. Although as an "American Commercial Agent" he was clearly a representative of foreign interests,[23] Collins was nonetheless regarded in Russia as something very akin to a native son. Murav′ev welcomed him into his inner circle and provided every assistance so that he could carry out his journey as thoroughly as possible.[24] Moreover, there was an an immediate and enthusiastic response to his prognoses for the future of Russian Pacific commerce, as soon as they began to appear in print in the United States.[25] They were restated at length in the Russian press, and coming from his – an American – pen lent an entirely special authenticity to the vision of a far-eastern commercial civilization which the Russians themselves were propagating.[26] The euphoria of this brief period was of such an intensity, and the desire to believe in a community of mutual international interest on the Pacific so strong, that Russia's endemic xenophobia about foreign interest preying on the country's riches was suppressed for the moment, at least as far as the Amur was concerned. Foreigners, and above all Americans, were welcomed into the Far East, and their activities there seen as a great boon to the development of the region.[27]

As sort of logical completion to the vision of international commerce evoked by the acquisition of the Amur was its potential role as part of a grand system of railways across European and Asiatic Russia. The era which had begun in Russia with the accession of Alexander II witnessed the country's first large-scale railway construction, based on foreign capital and technology.[28] Russian observers familiar with the territories east of the Urals imme-

[23] American profits were certainly uppermost in Collins' own mind. Calculating on the basis of a Siberian population of 4 million individuals, each of whom would spend five dollars yearly on foreign merchandise, he claimed to Secretary of State Lewis Cass that the "Siberian market" could ultimately be worth as much as 20 million dollars to American enterprise. Letter of 6 March 1858, in *Siberian Journey*, p. 184. At a maximum salary of $1,000, though, Collins himself was not exactly enriching himself, although the terms of his position allowed him to engage in private trade as well. Griffin, *Clippers*, p. 238n.

[24] Barsukov, *Graf . . . Amurskii*, II, p. 144; Griffin, *Clippers*, p. 341.

[25] The first edition of Collins' book *Siberian Journey* was not published until 1860, but he had presented the essence of his views in a number of letters and reports to the government, some of which had been printed in the *Congressional Record* in 1858. The fact that Collins had such powerful patrons in Russia allowed him to travel widely and visit virtually every locality in Siberia and the Far East of importance to his mission. This fact, in turn, along with his openness and peculiar eye for detail – he commented incessantly, for example, on the quality of the teeth of the indigenous women he encountered in Asiatic Russia (by whom he was often quite taken) – make his account one of the most engaging and useful travellers' reports from nineteenth-century Russia.

[26] [Lamanskii], "Mnenie anglichan," pp. 55–56; "Zametki ob Amure," pp. 99–101; Eima, "Issledovaniia," *passim*, and "Donesenie Kollinsa," *passim*.

[27] The strong interest on the part of various Western countries in Russian activities on the Amur was indicated, among other things, by the string of books about the Amur region published at this time in France, Germany, and England. Sabir, *Le fleuve Amour*; von Etzel and Wagner, *Reisen*; Andree, *Das Amur-Gebiet*; Ravenstein, *Russians*; Atkinson, *Travels*.

[28] Von Laue, *Sergei Witte*, pp. 5–7.

diately pointed to the desirability of connecting Siberia to European Russia by rail.[29] Now, with Russian steam navigation on the Amur, the centuries-old dream of a grand trading route across Russia which would connect the population centers of Europe with those of Asia actually appeared to be within grasp. Collins, true to form, did not miss the opportunity to assure the Russians that, if only they would "assist Nature a little" by building a rail line connecting the headwaters of the Amur to Lake Baikal, they would thereby "open up a system of inland navigation, wonderful in extent, and absolutely past calculation in its commercial results."[30] An article in the local newspaper in Irkutsk proposed a rail connection from Nizhnii Novogorod to Kiakhta, with a branch line to the Amur, pointing out that this would "connect the center of Russia with China, Manchuria, Mongolia, and in particular the Pacific Ocean" and allow Russia to exert an influence "on all of the trade and international relations of Europe with the peoples of the East and with China."[31] The point was not lost in European Russia, where an article in a St. Petersburg journal observed in a similar spirit that by linking the Volga basin by rail with the basins of the Irtysh, Enisei, and Amur rivers,

> all of the products of Northern and Central Asia, the [Indonesian] Archipelago, the Pacific and the western shores of America will be drawn into the full network of Russian, and with them European railroads, and to the ports of the Baltic, the Black and White seas. This will [moreover] spread the influence of the European continent and European civilization over three-quarters of the entire population of the earth, and having created a new epoch in the alteration of world trade routes, will weaken the [current] predominance of the sea powers.

From this perspective, the Amur assumed an importance going beyond its function in simply connecting Russia to the Pacific, for it would become a link between East and West which would revolutionize the world's trade routes. This development in turn, the author reasoned, would have the potential of altering the geopolitical balance between land-based Russia and the so-called naval powers – Britain, in the first instance, and France. The article ended by depicting how the Amur region itself would become the home of a great commercial civilization:

> From Kiakhta and heading toward Nerchinsk, [a railroad from European Russia] will revive all of Dauria . . . and have its terminus at the Amur river, where the *seventh* most outstanding international trading center in the world will arise. From the railroad network of the European continent and Siberia . . ., from the Pacific Ocean and up the Amur on steam fleets, the products of three parts of the earth, Europe, Asia, and America, will flow into this center for mutual exchange. All this great circulation will

[29] Shatrova, *Dekabristy*, p. 124; Pasetskii, *Geograficheskie issledovaniia*, p. 112.
[30] Collins, *Siberian Journey*, pp. 130, 112 (quotes), 55, 326; Babey, *Americans*, pp. 106–107.
[31] I. N., "O sibirskoi zheleznoi doroge," pp. 1, 4.

take place peacefully and safely, under the aegis of the Russian eagle [which, as] the guardian of justice and law, . . . will protect it from the self-seeking interference of those who [currently] monopolize oceanic trade.[32]

It is fair to say that, after the painful experience of the Crimean War, no prospect could have been more appealing to the Russians.

"More a cruel step-mother than a mother"

If the vision of the Amur as a commercial link to the Pacific was understood in the first instance in terms of its broad implications for the country as a whole, and indeed for all of Europe, it carried at the same time a special significance for the Russian territories east of the Urals. As we have noted at an earlier point, Siberia was perceived in very different ways in the period under consideration, but concerning a few basic facts it was difficult for anyone to disagree. Above all, Siberia was very big, it was moreover very cold, and finally – due its peculiar physical–geographical configuration – it was an extremely isolated region. Its isolation came from three physical–geographical endowments, which in combination worked to cut it off from the surrounding world. To begin with, although Siberia was endowed with some of the mightiest rivers on the face of the earth, these arteries were not very useful in facilitating commercial and cultural contacts with neighboring peoples by virtue of the fact that they flowed in the "wrong" direction – that is, to the north – and emptied into dark and frozen bays along the Arctic coast. The second factor was the Arctic ocean itself, the impenetrability of which insured the absolute desolation of large portions of northern Siberia. Finally, its southern, eastern, and southeastern borders were rimmed either with high mountains or vast deserts, cutting it off not only from other peoples to the south but from the more temperate and promising waters of the Pacific.[33] Yet while it was clear that such a disadvantageous geographical disposition would naturally inhibit Siberia's settlement and its economic development, not everyone found this circumstance particularly disturbing. Quite to the contrary, as far as Nicholas I's conservative government was concerned, the region's isolation and ensuing retarded level of development seemed not only entirely appropriate but indeed positively desirable. Setting forth this perspective with scholarly authority, the geographer Konstantin Arsen′ev identified a definite benefit in Siberia's unhappy natural configuration, arguing that without such natural obstacles to inhibit its development, the tasks of maintaining and administering such an enormous and unwieldly region would be quite "beyond human forces." Arsen′ev spoke particularly favorably about the border that had been estab-

[32] Safronov, "O sibirskoi zheleznoi doroge," pp. 593, 594 (emphasis in original).
[33] See for example Nevel′skoi's discussion. *Podvigi*, p. 18.

lished between Russia and China by the Treaty of Nerchinsk, which was intended to take advantage of the "natural" boundary of the Stanovoi mountain range to separate the two powers. The political boundary that resulted was so effective and secure, he affirmed, that it required no more than a few scattered cossack divisions "to maintain peace and security in these remote and sparsely populated lands."[34]

It may be readily appreciated that the enthusiasts of a forward Russian policy in the Far East would read the geographical configuration of the region very differently. In their view, Eastern Siberia was a not a region protected but rather a region imprisoned, for although Nature had generously endowed the area with abundant natural resources and industrial potential, there were practically no natural means by which these might be developed and exploited. As was the case in most other regards, here as well the scope of the problem was immediately clear to the ever-prescient American Collins. "The necessities incident to Asiatic Russia are quite reversed by the laws of nature," he explained, "having within her borders the noblest system of rivers in the world; they are yet, in a great degree, unavailable to her necessities, from the fact that . . . the ocean into which they flow, due to its high northern latitude, is nearly useless in a commercial point of view." Accordingly, this vast country, "seeking an outlet for its commerce, as well as an inlet to its wants, . . . [and] a channel of communication with other countries and nations, must, of necessity, seek the sources of new rivers for an outlet."[35] Regarding the latter there was, of course, only one alternative, which unsurprisingly was precisely the river that our author had travelled halfway around the world to investigate. "Look upon the map," he commanded, "and regard the peculiar situation of Siberia, and you must at once perceive that it is only through this river that you can have communication with the great ocean, and thus an extended commerce with all the nations of the world."[36]

Here as elsewhere, Collins was merely giving voice to a sentiment that many Russians already deeply shared. As they expressed it, however, his point about the Amur as Siberia's single opportunity for contact and intercourse with the rest of civilization was embellished with a sort of geographical teleology. Arguing in practically the same terms, numerous observers spoke about the need to allow the river to fulfill that function "to which it was predestined (prednaznachena) by Nature itself, namely to be a great commercial route for Eastern Siberia."[37] Peter Semenov, whose own scientific authority was an easy balance for that of the sceptical Arsen′ev , expressed himself in this spirit as well. "Happily for the Russian population of Siberia, after a century-and-a-half of striving for the ocean Russia has finally taken control of the most

[34] Arsen′ev, *Statisticheskii ocherk*, pp. 24, 17. Also see [Baten′kov], "Obshchii vzgliad," p. 156.
[35] Collins, *Siberian Journey*, p. 93. [36] Collins, *Siberian Journey*, p. 111.
[37] "Ekspeditsii russkikh," p. 26 (quote); *Puteshestvie po Amuru*, p. 10.

convenient route from the internal regions of Siberia, [a route] which was provided by Nature itself."[38] This natural teleology was brought out with full and emphatic eloquence in an article by Nikolai Speshnev, one of the three former members of the Petrashevskii *kruzhok* living in exile in Eastern Siberia in the late 1850s. Prior to the arrest of the *petrashevtsy*, he had been one of its more outspokenly radical members,[39] and the police committee investigating the circle concluded that he was the most "dangerous revolutionary" of the lot. Indeed, Speshnev – who by all accounts was dazzlingly handsome – appears to have possessed a sort of diabolical charisma, which impressed his fellow *petrashevets* Fedor Dostoevskii sufficiently to use him as a model for the fiendish Stavrogin in *The Possessed*.[40] Despite this, he was able to gain the full confidence of Murav′ev during his banishment in Irkutsk. Speshnev became the governor-general's loyal and valuable assistant, accompanied him on a state visit to Japan and China in 1859, and was even able to convince Murav′ev to agitate for his pardon in St. Petersburg.[41] He also served briefly as editor of the local newspaper *Irkutskie Gubernskie Vedomosti* (Irkutsk Gubernia News), in which the essay in question appeared anonymously in 1857.[42]

Speshnev began by pointing out the key significance which a river system has for any region, at once guaranteeing its welfare and offering the basis for its material and industrial life. This general principle, however, was especially true in regard to the significance of the Amur to Eastern Siberia, which owing to its natural endowments (or lack of them) had virtually nothing else on which it could rely, nothing else with which to secure its existence. Nature itself turned out to be the chief villain in this sad state of affairs.

> Nature was harsh in its treatment of that enormous part of Russia which we call Eastern Siberia! It was not overly generous with the rest of the country, but here, in Eastern Siberia, she resembles more a cruel step-mother (machekha) than a mother.

It was only in regard to raw geographical space that Nature did not scrimp in Siberia, for its enormous expanses of tundra, taiga, and frozen wastes were large enough to hold all of Europe. This was a faint blessing, though, for these lands were largely unproductive, and moreover served to isolate the region almost entirely.

In specifying the natural handicaps which had been imposed on Eastern Siberia, Speshnev rehearsed the tripartite litany noted above. First of all was the problem of climate. Nature, he asserted, "awarded Eastern Siberia's southernmost [i.e. most temperate] parts with seven months of harsh winter, and in the rest of the region frost reigns for eight and nine months, a winter the likes of which Captain Ross did not experience near the North Pole." Only

[38] Semenov, "Amur," p. 190. [39] See above, p. 92. [40] Berdiaev, *Origins*, pp. 32–33.
[41] Evans, *Petrasevskij Circle*, pp. 45, 83–86; Semevskii, "M. V. Butashevich-Petrashevskii," no. 3, p. 21; Leikina, *Petrashevtsy*, pp. 59–60, 125.
[42] Koz'min, "M. B. Zagoskin," p. 194. On Speshnev's authorship, see Leikina, *Petrashevtsy*, p. 58.

Antarctica was less favorably endowed climatically than Siberia, but with this exception "there is hardly another place on the earth which can compare with harsh and inhospitable Eastern Siberia." This state of affairs might even lead one to suspect Mother Nature's reputed good sense, a suspicion which he indeed voiced in discussing the next point, the problem of the Siberian rivers.

Here you involuntarily begin to doubt the much-celebrated good sense and expediency of Nature: take a look, if you please, at the giant river systems, at these titan rivers which are called the Enisei and the Lena. What are they here for? In another place, in another climate, under different physical conditions – each of them would be a blessing of God Himself on the earth, they would feed and nurture not millions, but of tens, hundreds of millions of people. What life, what trade would seethe on these rivers, the vastness of which can compete with the Mississippi! But here? . . . they flow uselessly far to the north, and empty senselessly into the Arctic, that cake of ice which is jokingly honored as an "open sea."[43]

The third handicap which Nature hung around the neck of Eastern Siberia was its enforced isolation from the rest of the world. Although its eastern coasts were washed by an ocean which was not at all as hopeless as the one to the north, still

Nature zealously bordered [the eastern reaches of the continent] with a broad rocky mountain range, passable only at certain points and there with difficulty. It is as if [Nature] was really afraid that some significant river would take it into its head to flow from Siberial into [the Pacific], one of those rivers which bring activity, trade, and life from the sea far inland, because they open an entry way for the products of other, more favored regions.

Nature shielded Siberia no less carefully from the warm south by means of several parallel chains of mountains, and "to complete the picture, not being content with these mountains, She partitioned off the abundance of China with a wide desert of rock and sand."

With this, the bleak picture of Eastern Siberia's natural endowments was complete. Implictly, Speshnev argued the teleological position that the physical environment was created for a certain purpose – that is, to provide an arena for the development of the economic and cultural activities of humankind. Eastern Siberia, however, had not been granted the natural conditions necessary for even the most rudimentary form of agriculture, and at the same time was denied convenient access to other centers of population from which could be obtained the necessities of life. "Make the best of it you can" (zhivi sebe kak znaesh′), he challenged the local readership ominously, "in total isolation from the rest of the world."[44] This compelling image of a region cut off and starving left his audience – natives all of Eastern Siberia – waiting anxiously for the solution which might relieve their homeland's plight. And the solution, of course, was obvious. It was the Amur, the single route that

[43] [Speshnev], Untitled lead article, p. 2. [44] [Speshnev], Untitled lead article, p. 3.

led out of Siberia's natural encasement. "At only one place in this country did Nature diverge from her inexorable logic, as if She made an error in Her calculations and forgot to block off the final route for the convenient intercourse between Eastern Siberia and the rest of the world." The Amur "is the only compensation which Nature gave Eastern Siberia for all of the disadvantages she apportioned to that region; it is Siberia's vital, living artery,"[45] a fact had long been appreciated by those who knew anything about Eastern Siberia.

The fact that Speshnev's florid imagery was so reminiscent of Collins' own exhortations suggests that the two might have exchanged views during the latter's sojourn in Irkutsk, which indeed took place before Speshnev published his editorial. However that may be, the promotional intent of this essay is clear. In the best spirit of the American entrepreneur, and writing in all likelihood at the behest of Murav′ev , the author was trying as hard as he could to praise both the annexation of the Amur and the development of private trading companies on it. At the same time, however, he chose a particularly effective way to argue the point, and we may readily assume that his exposition of Eastern Siberia's peculiar geographical fate had a definite appeal for those Siberians who read it. Despite the enormous distances separating it from Russia's European capitals, educated society east of the Urals was nonetheless inspired by the same optimistic mood of excitement at the prospect of rejuvenation, reform, and reconstruction that animated the rest of the country.[46] The Irkutsk intellectual M.P. Shestunov expressed these sentiments in terms that would have sounded not at all out of place in St. Petersburg or Moscow, criticizing "the inadequacy of our previous life, of what we might call the apathy" in which Siberia up to that point had languished. Now, however, everything had changed.

> The lethargy has now passed, and society, casting off its shackles, naturally sensed a need to change those principles that formed that basis of the previous life. People have now . . . begun to question authority and, animated by the desire to become conscious of themselves and to have confidence in their own strengths, have moved forward on the path to self-perfection and self-awareness.[47]

Effectively, Speshnev's argument can be seen as a geographical formulation of this perspective, in which the region's natural–geographical isolation itself represented the old order, which had operated effectively as a "shackle" to retard its development, and which Eastern Siberia would overcome and "cast off" by establishing a link to the Pacific.

That Siberian society understood the annexation of the Amur in precisely this sense can be seen in numerous examples. Rafail Chernosvitov, for example, who after his arrest and interrogation in the Petrashevskii affair had

[45] [Speshnev], Untitled lead article, pp. 3–4. [46] Koz′min, "M. B. Zagoskin." p. 194.
[47] *Irkutskie Gubernskie Vedomosti*, no. 29 (1857), quoted in Koz′min, "M. B. Zagoskin," p. 195.

returned to his commercial ventures in Eastern Siberia, wrote in 1860 that Siberia had from the earliest times been locked in on all sides by insurmountable geographical obstacles. Consequently, lacking a "window" as he put it for communication with the rest of the world, the region had "vegetated in apathy, unaware of its own strengths." "But then an architect appeared," he went on, referring to Murav′ev, who instead of a window opened a broad gate and marched through it as a triumphant conqueror.

With the opening of the Amur region a new era should start for Siberia. Siberia should cast off its thick hide of prejudice and sloth. It has been asleep for long enough, it is now time for it to wake up . . . It is time to drop the narrow path of routine and follow the broad avenue of science and progress. The gates have opened for a way out, and the human race is calling to it to take part in the general arena of human activity.[48]

For Russia east of the Urals, this revival would be based above all on the acquisition of the Amur river, in its capacity as a life-giving connection to the outside world. Indeed, others on the scene described how this summons to "follow the broad avenue" of progress was already being heeded. One local resident raised his voice in celebration of the arrival in Nerchinsk of the first steamboat up the Amur from Nikolaevsk, and used to occasion to confirm the multifarious benefits already accruing to an Eastern Siberia newly blessed with the Amur river.

From year to year navigation expands on the Amur, commerce intensifies, and our region is being revived . . . Bells are ringing now day and night, coaches and postal convoys are rumbling, couriers rush about: it is in a word as if everything has arisen out of its slumber and come to life, raised itself up on its own legs, was renovated, began to speak, to look after itself, to hurry about.[49]

If Chernosvitov's was a call for the revival of civic life in Siberia, it was quite identical to those appeals from European Russia for the country as a whole to stir itself like Ilia Muromets and embark upon a new period of progressive activity. On the pages of the first issue of the newspaper *Amur*, Russia's first private newspaper which began publication in Irkutsk on New Year's Day 1860, Mikhail Petrashevskii himself heralded the new decade by making this connection very strongly.

The wave of social movement originating in European Russia has spread out broadly and reached even into Siberia, where it is felt on all rungs of the social ladder. Everyone has responded to this call to life, [and] voices which up to this point kept their silence have spoken out. Thanks to the acquisition of the Amur river, the common interest has been stirred . . . The epoch during which Siberia's society stagnated is passing, and a rudimentary striving toward consciousness is becoming apparent. In a social sense, this fact is as important to us as the gift of speech is to a mute.[50]

[48] Chernosvitov, "Neskol′ko slov," p. 1. [49] Zenzinov, Untitled communication, pp. 529–530.
[50] "Mestnoe obozrenie" (1 January 1860), p. 1; Dulov, "Sibirskaia publitsistika," p. 10.

In a letter to Herzen, Bakunin emphasized the significance for Siberia of the acquisition of the Amur and the region's revolutionary new juxtaposition to the ocean. "Siberia is now connected to the ocean, it has ceased to be a locked-in desert, [it has ceased to be] Siberia." We already feel the influence of the ocean, he reported, and in Irkutsk "we are closer to Europe than is Tomsk [in Western Siberia]. With the Amur, Siberia has for the first time come to its senses (osmyslilas′). Is this not a great affair, and who can enumerate all of its results!"[51]

The primary importance of the Amur specifically for Siberia as a link to the ocean was stressed again and again.[52] It was highlighted even in the official rescript from St. Petersburg in August 1858, which conferred upon Murav′ev the title of Count Amurskii: "The civic resurrection of this region is a result of your enlightened activities . . . The treaty which you have concluded gives Siberia a new commercial route along the Amur river."[53] For his part, Dmitrii Romanov brought the point home by means of a parallel with another period in Russian history. Pre-Petrine Russia had bordered on Europe, he noted, but "like Siberia, it remained closed off because it was cut off from the Baltic and Black seas." Stubbornly, in the course of the centuries, Russia broke through to establish itself on these seas, and the great benefits that the nation reaped through these advances were more than obvious. In Siberia a similar geohistorical imperative – the need for a link to the ocean – had long existed, but only in the author's day was being realized.

> The western coast of the Sea of Japan represents the very same type of natural supplement for Siberia as was the Black Sea coast for pre-Petrine Russia . . . The Amur river serves as the natural link between Siberia and this coast in the same way that the Dnieper links the internal regions of Russia with the Black Sea.

This rather overstretched comparison was also intended to dispel any concern about the lack of development, or even of population in Russia's newly acquired far eastern regions, for

> did not the tundra of the Gulf of Finland or the arid steppes of the Black Sea coast present the very same picture 150 years ago for the Russia of that time? The population of Russia then, as in Siberia today, was not particularly large, but persistent effort resulted in the course of time in the solid colonization of these coasts, and the seas which washed them became the bases for the brilliant development of Petersburg and Odessa.[54]

The vision of the Amur as the connection between Siberia's landlocked continental interior and the open sea suggested yet another comparison, not his-

[51] [Bakunin], *Pis′ma*, p. 123 (quote), 163.
[52] Middendorf, *Puteshestvie*, I, p. 169; Kpt, "Gde budet?" p. 395; A. K. "Amurskii krai," pp. 132–133, 142.
[53] "Mestnoe obozrenie," (1 January 1860), p. 2; Romanov, "Prisoedinenie," p. 164.
[54] Romanov, *Poslednie sobytiia*, p. 33 (quote); Romanov, "Prisoedinenie," p. 332.

torical like Romanov's but rather geographical. The example of the grand effects of the Mississippi river in opening up the Louisiana territories and the vast heartland of an entire continent captured the imaginations of many Russians, who fancied that the Amur, as their Mississippi, would lead to the same blossoming of agriculture, settlement, and commerce which could be witnessed in North America. Although a number of Russians in the region had spoken in very general terms of the similarities between the early exploration of North America and their own activities in the Russian Far East – Kropotkin, for example, specifically compared Russian cossacks in the Far East to the "Canadian *voyageurs*" colonizing "the banks of the Mississippi"[55] – it was apparently Collins who had first insisted on the parallel between the two rivers themselves. It was then taken up quickly and eagerly by Russian commentators.[56] Much was made of this comparison for the purpose of allaying suspicions that the Amur was not a navigable river – the American had insisted that the Amur "will not prove as difficult or dangerous to navigate as the Mississippi"[57] – but the ultimate import of the association went far beyond this. Indeed, this was an enormously powerful image at the time, particularly for European Russia. While very few Russians west of the Urals had so much as heard of the Amur prior to its annexation, we may assume that all who possessed any worldly education at all were familiar with the Mississippi, if no more closely than as a river in frontier America watering a region formerly inhabited by wild Indians but now the artery of a great advance of modern civilization. This comparison both enhanced and was enhanced by the general feeling of affinity on the part of the Russians for North America which was at a pitch in this period, and they could reflect with satisfaction that with their own Mississippi they would be able to recreate the experience of their sister nation across the Pacific.

"A blessed Russian Kentucky"

Russian attitudes toward the United States in the nineteenth century diverged dramatically.[58] To some extent, the different evaluations followed political lines. Conservatives were for the most part negative, castigating the young republic for its degenerate materialism, utilitarianism, and – ironically enough – the inhumanity of the insititution of slavery.[59] America "displays all the faults of its illegitimate birth," sermonized Mikhail Pogodin in 1837. "This is not a government but rather a trading company . . . which thinks only about

[55] Kropotkin, *Memoirs*, p. 205 (quote); Anisimov, *Puteshestviia*, p. 47; Veniukov, "Vospominaniia," p. 97.

[56] E.g. [Lamanskii], "Mnenie anglichan," p. 55; "Russkie na Amure," p. 270; D. R., "Pis′ma," no. 17, p. 19.

[57] Collins, *Siberian Journey*, p. 261.

[58] See especially Laserson, *American Impact*; Hecht, *Russian Radicals*; Boden, *Das Amerikabild*; Allen, *Russia*; Rogger, "America."

[59] E.g. "Vzgliad," pp. 178, 192; Laserson, *American Impact*, pp. 32, 146–149.

profits . . . and will hardly be able to produce anything of major significance for statesmanship or humanity."[60] Radicals on the left could be highly critical as well, lambasting the cold competition and unrelenting individualism upon which this "Kingdom of the Dollar" was founded.[61] At the same time, however, other progressive and reform-minded thinkers were more appreciative, admiring not only the success of America's revolution against imperial tyranny but also the egalitarianism, constitutionally assured freedoms, and democratic federal structure that replaced it.[62] By the time Collins arrived in the mid-1850s, sentiments on the part of many Russians were particularly favorable, as was obvious in the singular cordiality and openness extended to him on his travels across Siberia.

There were good reasons for this partiality. During the Crimean War, the United States acted, if not as Russia's ally in a strict sense, then at least as a extremely good friend. The American consul in Honolulu supplied valuable information about an impending attack by British and French flotillas on Petropavlosk, and the successful defense which this enabled the Russians to prepare was one of the brightest of the few military successes they were able to score in the war.[63] Support for Russia among the American public during the war was extremely strong, as indicated by the example of a group of several hundred Kentucky riflemen who contacted the Russian legation in Washington with the offer to fight for the Russian cause.[64] For their part, the Russians did not hesitate to return this support a few years later during the Civil War, in which they staunchly supported the Union forces, and some Russian officers actually did service on American battlefields.[65] "There is no capital in Europe," wrote the American ambassador in Russia to Secretary of State Seward in the middle of the war, "where the loyal [i.e. unionist] American meets with such universal sympathy as St. Petersburg."[66]

All of this mutual good will, which struck even Herzen in London,[67] culminated in a mission, dispatched to St. Petersburg in 1866 and led by former Assistant Secretary of the Navy Gustavus Vasa Fox, to offer the special condolences of the United States' government over an assasination attempt on

[60] Letter to Tsarevich Alexander, quoted in Barsukov, *Zhizn'*, V, p. 173.

[61] The characterization was Peter Lavrov's. Cited in Gleason, "Republic," p. 12.

[62] Kohn, "Introduction," p. 15; Laserson, *American Impact*, pp. 119–121, 139. Mikhail Petrashevskii in particular was an admirer of American federalism, and felt that it offered a model for the reorganization of Russia itself. Shchegolev, *Dekabristy*, III, pp. 18, 43; Desnitskii, *Delo*, I, pp. 546–547; Boden, *Das Amerikabild*, p. 143.

[63] Adamov, "Russia," p. 592; "Soedinennye Shtaty v epokhu," p. 152; Golder, "Russian–American Relations," p. 474; Laserson, *American Impact*, p. 164.

[64] Bailey, *Americans*, pp. 63–64.

[65] Bailey, *Americans*, pp. 70–72; Adamov, "Russia," p. 586. Not all of these officers served with equal distinction. Perhaps the most notorious among them was one John B. Turchin, the so-called "Mad Cossack," who was actually reprimanded with a court-martial after his men sacked Athens, Georgia.

[66] Letter of 23 July 1862, in *Papers relating to Foreign Affairs* (1862), p. 450.

[67] [Gertsen], "Amerika," pp. 1861–1862. Also see *Amerikantsy v Rossii*, *passim*.

Alexander II.[68] In a poem composed for the occasion, Oliver Wendell Holmes gave lyrical expression to American sentiments toward their distant friends.

> Though watery deserts hold apart
> The worlds of East and West,
> Still beats the self-same human heart
> In each proud nation's breast!
> . . .
> A nation's love in tears and smiles
> We bear across the sea;
> O Neva of the hundred isles
> We moor our hearts in thee![69]

The Russian response to the mission made clear the degree to which the strength of their friendship with the United States was inextricably related to the trauma of their relations with Europe. Indeed, it was precisely the shared identity of the two countries as "New Worlds" – offshoots of Europe, that is, which stood apart from and to a significant extent in opposition to the Old World – that secured their own closeness. This particular commonality was appreciated on both sides, to be sure, but the Russians were inclined to be far more emphatic and outspoken about it. The historian Pogodin, for example – having apparently put his dim views of the United States as a bastard nation well behind him[70] – delivered a speech at a reception for the American delegates in which he affirmed that the two countries "love each other, wish each other well, and value each other *without any ulterior motives*." The European Old World, on the other hand, looks askance at the United States with an apprehension born of suspicion and envy, and "[i]t regards the *other New World*, that is, Russia, in precisely the same way." And where was the source of this antipathy to be found? "Perhaps in [Europe's] senile jealousy, in the general and involuntary conviction that the *future* belongs primarily to Russia and America, while to Europe belongs the past and the distant past. Yes, . . . a great future awaits Russia and America, a future which we are now approaching *with faith and hope*."[71]

In the remote reaches of eastern Siberia, American and Russian activities and ambitions came together on a common geographical arena, and the two countries had an opportunity of sorts to consummate in practice the mutual commitment about which Pogodin so determinedly spoke. For their part, the Americans had as early as the 1840s displayed a keen interest in Russia's Pacific littoral, and the Amur region in particular. Even as the occupation of Oregon and California was taking place, they had begun to direct their attention yet

[68] Williams, *American–Russian Relations*, pp. 20–21. On the Fox mission see Loubat, *Narrative*; Allen, *Russia*, pp. 229–233.

[69] Quoted in Loubat, *Narrative*, pp. 180–181; Jensen, *Alaska Purchase*, pp. 40–43.

[70] Boden, *Das Amerikabild*, p. 132.

[71] Pogodin, "Rech′," pp. 262, 263 (emphasis in original); *idem*, "O russkoi politike," p. 233.

further westward, and a trans-oceanic vision began to take shape. American activity and influence, it was imagined, was naturally destined to spill far beyond the shores of North America and across the Pacific basin to the countries of East Asia. As early as 1848, an elaborate "memoir" was submitted to Congress by one Aaron Haight Palmer, who spoke at length about the importance of northeast Asia to American interests, and he laid out a plan for the expansion of American influence on Siberia's Pacific coast. Setting forth in prodigious detail the alluring prospect of convenient maritime access to Manchuria, Mongolia, and northern China, the author specifically urged his government to demand of China the rights of "navigating the great Manchurian river Amur and its affluents, and of trading with the colonial dependencies of China, upon the same footing as the Russians." An American settlement should be established, moreover, "at or near its embouchure, [which] would open a new and most profitable trade with Manchuria, Central Asia, Siberia, Japan, Corea, etc."[72] All of this would then contribute toward securing what Palmer referred to as "the permanency of our commercial and maritime supremacy" on the Pacific. In addition to a strong American presence on the Amur, this supremacy would require as well the construction of a trans-continental railway across the United States and a canal to connect the Atlantic and Pacific.[73] Some years later, as part of the Perry mission to Japan, an American ship was actually dispatched to conduct reconnaisance in the estuary of the Amur in order to determine its practical navigability. By this time, however, it was already too late, and whatever opportunity might have been available earlier for fully independent activity on the part of the Americans had passed. Although the head of the expedition was convinced that "at some future day a vast commerce will doubtless be borne" upon the waters of the "River Amour," the Russians had already taken decisive steps to establish their own predominance in the region, and the American mission remained unfulfilled.[74]

The Palmer memorandum and American exploratory interest in the Amur were duly noted in Russia,[75] but whatever concern this might have stirred among Russians wary of yet another competitor nation had evaporated by the mid-1850s. When Collins arrived in Siberia for a closer investigation of the Amur region, it was clear that whatever commercial value the river might represent for the United States would be available only with full Russian cooperation, and the American "Commercial Agent to the Amoor River" was

[72] Palmer, *Memoir*, pp. 2, 34, 46–47. Palmer's essay was accompanied by a remarkably detailed map of the river valley, which gave a considerably more accurate picture of the region than would have been commonly available in Russia at the time.

[73] Palmer, *Memoir*, p. 1. For another American view which argued a direct connection between American interests on the Amur and a trans-continental North American railway, also see J. G. S., "Explorations," p. 181.

[74] Cole, *Yankee Surveyors*, p. 56.

[75] Struve, *Vospominanii*, pp. 122, 124. Palmer's lengthy memorandum, however, was not translated and published until 1906. Pal'mer, *Zapiska*.

entirely prepared to proceed on this basis. Accordingly, his mission became one of convincing his hosts that the benefits to come from the development of the region would accrue equally to both sides. In making this point he was well served by his considerable diplomatic acumen, but the message was one to which the Russians were in any event prepared to respond eagerly. At this point governor-general's Murav′ev 's support for American activities in the Far East was, if not unconditional, than in any event very strong,[76] and he was by no means alone in his enthusiasm. Immediately upon the conclusion of the Crimean War, Russia's ambassador to Washington de Stoeckl assured American officials confidentially that their merchants would be given a particularly warm welcome in the Far East and receive special treatment there.[77] His assurances were echoed at an even higher level by the Russian foreign minister Gorchakov, who spoke enthusiastically to de Stoeckl's American counterpart in St. Petersburg about the Russian advances on the Amur and indicated his willingness to accord the United States special trading privileges in the region.[78]

In the event, the Americans did not fail take advantage of Russia's favorable disposition. Collins' mission was expanded with the appointment of two "vice-commercial agents," and American traders easily outnumbered all other foreigner merchants active in Russia's newly acquired far-eastern territories.[79] Indeed, they so dominated activity there that, according to Peter Kropotkin, all foreigners were referred to as "Amerikantsy," regardless of what their nationality and provenance may actually have been.[80] Despite the fact that the value of American interests already in place on the Amur was considerable – Collins estimated that it amounted to no less than half-a-million dollars[81] – any fears that the Americans might take advantage of their trading privileges at the expense of their hosts were largely dismissed. An article appearing in the official organ of the Ministry of Trade, for example, mentioned the potential problem of foreign encroachment, but with an uncharacteristically stoic reference to "that fateful law . . . according to which major trade routes and markets cannot escape their fate, and some day will have to be opened to the needs and products of all nations" it concluded that free trade in Siberia simply could not and ought not be avoided.[82]

Indeed, no one seemed to want to at the time. The desire to enhance and facilitate relations between Siberia and the United States was extremely strong, indeed so strong that (again according to Collins) Murav′ev ordered that in the future "the American language" was to be taught in schools in

[76] Barsukov, *Graf . . . Amurskii*, I, pp. 355–356, 486, II, pp. 275–276; Cole, *Yankee Surveyors*, pp. 132–135; Vevier, "Introduction," pp. 19–20, 27.
[77] Golder, "Russian–American Relations," p. 475. [78] Vevier, "Introduction," pp. 26–27.
[79] Griffin, *Clippers*, pp. 340, 359.
[80] Kropotkin, "Iz Vostochnoi Sibiri" (no. 45, December 1863), p. 4n.
[81] Collins, *Siberian Journey*, pp. 299, 303–304; "Otto Esche's Expedition," p. 162n.
[82] Eima, "Issledovaniia," p. 50.

Eastern Siberia, instead of German as in European Russia.[83] The most significant indication of Russian interest at the highest levels for far-reaching cooperation with the Americans in the Far East, however, was their agreement to participate in the construction of an intercontinental telegraph line from the Amur across the Bering strait and down the North American coast to San Francisco. This undertaking, first conceived by the Russians and Collins in the late 1850s,[84] received the enthusiatic endorsement of President Abraham Lincoln himself, who deemed it important enough to merit reference in his State of the Union addresses for several years running.[85] Western Union supported the venture on the American side, and after Collins secured an imperial promise of cooperation from the tsar in 1863, the Russians indulged their imaginations about the "degree of influence on political events and on the commercial affairs of the entire world that Russia stands to gain" if the first telegraph line between Europe and America were to traverse its territory.[86] Reconnaissance work – carried out, among others, by the young George Kennan – began in 1865, and over 850 miles of line were in place by the time the project was halted in 1867, thwarted by Cyrus Field's success in laying a trans-Atlantic cable the year before.[87]

The commonality of Russian and American interests in the Far East was not limited to the anticipated material benefits which would come with the growth of commercial activity in the region. Both sides were prepared to recognize a more profound national–political significance in the acquisition and development of a region that would enable them at once to move closer together and turn more definitively away from Europe. On 8 October 1858, The Philadelphia newspaper *Daily Evening Bulletin* carried an article entitled "Our Western Neighbor," which reported on the conclusion of the Aigun treaty earlier that year. Noting that the treaty had caused "great rejoicing" in Siberia, the article asserted that if only its potential significance for the United

[83] Letter of November 1858, to Lewis B. Cass, cited in Quested, *Expansion*, p. 190. Swan describes this decision (no doubt mistakenly) as an imperial edict from Alexander II himself. J. G. S., "Explorations," p. 179.

[84] Romanov, "Proekt . . . telegrafa," *passim*. For earlier schemes to construct an international telegraph connecting Russia and American in the Far East, see *idem*, *Prizabytyi vopros*, pp. 4–5; S., Untitled article, p. 3.

[85] Lincoln, "Second Annual Message," p. 3329; *idem*, "Third Annual Message," p. 3382; *idem*, "Fourth Annual Message," p. 3445.

[86] A. K., "Amurskii krai," p. 144 (quote); Romanov, *Prizabytyi vopros*, *passim*. On Western Union, see letters of Cassius Clay to Seward, 17 June, 1863, *Papers relating to Foreign Affairs* (1863), p. 311, and 14 November 1864, *Papers relating to Foreign Affairs* (1865), III, p. 363.

[87] Letter of William Orton (vice-president of Western Union) to Seward, 25 March 1867, *Papers relating to Foreign Affairs* (1867), I, p. 385; Vevier, " Collins Overland," pp. 250–251; Babey, *Americans*, pp. 17–18. On Kennan see Melamed, *Russkie universitety*, pp. 18–30. The efforts of the Americans were not in vain, however, for the reconnaissance team to Alaska, led by Robert Kennicott, produced the first scientific survey of Alaska, which proved useful to the United States' purchase of the region from Russia two years later. James, *First Scientific Exploration*, pp. 13–20 and *passim*.

States could be fully appreciated, "we should have good cause to celebrate it here, too, . . . as the history of the next half century must inevitably show." The United States, after all, was Eastern Siberia's "only civilized neighbor," and the development of the American West would be heavily influenced by all positive developments on its opposite shore.

When the Pacific Railroad is finished, and when Russia has an open sea coast, the American and the Muscovite can afford to look the one west, the other east over the Pacific – name of good omen – and turn their backs on Europe . . . The Russians know the value of America and respect it; John Bull snubs us and France despises us. But the inevitable laws of industrial progress, as conditioned by geography and climate, will force their way. There is a manifest destiny for nations![88]

The publishers of the newspaper demonstrated their own committment to this prospect a short time later by having a English-language leaflet, addressed "To the Directors of the Society for the Colonization of the Amoor in St. Petersburg," inserted into an issue of Herzen's *Kolokol.* Characterizing the occupation of the Amur basin as "one of the greatest enterprises of our century" and predicting that the Russians "have all reasons to be sure that in a few years Eastern Asiatic Russia will become much more important than California," the leaflet assured them that "for us in America it is extremely important that your colonization of the Amoor region shall develop quickly, without wasting time." It ended by mentioning the possibility that Americans might even immigrate into the region as settlers.[89]

It was hard indeed to miss the connection between the authors' rather emphatic interest in the Russian Far East and their optimistic (if misguided) belief that "the Amoor and its arms irrigate a country as abundantly rich in gold as no other country in the world." Nonetheless, *Kolokol*'s editor was deeply moved. Ever since the failed European revolutions of 1848, Herzen was increasingly prepared to acknowledge a special positive affinity between Russia and the United States setting them both off from the Old World,[90] and the stirring summons issued by the editors of the *Daily Evening Bulletin* provided an ideal opportunity for him to expand on the subject. On 1 December 1858, he responded with his own editorial on the signing of the Aigun treaty, entitled "America and Siberia." The Crimean War, he suggested, had demonstrated quite conclusively just how pointless it was for Russia to fight against the West. Yet as it lost the war, a process of moral liberation from Europe had

[88] Cited in Laserson, *American Impact*, pp. 216–217.
[89] Cited in Laserson, *American Impact*, pp. 217–218.
[90] E.g. Gertsen, "La Russie," p. 169; *idem*, "Pis′mo k Dzhuzeppe Matstsini," p. 349. As was the case with Pogodin, Herzen's earlier views of North America had been far more critical, and he quite pointedly rejected de Tocqueville's comparison of the two countries. Letter to K. Kh. Ketcher (20 August 1838), p. 386; see also his 1839 play *William Penn* about early Pennsylvania. Gertsen, "Vil′iam Pen." On the evolution of his views, see Bassin, "Inventing Siberia," p. 788; Boden, *Das Amerikabild*, pp. 133–142.

begun, which in turn would begin to enable Russia to overcome the "Petersburg tradition" that had denigrated everything natively Russian and slavishly venerated everything foreign. "As long as we imitated the West, we didn't know our own soil under our feet." All of this has now changed, and out of the desperation of defeat a new Russia, intensely aware of its "moral individuality," has lifted its head.

Events are revealing an embryo which is strong and powerful. Not in St. Petersburg – old [Official] Russia, irreverent and losing its head at the first misfortune, perished there . . . Some sort of craziness has possessed people. Rather than becoming desperate for themselves and Russia, Russian thought is daring *to doubt* in Europe, and is searching in the rude principles of its own existence for elements for the future.

Herzen's reference to the imperial capital is significant. For him this capital, and the system which emanated out of it, represented nothing more than the artificial grafting onto Russia of something foreign and essentially European, which then effectively repressed and smothered the real Russia. Everything that was hateful in Europe was thus also present in this system, with the added repugnant feature that, while in the West it was at least a natural product of indigenous historical evolution, in Russia it was totally out of place, and maintained only by the bayonets of the imperial *gendarmes*. It was in terms of all these points that Herzen understood the importance of the Russian advance on the Amur, which he saw as vital evidence of the on-going break with the St. Petersburg tradition. In the most literal sense it was a *geographical* break that shifted the center of activity in Russia away from the established seat of traditional authority into a region which had had no association with the old system. By the same token, the Amur represented a direct link to America. "Between Russia and America there is a great salt ocean, but there is not an entire world of old prejudices, frozen conceptions, a jealous system of seniority, and a stagnant civilization." Russia had only to free itself from its own prejudices and from the "leaden atmosphere of Petersburg," to take a fresh and independent look at the world, in order to realize that "It is now clearly Russia's and America's turn."

The union between the two countries was natural in view of their inherent affinities, which Herzen stressed very strongly.

Both countries abound in strengths, flexibility, a spirit of organization, and a persistence which knows no obstacles. Both are poor in their historical experience, both begin with a complete break with tradition, [and] both swim through endless valleys searching for their borders. From different sides, both have traversed awesome expanses – everywhere marking their path with cities, villages, and colonies – right up to the shores of the Pacific Ocean, that "Mediterranean of the future."

[91] Iskander "Amerika," pp. 233–235 (emphasis in original).

Russia's advance in the Far East, heralded by the Treaty of Aigun, was a profound confirmation of the enduring youthfulness and vigor with which the country emerged from under the domination of Official Russia. At the same time, it provided a means through which Russia could effect a solid union with the United States, Russia's only natural partner in the world. "If Russia succeeds in liberating itself from the Petersburg tradition, then it has but one ally: the North American States." Together, the two countries were the heirs to a future civilization. Europe was now left in the background, he argued, and all eyes in Russia were cast east to the Pacific, which would be the scene of this civilization. The essay ended with a stirring pronouncement.

> The names of Murav′ev , Putiatin, and their comrades have been entered into history, they have laid the foundation for a long bridge . . . across an entire ocean. At a time when Europe is going through its gloomy sepulture and everyone has something to bewail, they from one side – and the Americans from the other – have put the cradle back together![91]

For all of Herzen's gushing optimism, however, it may be noted that his appealing image of a "long bridge" across the Pacific was fundamentally ambivalent in its significance. In the same way that the vision of the Amur as a commercial link to the Pacific implied something rather different for the country as a whole than it did for Siberia, so the prospect that the river would facilitate a new closeness with the United States was similarly bifurcated. Indeed, these nuances were even more important when relations with North America were at issue, and they pointed to a divergence in political implications that was in the final analysis profound. At the same time that European Russia, as we have seen in the statements of Pogodin and others, counterposed its relations with the United States as a positive alternative to Western Europe, Siberians counterposed the American connection to their relations with European Russia itself. Their expectation was that to the extent the Amur would foster a bond with their neighbor across the ocean, it would act as a conduit of sorts for the penetration of American political ideas deep into Siberia. This in turn would not only stimulate the spread of democracy and federalism in their Siberian homeland, but beyond this would promote its eventual political separation from Russia. Siberian separatists had traditionally entertained a special sense of affinity with the United States – indicated among other things by their frequent reference to the whole of Siberia as "Russian America"[92] – for the North American contest with Britain was seen as a brilliant example of the very sort of colonial struggle they would eventually wage with Russia.[93] In the pervasive

[92] The first to point out the similarities between Siberia and the United States were in fact not Siberians themselves but rather the exiled Decembrists. See Rozen, *Zapiski*, p. 213; Basargin, *Zapiski*, p. 198; Pushchin, *Zapiski*, pp. 196–197; Gurevich, *Vostochnaia Sibir′*, p. xi.

[93] Svatikov, *Rossiia*, p. 47.

atmosphere of liberalization and reform in the late 1850s and early 1860s, regionalist–separatist sentiment, or *oblastnichestvo* as it was called, grew intense in Siberia, where it was much stimulated by the activity and excitement associated with the annexation of the Amur.[94]

The most highly placed proponent of separatist ideas was perhaps Murav′ev himself, who according to Kropotkin gathered in his study with his officers and Bakunin to discuss "the chances of creating the United States of Siberia, federated across the Pacific Ocean with the United States of America."[95] While the degree of Murav′ev's genuine commitment to such a program has with good reason been vigorously disputed,[96] it was enthusiastically endorsed by others whose sincerity there can be no reason to doubt. An unsigned article from Eastern Siberia appearing in *Kolokol* in 1862, for example, chastised officials in St. Petersburg for refusing to consider Collins' plan for an Amur–Baikal railroad.

> We wish it [the government] to sit firmly on the throne of its fathers and grandfathers, but let it also understand that the breeze of civic freedom, which is hostile to it, will penetrate up the Amur into all of Siberia, and then it will be compelled to part with its territories east of the Urals even more surely than it is today parting with Poland.[97]

Two years earlier, an officer on Murav′ev's general staff had written back to a friend in St. Petersburg describing the Amur as the "means of escape from the tsarist embrace" which had at last become available to Siberia.

> If the western breeze [of liberalism] is not allowed to pass through Tsarist customs [on the European frontier], then the breeze from the east will bring with it everything necessary for the Siberian. The conduit (provodnik) will be the Amur and trade with America.

This envigorating and liberating influence will help Siberia to "break out of its chains" and "light the torch of freedom" for all of Asia.[98]

[94] Stephan, "Far Eastern Conspiracies?," pp. 137–138; Sokolov, *Dekabristy*, p. 206.

[95] Kropotkin, *Memoirs*, p. 169; Stanton, "Foundations," p. 132.

[96] E.g Barsukov, *Graf . . . Amurskii*, I, p. 207; Steklov, *Mikhail Aleksandrovich Bakunin*, I, pp. 503–504; Sokolov, *Dekabristy*, p. 207; Lemke, "Krest′ianskie volneniia," pp. 152–153. Murav′ev's loyal assistant Struve admitted that the governor-general did not really believe that Siberia would break away, and merely used this specter to further his cause in St. Petersburg. Barsukov, *Graf . . . Amurskii*, I, p. 207; Sokolov, *Dekabristy*, p. 207. If Murav′ev's duplicity on this issue appears troubling, there might be some satisfaction in the fact that it did not escape the attention of the Siberian regionalists at the time. See Svatikov, *Rossiia*, pp. 36, 39.

[97] "Iz Sibiri," p. 1092. For its part, the Russian government was more sensitive to the dangers of the ominous "breeze of civic freedom" than this author realized. It later came to light that the potential dangers of American–Siberian intercourse were the very reason Collins' plan of 1858 for a railroad constructed with American capital was rejected. B. P. Butkov, the head of the Siberian Committee in St. Petersburg, sent the following unofficial explanation back to Murav′ev in Irkutsk: "It is impossible to allow the republicans [the Americans] into Siberia: they will spread their spirit there, and Siberia will break away from us." Quoted in [Veniukov], "Primechanie," p. 4; *idem*, "Postupatel′noe dvizhenie," p. 84; *idem*, "Vospominaniia," p. 111; Svatikov, *Rossiia*, pp. 30–31.

[98] Quoted in Svatikov, *Rossiia*, pp. 38, 61–62.

Whatever may or may not have been discussed in Murav'ev's study, Mikhail Bakunin expanded energetically upon these themes in one of his letters to Herzen. Must I tell you, he asked, about he political significance of this gigantic region, with its blessed (blagodatnyi) climate, fine soils, and the two great navigable rivers that join it to the Pacific? Emphasizing the monumental changes which the annexation of the Amur had already wrought, he stated flatly that "[t]his is a new Siberia, but [one which is] blessed, enlightened, maritime." Thanks to the Amur, the Russian realm has placed a firm foot on the Pacific, and "a union with the United States, up to now Platonic, has from this point on become real." There was no question, he assured Herzen, that the Americans will soon take full control of navigation and trade on the Amur, and he greeted this prospect with the same optimistic enthusiasm we have noted above. The most important point, however, was the new spiritual union between the two countries, a union which would allow Siberia to break from the St. Petersburg tradition and begin a new, democratic life.

There is no doubt that with time the Amur will draw Siberia away from Russia, and give it independence and autonomy. This is much feared in St. Petersburg, where they were even worried that Murav'ev [himself] might proclaim Siberia's independence. But is this independence, which is impossible today but necessary perhaps in the near future, really bad? Can Russia really long remain an awkward monarchy, held together by ugly force? Should monarchial centralization really not dissolve into a Slavic federation?[99]

In making these points, and in particular by taking his stand in support of Siberian independence, Bakunin betrayed a significant difference with Herzen's 1858 position. Both of them related Russia's advance on the Amur to the collapse of the "St. Petersburg tradition," but they did so in different ways. For Herzen, official Russia was destroyed, but the positive forces in the country lived on and manifested themselves valiantly in the Far East. In this very important sense, he saw a genuine and progressive *Russia* at work in the Far East, and it did not even occur to him that the benefits that were to come from the activities on the Amur would be shared by anything less than the entire country. Bakunin, on the other hand, pressed the implications of the collapse of the St. Petersburg system rather further. For him, the spiritual destruction of Official Russia involved the practical destruction of imperial Russia – that is to say, the political liberation of its numerous captive borderlands. From this standpoint, while the annexation of the Amur could perhaps be seen as a benefit to all of Russia in the negative sense that it contributed to the collapse of an oppressive order, it was of positive significance to the future of Siberia alone, for it was above all a harbinger of Siberian independence. To

[99] Letter of 7 November 1860, in [Bakunin], *Pis'ma*, pp. 123–125. Interestingly, Frederick Jackson Turner described an essentially identical geopolitical situation in the Mississippi valley, arguing that prior to the railroad it formed "potentially the basis for an independent empire" whose "natural outlet was down the current to the Gulf." Turner, "Significance," p. 187.

be sure, Herzen eventually came to appreciate this point and essentially adopted Bakunin's position. Nevertheless, the very fact that they expressed these constrasting perspectives at all is significant, for it was an expression of a deep and extremely characteristic ambivalence in the perception of Russia's political–geographical identity. Simply put, for one of them the Amur region and Siberia formed an integral and natural part of that Russia whose center lay west of the Urals, while for the other they did not.

This ambivalence emerges yet more clearly if we consider precisely how the Russians understood the term "America" and the affinities between it and the Amur region. Herzen's enthusiastic characterization of Siberia as an "America *sui generis*," Goncharov's thoroughly laudatory description of Murav′ev as a "courageous, enterprising Yankee," Kropotkin's easy comparison of the Amur to the Mississippi, Murav′ev's predeliction to refer to Nikolaevsk as "San Francisco"[100] – all of this suggested that the notion of "America" was imbued with shades of significance going beyond the simple designation of a friendly and supportive neighbor. In fact, "America" had come to represent two rather distinct qualities. The first was the United States itself: a progressive, dynamic, and thoroughly non-European society located in North America, with which as we have seen many Russians of the period declared their solidarity and yearned to materialize some sort of lasting bond or even alliance. At the same time, however, "America" represented not so much a specific country as the process by which that country had come into existence. In effect, the term described a model or a vision of a particular pattern of vigorous national development, one which involved the hewing of civilization out of unsettled and savage realms, the rapid colonization and "taming" of virgin territories, and the dissemination of agriculture, urban life, and all the other associated amenities which came with enlightenment. The end result of this process would be the creation of a new world: the forging of a single nation out of highly mixed streams of population, founded on self-reliance, personal initiative, and an all-inclusive popular democracy. While the experience of the United States offered the most obvious prototype for this pattern, the model was not tied geographically to North America, and was instead conceivable effectively in any place where the appropriate conditions seemed to present themselves. Here, then, was an "America" with which the Russians did not want merely to ally, but which they actively sought to appropriate for their own purposes and to create within their own borders.

This sort of quasi-metaphorical sense of the term "America" was alluded to by Herzen when in an 1857 letter to Guiseppe Mazzini he pressed the parallels between the westward movement of the American frontier and the settlement of Siberia. He likened the first colonists of the Urals to the North American homesteaders "on the virgin lands of Wisconsin or Illinois," and

[100] Butsinskii, *Graf . . . Amurskii*, p. 44.

maintained that the Siberian experience was "right out of one of Fennimore Cooper's novels." And just like the Americans, he affirmed, the Russians had accomplished their own "miracle," for they had planted modern civilization in empty territories.[101] The original settlement of Siberia, however, was a chapter out of remote Russian history. The altogether unique appeal of Russia's new lands along the Amur and Ussuri lay in the fact that they offered the Russians a palpable contemporary locus for this type of thinking. It was in this sense that the putative similarities between the Amur and America's own Mississippi were so very critical, for they appeared to confirm the quintessential appropriateness of the parallel in a manner that no other part of Siberia – indeed no other part of the country – could. An observer from European Russia, who ventured out to the Far East in the early 1860s, later recalled his initial impressions.

There on the Amur, it seemed to me, an interesting and instructive process is taking place which will last long into the future. This is the transformation (pererozhdenie) of various nationalities in the form of a new country (vo imia novoi strany), in accordance with the influence of climate, physical geography, and local administration. This process has up to now escaped the attention of history but can be followed with difficulty on the opposite shores of the ocean, in the United States and New Holland.[102]

And perhaps the clearest prospect of what this process might yield was offered by the English traveller Atkinson, who in 1860 described the anticipations and expectations of the Russians in the following terms.

Ten years hence the aspect of this region will be materially changed, flourishing towns will be seen on the banks of the Amoor, the vessels moored on the shore will show that the people are actively engaged in commerce and other industrial pursuits, while the white churches with their numerous turrets and green domes will prove that religion and civilization have taken the place of idolatry and superstition. A country like this, where . . . all the necessities of life can be easily produced, must prosper. [T]his country is destined to have a great future.[103]

It was the brilliant and understandably irresistible prospect of capturing and reproducing an America-like experience for themselves that fired the imagination of the Russians in the early years of Alexander II's reign, enthused as they were by the spirit of reform and rejuvenation that was reverberating throughout the country. Their efforts would amount to nothing less than the recreation of America on the Amur, a seemingly bizarre notion but one which was nevertheless articulated quite explicitly at the time, indeed by even such a crusty local nationalist as Mikhail Zenzinov. Although two decades earlier, Zenzinov had written to Pogodin's *Moskvitianin* in a blaze of patriotic indignation to criticize his compatriots for preferring to admire

[101] Gertsen, "Pis'mo k Dzhuzeppe Matstsini," p. 350; Bassin, "Inventing Siberia," p. 790; Kucherov, "Alexander Herzen's Parallel," p. 36; Hecht, *Russian Radicals*, pp. 36–37.
[102] Maksimov, *Na Vostoke*, p. 117. [103] Atkinson, *Travels*, p. 453.

foreign venues rather than their native Russian provinces,[104] a letter to an Irkutsk newspaper in 1860 indicated that now even he was caught up in the common enthusiasm for creating America in Russia. He began his missive with characteristic nationalist bravado, echoing his idol Gogol to the effect that "The [Amur's] vast territories and the abundance of their natural endowments will provide an opportunity to the steadfast and firm Russian knight to display all of his strengths." By the end, however, his tone had changed markedly, and he concluded in the most un-Gogolian manner imaginable: "and God willing . . . the Amur will eventually become a blessed Russian Kentucky."[105]

It will become clear by the end of this study just how quickly the Russians were to appreciate the fanciful and even frivolous quality of these sorts of expectations. For the moment, however, we might relate them back to the points made earlier regarding Herzen's and Bakunin's perspectives on Siberia. The ease and satisfaction with which the metaphor of "America" could be appropriated for the Amur region was an indication of the same ambivalence as to how "Russian" the region actually was. Indeed, in the case of the Amur the point may be put rather more strongly: it was only by virtue of their essential foreignness of these territories that such exaggerated anticipations could be foisted upon them in the first place. This sense of foreignness was underscored by the Orientalist Vasilii Vasil′ev, whose influential writings on Russia in East Asia will be considered at a later point in this study. In an essay on Russo-Chinese relations written in 1859, Vasil′ev felt the need to make a special plea for Russians to appreciate their presence in the Far East.

> Out there, on the Pacific, is also our fatherland (otchizna) . . . Our remote possessions in the East should be dear to the heart of each Russian, and every true Russian can regard them with the same pride with which Europe regards America.[106]

Vasil′ev may well have been expected to realize that at this late date Europe no longer took paternal "pride" in the existence of the obstreperous North American states, but the real significance of his comparison was what it said about the disengaged and indeterminate attitude of Russians towards their own realm. To the extent that the Russian Far East could be perceptually integrated into the broader matrix of national and civic preoccupations of the day – namely as an arena for national reform and reconstruction – its "Russianness" was relatively unproblematic; indeed the region appeared to be a vital national location. This, for example, was Herzen's position in 1858. It was not however an enduring perception, and the moment the basis for the region's integration came into question, there was nothing, or at least very little to bond it not only to the Russian heartland west of the Urals, but even

[104] See above, p. 59.
[105] Zenzinov, Untitled communication, p. 530. On his appreciation of Gogol's patriotism see his letter to Pogodin (30 January 1843), p. 650.
[106] Vasil′ev, "Otkrytie Kitaia," p. 19.

to Siberia itself. This point will be of considerable significance later in this study, for it will help explain the enigma of how the intense, practically euphoric expectations initially associated with the Amur annexation could dissipate so completely in the space of only a few years.

6

Civilizing a savage realm

"For the good of all Slavdom"

In the seventeenth century, Russian interest in the Amur region was fixed above all else upon its agricultural potential. Although the area had come to represent much more than this by the time it was incorporated into the Russian empire some two centuries later, the vision of a Siberian river valley that was uniquely endowed with a moderate climate and splendidly fertile soils still retained its essential appeal, and the hopeful anticipation persisted that it would become a thriving center of agricultural production. Those who reported back to European Russia about the region in the mid-nineteenth century often spoke to this point, describing at length the auspicious physical–geographical conditions on the Amur and the bright prospects for agricultural settlement there. The climate was "warm and healthy" and the soils were "extraordinarily fertile," noted one observer, such that not only grains of all sorts but orchard fruits, berries, and even extremely delicate plants such as tobacco and grapes could be cultivated there.[1] In the best locations along the Amur, suggested another, agriculture would be free from the frosts, droughts, and other climatic intemperances that plagued European Russia. To convince the most stubbornly sceptical readers of the potential of this cornucopia, he described the huge *arbuzy* or watermelons that grew wild for the taking along the river.[2] Indeed, the British traveller Atkinson reckoned that not only foodstuffs but valuable industrial crops could be grown here as well, and he concluded somewhat ruefully that with the Amur valley Russia obtained lands "more valuable than all the supposed cotton districts of Africa."[3]

In view of such apparently outstanding natural endowments, then, there appeared to be no question that the Amur region would yet become the *zhitnitsa* or granary that Khabarov and others had dreamed about so many years

[1] *Puteshestvie po Amuru*, p. 30; Romanov, "S ust′ia Amura," p. 107.
[2] Noskov, "Amurskii krai," no. 34:5, pp. 1–2, 4; Maak, *Puteshestvie na Amur*, pp. 103–104.
[3] Atkinson, *Travels*, pp. 463–464; Potanin, Review of R. Maak, *Puteshestvie na Amur*, p. 91.

earlier. In fact, after annexation by the Russians this vision became more grandiose than ever, and the hope of the early nineteenth century that this "Crimea of the Far East" would provide sustenance only for other Russian settlements in the North Pacific now seemed to some to be altogether too modest.[4] Bakunin, for example, whose exuberance in regard to Russia's newly acquired territories has already been noted, described the Ussuri valley to Herzen as a region "endowed with luxuriant soils, and a blessed, practically southern climate – with everything that the soul could desire." One could reckon with certainty, he confidently assured his correspondent, that "in 10 years or so" the region was sure to become the "breadbasket of the Pacific Ocean."[5] The scope of its potential, indeed, knew no bounds. A popular description of these territories published the same year speculated that beyond northeast Siberia and Russia's American colonies, the Amur region would become an important agricultural supply market for all of Japan and China as well. The current low level of agricultural development was no indication of what was to come, for "a beneficient Nature will more than reward the work of local tillers," and there was every reason to expect that with time the region "will be one of the very richest in the entire world."[6]

In addition to sharing this anticipated surfeit of agricultural bounty with hungry regions near and far, there was a related service for the country as a whole which the Amur region could fulfill. This was its potential as an area of resettlement for excess agricultural population. It was precisely as such a geographical receptacle that European Russia came to regard Siberia as a whole in the late nineteenth century, when the resettlement of peasants beyond the Urals – much facilitated by the construction of the Trans-Siberian Railway – assumed massive proportions.[7] This development of what might be called a free or voluntary migration was essentially unprecedented in modern Russia, where since the emergence of serfdom the population had been bonded to a certain locale and could not legally move at will. The massive eastward migrations of the turn of the century, therefore, were possible only after serfdom had been abolished and restrictions on peasant mobility significantly relaxed. The idea of peasant resettlement, however, was much older, as indicated among other things by scattered references to the value of the Amur region in this regard already in the 1850s. Murav′ev himself drew attention to this possibility as early as 1853, when in one of his many memoranda to the capital he referred to the "empty spaces" of Eastern Siberia which "are important to us, for they can hold the entire excess of the agricultural population of European Russia for an entire century."[8]

[4] *Istoriia reki Amura*, p. 147 (quote); [Speshnev], Untitled lead article, p. 4.
[5] [Bakunin], *Pis′ma*, pp. 124–125.
[6] *Puteshestvie po Amuru*, pp. 11–12, 33 (quotes); "Ekspeditsii russkikh," p. 26.
[7] Treadgold, *Great Siberian Migration*, *passim*.
[8] Barsukov, *Graf . . . Amurskii*, II, pp. 104–105 (quote); Veniukov, "Ob uspekhakh," pp. 57–58; and Potanin, Review of R. Maak, *Puteshestvie na Amur*, pp. 82, 98–99.

The point must have been appreciated in St. Petersburg, for the very first, and for many years the only legislation in Russia to promote peasant resettlement was directed at the Far East. Offering substantial incentives, which included extended exemptions from taxes and military conscription, the government opened the Amur region to Russian and foreign colonists in 1861, the same year that it emancipated the serfs.[9] The publicists of developments on the Amur, for their part, did not fail to call attention to the suitability of the region for colonization. On the first page of one of the first books to be published about the region, the naturalist Richard Maak wrote that Russia's new Far-Eastern territories "present the most advantageous conditions for colonization. The region possesses a healthy climate, convenience of internal movement over land and water, highly fertile soil, and close proximity to the administrative center of Eastern Siberia." And finally, to round out this appealing picture, he assured his readers that these splendid territories were "almost entirely empty" of indigenous population. Peter Semenov claimed that the capacity of the Amur region for absorbing colonists was "extraordinarily high," surpassing that of all the other parts of Siberia taken together. In a review of Maak's work, the Siberian ethnographer and historian Grigorii Potanin pointed to the need to disseminate information for potential colonists in the form of resettlement manuals, and an attempt to present some such material in a rather summary form was made in an article in the St. Petersburg journal *The Northern Bee* (*Severnaia Pchela*).[10] It is significant to note that all of these commentators stressed the imperative that colonization must be entirely free, voluntary, and carefully organized by the government, for in so doing they implicitly associated this aspect of activity in the Far East as well with the overall movement for progressive reform in Russia.

The willingness on the part of the Russian government to open up the Amur region to non-Russian immigration was unusual but not unprecedented. Since the eighteenth century, communities of foreigners had been solicited on occasion to undertake the colonization of newly acquired agricultural lands in Russia's various frontier regions. The most famous and important example was that of the Germans, whom Catherine II invited to settle and cultivate the Volga region and the southern steppes of so-called "New Russia." This endeavor proved to be highly successful, for which reason it was perhaps natural that those convinced of the need for the rapid colonization of the Amur region in the mid-nineteenth century should have been tempted by the notion that the earlier experience might be easily repeated in the Far East.

[9] Kaufman, "Pereselenie," p. 275; Treadgold, *Great Siberian Migration*, pp. 69–71; Malozemoff, *Russian Far Eastern Policy*, p. 9.

[10] Maak, *Puteshestvie na Amur,* p. v; Semenov, "Amur," p. 189; Potanin, Review of R. Maak, *Puteshestvie na Amur*, pp. 98–99; Veniukov, "Kolonizatsiia," *passim*. As it turned out, the sort of brochure that Potanin had in mind, directed at prospective peasant colonists to the Russian Far East, began to be produced only some 40 years later. Cf. Iakovlev, *Rasskazy*.

Suggestions were made for trying to attract this "deutsche Element" directly from Germany itself,[11] but most attention fell on those groups who had already made the move once. In the late 1850s, German Mennonite communities in southern Russia were approached with offers of land and special privileges in the Amur valley, in order to found new colonies. The offer was taken very seriously, several agents were dispatched to inspect potential sites, and Murav′ev was extremely enthusiastic about the possibility.[12] Despite the initial interest, however, and the eschatological conviction of at least one pious agent that the invitation of the Russian government was a sacred signal "directed to the children of God . . . to relocate to the Amur region," the Mennonites came rather quickly to the conclusion that the Far East was simply too remote and too unknown to be very attractive.[13]

The Mennonites might have seemed a particularly appropriate choice to provide the sort of "thrifty colonists" that in Ravenstein's estimation the Amur region was in such desperate need of, but they were not the only one.[14] There was one other alternative, a group who unlike the Germans had no previous experience with colonization in Russia but who were far more attractive, at least for some, in that they were a fraternal Slavic people. We have seen earlier in this study that Asia was only one of the geographical arenas to which the Russians felt called in order to realize their messianic destiny. No less important, and certainly more widely endorsed, was the belief in a Russian mission of salvation directed at the other Slavic nations, a belief which served to focus attention in a very different direction, namely westward to Eastern Europe, and to a lesser extent to the Near East. This amounted to a geographical bifurcation of sorts in the Russian national mission, but it was one which did not necessarily involve a corresponding ideological dichotomy or ambivalence of any significance, at least in the period under consideration. In whichever geographical realm Russia might choose to concentrate its energies, that is to say, it was felt that in some way the same essential impulse was being satisfied. Indeed, rather than an antagoism between these eastward and westward biases, there was more than anything an entirely positive affinity between them, a point which is unmistakably indicated by the fact that many if not most of the individuals responsible for urging Russia's expansion in the Far East in the second half of the nineteenth century were at the same time very strongly committed to the Pan-Slav movement. In the case of the Amur, this was true of such major actors as Yegor Kovalevskii, Nikolai Ignat′ev, Putiatin, Konstantin Nikolaevich, and Murav′ev himself. An identical pattern can be

[11] "Gustav Radde's Vorlesungen," 1860, p. 267.

[12] Barsukov, *Graf . . . Amurskii*, II, pp. 296–297.

[13] Quoted in Friesen, *Mennonite Brotherhood*, p. 102n, 248; Urry, *None but Saints*, p. 197. Ravenstein's reports that 40 German families were to be brought from California and that 100 Mennonite families from Tavrida actually set out for the Amur in 1860 were in all likelihood apocryphal. *Russians*, p. 150.

[14] Ravenstein, *Russians*, p. 151n.

identified regarding Russian expansion in Central Asia and the Caucasus as well.[15]

It was on the Amur, however, that we encounter what was probably the only deliberate attempt literally to combine the two causes and realize the Pan-Slav vision on the shores of the Pacific. One of the most imaginative of all the undertakings in the Russian Far East during the period under consideration – and it was a period which, it should be quite clear by now, did not lack for imaginative endeavor – was a project advanced by some Russian Pan-Slavs for promoting the resettlement of Czech agriculturalists in the Amur valley. This undertaking, which in retrospect might seem to be little short of fantastic, actually appeared a good deal more reasonable in the late 1850s and early 1860s. This was a time when the heady atmosphere born of the first Pan-Slav congress in Prague in 1848 had not yet dissipated, and the Russian Pan-Slavs, fresh with the inspiration of their newly-discovered messianic role and still smarting from the wounds inflicted in the Crimea, were eager to cast themselves in the role of natural protector and saviour of the other Slavic peoples.[16] Their inclination toward such an endeavor, it should be noted, was endorsed by other Slavs, whose hopes were to be undermined only with the Russian repression of the Polish revolt of 1863.

As the Russians explained it, the notion of bringing Czech settlers to the Amur originated with the Czechs themselves. The Pan-Slav ethnographer and philologist Alexander Gil′ferding , who became one of the principal advocates of the project, reported that it had first been suggested to him by Frantiček Rieger, a prominent Czech Pan-Slav, during a visit to Prague in 1859.[17] In the following year, Gil′ferding presented his views on the enterprise in an article in the newly founded Irkutsk newspaper *Amur*. The Slavs of the Austrian and Ottoman empires, he wrote, have accepted Russia as their brother and friend, and look to it for their liberation. They understood just how much their national existence depended on Russia, and thus "it may be judged with what joy, with what exultation the Slavs of Austria and Turkey greet all news about the internal prospering and the external growth of Russia." Nowhere, he asserted, perhaps not even in Russia itself, was there such enthusiasm over the annexation of the Amur region as in these Slavic lands. Having recently visited

[15] Ritchie, "Asiatic Department," pp. 279–298; Hunczak, "Pan-Slavism," pp. 93–94; Kipp, "Grand Duke Konstantin Nikolaevich," pp. 103–104; Dahlmann, "Zwischen Europa," p. 61; Simon, "Russischer und sowjetischer Expansionismus," p. 102; Kazemzadeh, "Russia," p. 495. In Turkestan, the example of General M. G. Cherniaev might be mentioned, and in the Caucasus that of A. I. Bar′iatinskii. MacKenzie, *Lion*, pp. 118–121; Rieber, *Politics*, p. 61. On the strong presence of Pan-Slavs in the Asiatic Department of the Foreign Ministry, see Koot, "Asiatic Department," pp. 26–28. Not all Pan-Slavs supported Russian expansion in the Far East, however. See the discussion of Nikolai Danilevskii, below, p. 270.

[16] Petrovich, *Emergence*, pp. 31, 38.

[17] Gil′ferding, "Pis′mo," p. 287. On Gil′ferding, see Lapteva, "Gil′ferding," pp. 121–125; on Rieger see Kohn, *Pan-Slavism*, pp. 22–23.

Prague, Gil′ferding could report with authority that the Czechs had become infected with the Amur fever. They

> regard the annexation of the Amur not merely as a Russian but as a world event and an all-Slavic triumph. [They do so] because this acquisition, in their opinion, opens up the Pacific ocean – an ocean which previously was entirely in the hands of the Germanic tribe (plemia) represented by the English and the North Americans – to the activities of the Slavs.

For this reason, he went on, one could read "expressions of joy and sympathy" about the Amur river in Czech newspapers, the ardor of which matched that of the Russians themselves. There was already talk of how the Amur region would compete with the western United States, and how the Russians would build a railroad to connect Nikolaevsk and Irkutsk with Moscow.[18]

The Czechs were not only writing about the Amur, Gil′ferding went on, but wanted to talk about it at every opportunity, and thus it had been a favorite topic of conversation during his trip. His hosts spoke approvingly of the attempts to encourage Russian resettlement to the region, but they noted that Russia itself could probably not supply a large enough number of colonists without dangerously depopulating those regions out of which the colonists would emigrate. Because of this, they suggested, Russia would be constrained to turn "against its will" to the West, where there were available colonists who could be conveniently dispatched to the Far East by sea. Pressing the point further, Gil′ferding went on, the Czechs speculated that the Russians would turn first of all, as in the past, to the Germans, an option which was problematic insofar as the Germans tended to remain as a foreign element and did not as a rule assimilate. While such insularity was not inappropriate for such internal regions as the Volga and the southern steppes, where German settlements were tiny islands in a dense Slavic sea, the situation in the Far East – "a territory still so little penetrated by the Russian element" – was far more precarious. Should another war with Europe break out, there would always be the real danger that they would remain loyal to their original homeland. Gil′ferding concurred with his Czech hosts that this this proposition was a disturbing one, and he enthusiastically endorsed the solution they put forward, namely to try and attract fraternal Slavic peoples instead as settlers in these regions. In regions such as Moravia and Slovakia there were industrious agriculturalists who, in contrast to the Germans, "look upon Russia as their native land, and would travel to Russia more eagerly than anywhere else." These people would assimilate into their new environment quickly, so that "after one month the Czech, Moravian, Slovenian, and Slovak would be speaking Russian, and their children would be indistinguishable from the Russians."[19]

[18] Gil′ferding, "Mnenie," p. 372. [19] Gil′ferding, "Mnenie," pp. 372–373.

Although the project which Gil′ferding sketched out included measures to attract Czechs to the Amur directly from the Old World, his main hope rested on those who had already become immigrants and moved to the United States. While these settlers may have met with material success in their new home, they nevertheless felt uncomfortable and out of place in a non-Slavic environment, and they were highly apprehensive about the inevitable assimilation of their children into American culture and society. "Czechs [in Prague] who have kept up contact with them assured me," he explained, "that many of them would eagerly leave the New World for the Amur and would bring with them their capital and their activity." In the United States, he explained, they feel that they fade into the "Yankee and Germanic mass" and "are insulted that the Americans consider them to be Germans." Most distressing of all was the fact that their children were forgetting their native Slavic tongue. "This is why they prefer Russia and the Amur to America." And not only were they spiritually willing, but their experience in the New World had prepared them "for the struggle with wild, untouched Nature and with all the conditions of rudimentary colonization." In short, a more ideal group of settlers for the Russian Far East was hardly to be found. All that would be necessary would be to print an invitation in the Czech newspapers in the United States – Gil′ferding mentioned papers in St. Louis, Racine, Chicago, and Ohio – and indicate a departure date from San Francisco or Panama.[20]

In all likelihood, Gil′ferding , who was employed at the time in the Asiatic Department of the Foreign Ministry and thus had access to the most up-to-date information about developments in the Far East, had been a rather more proactive propagandist for the Amur during his sojourn in Prague than his own account would indicate. Nevertheless, there was at least some truth in his claims regarding Czech interest in these regions.[21] In October 1861, Ivan Aksakov's Pan-Slav journal *Den′* carried a report about a meeting of American Czechs in St. Louis that summer, during which the idea of relocating to the Amur was discussed at length. The article conveyed the considerable enthusiasm on the part of the group, which believed that on the Amur every "work-loving Czech would be able count with certainty on [finding] an abundant and noble field for the most fruitful activity."[22] Announcements of this project were published in various Czech newspapers in the United States, from which excerpts were reprinted later the same year in *Den′*. One passage, from an statement appearing in the weekly *Amerikanski Slovan* published in Racine, Wisconsin, clearly conveyed the Czechs' mood of discontent with their new home.

[20] Gil′ferding, "Mnenie," p. 374.

[21] Ritchie, "Asiatic Department," p. 261; Petrovich, *Emergence*, p. 65; Lapteva, "Gil′ferding," p. 121. [22] P. L., "Chekhi," no. 2, p. 12.

Our Slavic policy does not allow for any sort of American lies, deceptions, or fraud, and we despise all such actions which demean human dignity. The time is approaching when we shall have to extend our fraternal hands to our blood brothers, the Russian Slavs, for mutual unity and for the good of all Slavdom. After this we will not tarry in returning to the bosom of our mother, the Slavic world (Slavii) [i.e. to the Amur].[23]

To this the author of the article in *Den′* added his own thought that the desire of the Czechs to leave the United States was a natural product of that country's "moral collapse." A healthy Slavic future could be expected to develop "only where this people matured and where it put down roots, not as a transplant [i.e. in the New World] but by means of natural growth, that is, in Europe and Northern Asia."[24]

By the early 1860s, the affair had advanced to the stage where the American Czechs, like the Mennonites, sent agents to the Amur region to investigate its actual potential for agricultural colonization.[25] We have no information about the conclusions of these agents regarding conditions in Siberia, but in any event the entire enterprise was aborted in the aftermath of the Russian suppression of the Polish revolt in 1863. The Czech population and press in the United States sided firmly with the Poles, and their outspoken attitude was endlessly distressing for the Russian Pan-Slavs, who felt that this defiant stance violated the principles of Slavic fraternity and common interest. In a letter to Rieger in Prague, Gil′ferding expressed his bitter consternation over the attacks on Russia in Czech newspapers, and related them directly to the project of Czech resettlement on the Amur. "If the voice of contemporary Czech journalism is the true voice of Czech society and the Czech people," he wrote, "then I would despair of my involvement in this affair [resettlement], fearing that on the shores of the Pacific ocean would be born a new Poland."[26] The American Czechs, for their part, suggested in turn that the collapse of the resettlement project was responsible for destroying their "last hope . . . in old Russia as a savior nation for our posterity."[27]

Like so many other schemes and projects regarding the Russian Far East during this period, this flash of interest in settling Czech immigrants on the Amur was fanciful and even contrived. Indeed, some liberal critics at the time took indignant exception to it. Herzen's *Kolokol*, for example, which had expressed nothing but the warmest sympathy for Russian expansion in the Far East, printed a letter from Russia in 1862 which described the Pan-Slav project as the very apogee of official lunacy: "My God! My God! Where the devil are we? What are we? Are we seeing this dream in our delusions, or are we actually living it?!! Sire! The government is trying to attract American Czechs

[23] *Amerikanski Slovan*, no. 29 (18 July 1861), quoted in P. L., "Chekhi," no. 10, p. 15.
[24] P. L., "Chekhi," no. 10, p. 16. [25] Popov, "Vopros," p. 1. [26] Gil′ferding, "Pis′mo," p. 287.
[27] Popov, "Vopros," p. 2.

to the banks of the Amur."[28] Yet if the scheme had no practical outcome, it was still significant for what it indicated about perceptions of the Amur and of the sort of function this region might be able to fulfill for the country as a whole. By enabling the resettlement of a fraternal Slavic peoples back onto native Slavic soil – and for these purposes the notion of native Slavic soil was eminently transportable to the Far East – the annexation of the remote and obscure Amur region immediately became meaningful for one of the most important projects that animated the entire period, namely Russia's messianic task of salvation. The attempt to bring this significance to life in terms of the Pan-Slav preoccupation with Russia's brethren to the west was clumsy, and ended in a quick and unceremonious failure. Far more plausible and enduring, however, was the alternative view from the east, and it is to this perspective that we will now turn our attention.

"Who is transforming the soil and the climate?"

One of the most important elements animating the sense of Russian national identity that developed in the period under consideration was the belief in a special mission of salvation in regard to the peoples of Asia. The Russians did not necessarily feel that the task of bringing enlightenment and justice to these ossified societies was something they had wilfully elected to undertake, but understood it rather as a sort of collective national responsibility that had been thrust upon them, assumedly through divine agency. In this regard, as in many others, the Russian attitude was similar to the notion of a "white man's burden" to bring enlightenment and Christianity to the dark corners of the earth that was shared by all the colonial powers of the nineteenth century; indeed the essential unity of their cause with that of the West was a recurrent theme. The belief in a common mission was of course much enhanced when confirmed by representatives from the West itself, which it frequently was. Surveying the Russian position on the Amur, for example, Atkinson predicted that the superior "moral influence" of the Russians would "spread rapidly over Manjouria and destroy the power by which China holds the people in thraldom."[29] The ever-observant Collins was even more emphatic, and he put forward a grandiose vision of the civilizing mission that was Russia's: "[N]ow [that] the Russian finds himself master of the easternmost limits of the ancient dominions of Genghis," he mused, "[w]hat is left to be done in Asia?"

> Only to place the second son of the Sclavic Czar on the throne of Genghis. It would be a vast step in the progress of the Mongol race and of civilization and worthy of the great advance and the great epochs of the nineteenth century, and the only means by which nearly half of the inhabitants of the earth can be Christianized and brought

[28] Mart′ianov, "Pis′mo," p. 1096. *Kolokol* printed this letter a second time two years later: no. 178 (1 February 1864), pp. 1463–1467. [29] Atkinson, *Travels*, p. 460.

within the pale of commerce and modern civilization. May we not look to this as a solution of the Chinese riddle? For without Russian interposition, the Mongol race must go down in intestine [sic] religious wars, pestilence, and famine, pressed as they now are on all sides by the irresistible force of Christian powers . . . [T]wo contending forces, Sclavic and Tartar . . . have been wrestling for a thousand years. The blood of Japhet has triumphed over that of Sham. The curse of Noah is about to be accomplished, the prophecy fulfilled, and Asia Christianized.[30]

In regard to the specific situation confronting the Russians in the Far East, there were a number of different ways in which their civilizing obligations could be rendered. With his reference to the "Chinese riddle," Collins indicated the importance of Russia in his own mind to the grand historical project of opening the Middle Kingdom to outside influence and thereby reforming or transforming it along Western lines. This was an endeavor which the Russians, recently but by this point quite fully energized with a disdain for the stagnation and decay of this realm, were ready to pursue. While the dramatic opening of China to Western influence in the 1840s had been accomplished without their involvement, the expansion of their activities on the Amur in the following decade provided a timely opportunity for them finally to take their place among the ranks of those engaged in this most worthy enterprise. Russia's unity of purpose with the other European powers in China was endorsed at a relatively early date by the orientalist Valentin Korsh, writing in a major Moscow newspaper in 1851. Korsh contrasted the familiar image of China as a decayed culture wallowing in a "narrow, closed circle of long-dead concepts and morals" to the freshness of European imperial activity. No matter where the enterprising European settled, he demonstrated the ability to utilize the available resources and technology in order to "facilitate the improvement of the material welfare of all." Accordingly, the Europeans had "turned the steppes of America into a [civilized] country, and out of the Siberian desert created a rich province of the Russian realm." The Celestial Empire, on the other hand, "stubbornly clings to its fruitless and ossified ideas, on which it so pointlessly wastes its strengths and which will inevitably lead it to complete collapse."

In particular, Korsh continued, the backwardness of the Chinese rendered them ignorant about the environment in which they lived and incapable of utilizing it properly. The best evidence of this was their fear of the sea, about which they remained entirely ignorant despite an extensive coastline. "The enterprise of the European, calling him to the sea, is foreign to them: they make no use of the enormous advantages of their close proximity to America, and the countless islands of the Pacific." This then led to Korsh's major point, namely that the Chinese had never made use of the Amur river, never developed its potential, and thus never allowed it to attain the significance which,

[30] Collins, *Siberian Journey*, pp. 289–290.

by virtue of its natural position, it deserved. "In the hands of the Chinese the Amur loses all its importance" and remained an unutilized *terra incognita*, languishing without that "multifaceted historical significance" which it was destined sooner or later to achieve. This would moreover remain the case until such time as the river "crosses over into the hands of a more active and enterprising people," and although he left the point at that, there was no doubting that he had in mind the Russians themselves.[31]

In his article Korsh castigated the Chinese and placed his emphasis on what Russia would be able to accomplish in a remote border region where the Middle Kingdom had failed so miserably. No sooner had the occupation of the Amur valley actually got under way, however, than the Russian advance took on a broader significance, and began to be seen as a development which would foster the civilizing not merely of a heretofore neglected frontier zone, but indeed of China itself. It was with this prospect that Ksenofont Kandinskii, a merchant from Kiakhta, chose to celebrate Murav′ev's first successful voyage down the Amur in 1854, and at a banquet in honor of the governor-general he offered the following verse:

> Byt′ mozhet, nash orel dvuglavyi
> Probudit dremliushchii narod
> I, ozarivshis′ novoi slavoi
> Ego on k zhizni prizovet.
>
> Perhaps our two-headed eagle
> Will awaken the slumbering nation.
> And, radiant with a new glory,
> Will summon it back to life.[32]

The same sentiment was expressed in fuller detail in a speech delivered by Dmitrii Romanov in Irkutsk on the occasion of the signing of the Treaty of Peking in 1860. Inspired perhaps by the presence in the audience of N. P. Ignat′ev, the much-celebrated diplomat who had concluded the treaty, Romanov enthusiastically underscored the significance of Russia's new agreement with China as a means toward bringing the latter into the modern age.

[31] Korsh, "Bassein," p. 1327. An example of the remarkable symmetry in colonial attitudes can be seen in the contemporaneous reactions in the United States to Spanish domination of California in the period leading up to its annexation.

> Europeans – and now Anglo-Americans – who visited California and who lusted after the land described the present colonizers of the area as grossly unfit. Invariably the glowing descriptions of the California landscape – its salubrious climate, fertile soils, capacious harbors, and breath-taking natural beauty – were set against remarks on the unworthiness of its Spanish-speaking inhabitants. [Richard Henry] Dana expressed this view most fully. "In the hands of an enterprising people," he wrote, "what a country this might be!"

Korsh's own sentiments regarding the Chinese on the Amur and the prospects for Russian "enterprise" could not have been mirrored more precisely. Rawls, "California Mission," p. 347.

[32] Quoted in Barsukov, *Graf . . . Amurskii*, I, p. 360.

"A glorious and resounding event been achieved in the East!" he declared, for the Chinese empire has now been opened and made accessible to Western commerce, Western technology, and Christianity. "Fully one third of the human race, which up to this point remained as if it were non-existent for the rest of the world, is now entering into contact with the advanced nations, and is becoming accessible for European civilization." It was, Romanov claimed, the strength and vitality of what he called *evropeizm* or Europeanism that had finally been successful in breaking down the millennial stubborness of the Chinese, and with this success the West had gained a vast field for its progressive activities. "The consequences of this great and momentous world event" he concluded rather breathlessly, "are stupendous, innumerable, unfathomable!"[33] Like Korsh, Romanov felt that the unity of Russia's mission in the Far East with that the other Western powers was so self-evident that there was no need even to mention it. The prospect he laid out was a simple one of *evropeizm* in valiant struggle against the dark and resistant forces of oriental ignorance.

In addition to advancing the Westernization of China, the occupation of the Amur was meaningful to Russia's civilizing mission in terms of the indigenous population of the region itself. Indeed, these native groups formed a much more attractive object than the Chinese for the attention of those Russians anxious to exercise their new-found role as civilizers. In the first place, with their subsistence economies and paleolithic cultures, the natives of Russia's newly-acquired territories fit the stereotypical image of backward savages and children of nature far more closely than even the decrepit Chinese, and thus – as badly off as the latter may have been – necessarily stood in even more desperate and obvious need of salvation. Along with this, however, a rather more practical factor was also at work. Although Collins for one seemed to have been entirely comfortable with the prospect of letting the Russians "solve the Chinese riddle" by assuming exclusive control over the "throne of Genghis," there was no real question that the redemption and transformation of China would necessarily be a joint European enterprise, which meant that the glory would have to be shared as well. On the Amur, by contrast, the Russians could operate unencumbered by any competition, and thus could be secure in the confidence that whatever successes they would achieve would be theirs and theirs alone.

There was however a significant complicating factor in regard to the peoples of the Amur. China was a relatively familiar entity to the Russians, who had at their disposal numerous specialists and a variety of readily available sources, Russian as well as European. Indeed, even without recourse to this specialized knowledge, educated Russians were generally comfortable that their picture of Chinese society – which supported and justified their critical stance – was accurate. The Amur region on the other hand, as Korsh had quite correctly pointed

[33] Romanov, *Poslednie sobytiia*, p. 3.

out, was a vast unknown territory, not only for the Russians but for the West in general. While it was clear enough that the region was inhabited by Asiatic groups who existed on some sort of crude and primitive level of civilization, virtually nothing was known in any detail about who or what they were. This sort of information was urgently needed. It was necessary most obviously for the practical purposes of establishing Russian civil administration in the region, but it was also necessary for the subject we are discussing. Both the extent to which the Russians could exercise their mission of salvation in the Far East as well as the manner in which they exercised it depended upon the availability of what we might call suitable ethnographic material in the form of unenlightened indigenous populations. The Russian Far East promised to supply such material, but until a fuller picture of it could be compiled, its specific relevance and usefulness necessarily remained obscure. And in order to compile such a picture, the Amur region had to be "discovered" yet again.

The agents of this discovery, logically enough, were those individuals – military men and natural scientists, for the most part, but some merchants and incidental travelers as well – who were charged with the exploration and reconnaissance of the territories in question. Out of the accounts they presented, it is apparent that their exploration and discovery involved not one but at least two quite discrete processes. One of these corresponded to the practical concerns that were traditionally associated with the enterprise: navigating and describing unknown waters, surveying coastlines, determining meteorological conditions, natural resource endowment, and so on. The second process was more subtle, and corresponded not so much to the need for objective geographical information as to the needs of the new nationalist consciousness. For those intrepid Russians who were the first Westerners to penetrate the primeval wilds of the Amur region were not merely filling in a blank region on the map, they were at the same time identifying a population that clearly stood in need of the very civilization and enlightenment that they and their compatriots were desperately seeking to provide. From this standpoint, the entire project of exploration and discovery in the Russian Far East was transformed into an essentially psychological exercise in cultural perception, interpretation, and construction.[34]

Before considering the "discovery" of the Amur itself, a brief *excursus* would be useful in bringing out more fully the complex quality of this exercise. A particularly vivid example, which comes from the period we are considering but from a somewhat different region, can be seen in the reactions of the young Ivan Goncharov to his first encounter with Eastern Siberia. As a junior clerk in the foreign ministry, Goncharov had accompanied Evfimii Putiatin on the latter's mission to establish diplomatic relations with Japan in

[34] Among others, Paul Carter has recently examined this process in a splendid study of the "discovery" of Australia: *Road to Botany Bay*.

the mid-1850s. The goal of the mission having been accomplished, he continued with some other members of Putiatin's *entourage* northward, through the Tatar straits and north along the Sea of Okhotsk coast to the port of Okhotsk. Here Goncharov landed in 1854 and began the long overland journey back to St. Petersburg, the first leg of which involved the notoriously difficult passage across the coastal mountains to Yakutsk. Throughout his travels, he had kept a careful record of his impressions, and he continued to do so in northeastern Asia as well. At first, he was entirely overwhelmed by what seemed to him the utter desolation and wildness of the region.

Nobody lives here, from the Arctic Ocean to the Chinese border except the nomadic Tungus, scattered here and there across these enormous expanses . . . The heart is squeezed with grief when you traverse these mute deserts . . . There is nothing for a civilized (vyrabotannyi) person to do in these uncivilized (nevyrabotannye) deserts. After thousands of *versty*, one has to be a desperate poet in order to be able to take any pleasure in the enormity of either the desert silence or one's own boredom. One has to be a savage in order to consider these mountains, rocks, and trees to be furniture and domestic decoration, to consider the bears to be one's comrades, and to consider the wild game to be real food.[35]

When he finally reached Yakutsk, however, Goncharov's mood changed dramatically, for here the Russian "element" was a palpable presence, and he saw clear evidence of the positive benefits that modern civilization was capable of bringing to even the most helpless and backward of regions.

Despite the length of the winter and the severity of the frost, how everything is in motion in this region! I am now a living eye-witness to that chemical–historical process by which the desert is transformed into places where people live, and savages are promoted to the rank of [full] human beings. Religion and civilization struggle against savagery and call sleeping forces to life.

Indeed, this activity was so very impressive that it seemed to be the work of superhuman forces, who were constrained by no limitations on what they were able to accomplish.

The appearance and the form of the soil itself is transformed, the frost grows less severe, and from the ground warmth and plant growth are coaxed. In a word, something is being accomplished [under Russian influence] which, according to Humboldt, is [usually] accomplished with continents and islands by means of the hidden forces of Nature. Who then, it will be asked, is this titan, who tosses both dry land and water? Who is transforming the soil and climate? There are many titans, an entire legion of them, and they are all mixed together here in this laboratory: the nobility, the clergy, merchants, and exiles: all [Russians] are summoned to labor and work without stop.

"And when this region – once dark and unknown – has been entirely prepared, populated, and enlightened," he went on, it will present itself to an "aston-

[35] Goncharov, *Fregat "Pallada"* (1986), pp. 500–501.

ished humanity" and demand a proper name and rights for itself. History will someday have to take note of these titans, these bearers of progress who "erected pyramids in the deserts" and taught Yakuts, Aleuts, Tungus, and others "how to live and to pray: it was these very people who created, who thought up Siberia, who populated and enlightened it . . . And it is not as easy to create Siberia as it is to create something under a more blessed (blagoslovennyi: i.e. temperate) sky."[36]

Here then, would appear to be a splendid illustration of the supreme confidence on the part of Russians in the benefits which they brought with them to the debris of Eastern Siberia. In fact, however, Goncharov's flamboyantly overstated and seemingly unshakable self-confidence concealed a decidedly more vulnerable dimension of his imperial mentality. This latter aspect came to light in these very passages, as he expressed a rather peculiar irritation at the comments of an earlier author on the same subject. At one point, Goncharov paused in his narrative to deliver a stinging attack on Matvei Gedenshtrom, the same Arctic explorer whose 1830 work *Siberian Fragments* we have had occasion to cite earlier in this study.[37] Heavily influenced by the Romanticism of the early nineteenth century, Gedenshtrom understood the contrast between savagery and civilization in a rather different manner than the thoroughly modernist Goncharov. On the one hand, like other Romantics he found much to criticize in advanced and enlightened Western society. Moreover, he was prepared to recognize certain "primitive" *dobrodeteli* or virtues on the part of the indigenous tribes of Siberia, and while his depiction by no means presented their rude existence as the stereotypical ideal of the noble savage, he felt nonetheless that they would have little to gain from being "civilized" by the Russians. Consequently, in his work he urged that they be left undisturbed.[38] Gedenshtrom's conclusions in this regard were enormously offensive to Goncharov, who registered his objections most irately. First, he insisted on the importance of what the Russians could do for these peoples.

> The enlightenment of the Yakut consists for the moment in teaching him agriculture, herding, and trade; all of this is being done. It is not important that he lives in the desert: enlightenment will find a way to deal even with the desert. Earlier it was thought that grain will not grow here, but after [we] applied ourselves with knowledge and love to the cause, it turned out that it grows here after all. Now sheep are being raised here. Of course, we will be waiting for a long time yet before we will be wearing fabric from Yakutsk factories, but this is not necessary for the time being. Thank God, yes, thank God (and I don't want to insult our author) that the Yakuts are now eating bread, and not bark, that they are wearing Russian cloth and not musty animal skins.

[36] Goncharov, *Fregat "Pallada"* (1986), p. 525. [37] *Otryvki o Sibiri*. See above, pp. 70–72.

[38] For another emphatic statement of this view, see the comments of Karl Baer, a member of the Baltic German "Old Guard" at the Russian Geographical Society. Ber, "Ob etnograficheskikh issledovaniiakh," p. 104. For a fuller elaboration of Russian Romanticism's attitude toward Asia, see Susan Layton's most insightful discussion: *Russian Literature*, pp. 84–85 and *passim*.

What annoyed Goncharov most, however, was the suggestion that these indigenous groups might possess some sort of native virtues of their own, virtues that were not only autonomous from civilized society but might indeed stand in contradiction to it, such that civilization could actually be a threat to them. "Wild virtues and simple morals – what a treasure," he continued sarcastically, mocking Gedenshtrom:

> there really is something to get excited about here. They say that the savages don't drink, don't steal – this is true, as long as there is nothing for them to drink or take. They say that they don't lie – but this is only because there is no need for them to do so. This is all well and good, but after all, one can't remain in a condition of savagery forever. Enlightenment, like a conflagration (pozhar), will encompass the entire globe . . . What an idiot this author is![39]

This much having been said, the editors of earlier editions of Goncharov's work apparently considered that the author's point was made, and the text turns to another subject. From Goncharov's original manuscript, however, it is apparent that the text was cut, for the subject was one of inordinate importance to him, and he was only warming up to it. A recent complete edition of the work has restored the extracted passages,[40] and thus enables us to follow how he went on to press the point with a fury that was only barely concealed.

> If we don't go to them [Siberia's native population], then they will come to us themselves with their fur pelts and will want to trade with us for something else, and they would in any event not escape enlightenment, they would in any event learn to tell good from evil and, having passed through the fire of experience and learned to reject the latter, they would chose the former, which is the essence of true enlightenment. Precisely how would you like it to be [again addressing Gedenshtrom] that a savage jumps directly from Yakutsk oblast′ into the realm of virtue? As for savage virtues – God save us from them! After all, it means nothing at all for one savage to knife another and rob him – this is a most virtuous act. The Cherkess are also a virtuous people: as long as you are a guest in their home, they will not only not touch you but treat you to kumis and homemade wine. After you leave, though, they'll catch up to you and treat you to something else. Many of us are charmed by this. Once we too were virtuous, and the Normans as well, and even the medieval knights, but God delivered us from these virtuous groups![41]

We have cited Goncharov at length because his nervous outburst about the Yakuts affords a particularly revealing insight into psychological complexities of Russia's messianic consciousness, complexities which we will encounter again in the Amur valley. On the one hand, through their activities in Asia the Russians wanted to feel themselves to be capable of truly miraculous accomplishment: the transformation of an entire country, indeed – in Goncharov's

[39] Goncharov, *Fregat "Pallada"* (1986), pp. 538–539. It is impossible not to note the appositeness of Goncharov's simile.

[40] Compare the version published as part of Goncharov's collected works (1952–1955) with the annotated 1986 edition.

[41] Goncharov, *Fregat "Pallada"* (1986), p. 825n.

wonderfully telling phrase – the literal "creation" of Siberia. At the same time, however, his near-hysterical reaction to Gedenshtrom's obscure comments, obviously made many years earlier and in an entirely different context, was indicative of a very different side to this same frame of mind. Specifically, it exposed the elemental tension and uncertainty that twisted underneath the surface of Russia's apparently self-assured progressivism and loudly proclaimed belief in its own powers. By implicitly raising a question about the value of that "chemical–historical process" through which modern civilized society sought to recreate the rest of the world in its own image, and by casting even a shadow of a doubt upon the absolute beneficience of Russia's "legion of titans" in Siberia, Gedenshtrom unwittingly touched upon what in the intervening quarter-century had become an extraordinarily sensitive nerve. For Goncharov, and for the nationalist project to which he was unreservedly committed, it was absolutely imperative that the natives of Siberia be utterly virtueless wretches, as it was absolutely imperative that civilized (Russian) society be in unchallenged position to ameliorate and improve their lot. Things simply could not be otherwise. This imperative came not, or at least not primarily, from an altruistic desire to help the less fortunate, but rather from the fact that Russia's new and positive image of itself depended existentially upon these preconditions in order to be fully realized. After all, raising savagery up to the level of civilization could hardly be counted a virtue if there had been nothing particularly wrong with savagery in the first place, and the Russian nationalists now needed that virtue for themselves at all costs. Although poor Gedenshtrom had had not the slightest intention of doing so, by allowing the Siberian natives a few sundry *dobrodeteli* he was effectively denying the Russians themselves a critically important means of demonstrating the quality of their own civilization.

We may now return to the Amur. As part of his explorations in the early 1850s, Gennadi Nevel′skoi conducted a series of reconnaissance surveys in the territories around the river's mouth and south along the coast. The letters and diaries that have survived from these missions enable us to follow in some detail how a similar psychological process was at work from the very beginning of the occupation of the Amur valley. Indeed, it commenced as soon as Nevel′skoi disembarked from his ship for the first time, in the early days of August 1851. He related that he immediately encountered a group of Giliaks, a tribe indigenous to the lower Amur, who were being lorded over by an arrogant official from Manchuria. A brief exchange convinced him that the presence of the Manchu was resented, and that the natives indeed entertained a "concealed hostility" to their "oppressors." Nevel′skoi reported with satisfaction how he was able to excite the awe and barely concealed glee of the assembled Giliaks by drawing his revolver and threatening the offender, and their reaction convinced him that if it came to a confrontation, the natives were already sure to be on his – the Russian – side. As Nevel′skoi became more familiar with the natives of the Amur, he learned that the Manchus were not their only problem. For some years, British and American ships had been sailing up the Tartar straits. These "white men," the natives complained, would terrorize them, plundering their provisions and furs, and committing wanton atrocities. "The Giliaks, not knowing whom to turn to for aid and protection, and having not a single means of defending themselves, do not know how to repulse and punish the intruders." Nevel′skoi immediately recognized the opportunity to both announce and justify Russia's new presence in the region in a single breath, and he seized the occasion. Affirming that the Russians have always considered the river valley to be a part of their empire and its inhabitants to be imperial subjects, he informed them of his arrival as their saviour with the following proclamation:

> In order to protect you poor indigenous peoples who are his subjects from the offenses of foreigners, the Great Tsar has decided to erect military posts . . . on the estuary of the Amur, a decision which I, as the emissary of the Great Tsar, solemnly declare to you.[42]

Thus in the space of a single morning (as Nevel′skoi tells it, at any rate), an exploratory party began to implement one of the most precious principles of Russia's national mission of salvation.

Even more than Nevel′skoi himself, however, the "discovery" of the Amur as an arena for realizing Russia's mission was pursued by his wife Catherine, who in keeping with Russian military custom faithfully accompanied her spouse on all of his missions. While he was occupied with naval operations and the construction of the new settlements, she turned her full attention to the indigenous groups of the region. In this way, she was effectively the first Russian to have extended contact with them, and in her reactions we can see the outlines of an attitude taking shape that would become characteristic for those who followed. Her initial evaluation, of course, was uncompromisingly negative, for the groups she encountered seemed to her in every way to represent the lowest level of civilization imaginable. In a letter to her sister written a few days after her husband's speech, she depicted these first impressions.

> We often see the ugly faces of the Giliaks lurking about us. Despite their characteristic cowardice, the expression on their faces is ferocious and cunning. Their clothing consists of a shirt of dog-skin, and their shoes are made out of sealskin. Their black hair, rough as a horse's mane, they wear braided into pigtails, of which the men have many but the women only two, which they fasten together with a cord. The shape of their enormous hats is vaguely reminiscent of Tyrolean caps, crafted largely out of tree bark and decorated along the brim with awful designs. It is all atrociously ugly.[43]

In the weeks that followed, Catherine Nevel′skoi elaborated upon these impressions in a flow of letters. Her daughter assembled this correspondence

[42] Vend, *L'Amiral*, pp. 76–77.
[43] Cited in Vend, *L'Amiral*, p. 147; for a slightly different version, see Nevel′skoi, *Podvigi*, p. 436.

after Catherine's death and, making full use of her mother's vivid imagery, summarized the picture they depicted in the following manner.

Overcoming her disgust, [Catherine] feeds the filthy, stinking, bloodthirsty Giliaks in their hovels (for it is quite impossible to call them homes), and these savages, warmed by her goodheartedness, trustingly tell her everything she wants to know. She doesn't shun their ugly unwashed wives and children, but rather, stifling her fear and repugnance, combs their hair, sews their clothes, and teaches them how to sew . . . With them watching, she herself digs in the mud and plants potatoes, demonstrating in front of their very eyes that, contrary to their primitive fear, God does not punish the tiller of the soil with death, but rather rewards him for his labor.[44]

Here we have an early and unmistakable indication of the fact that with the Amur territories, the Russians were acquiring far more than a link to the Pacific and the resources of a river valley. Beyond this, the region represented a vast panaroma of precious ethnographic material, which could be easily identified as savage and uncivilized, and upon which the Russian nationalists accordingly had a full opportunity to exercise their mission as civilizers and enlighteners.

The high point of Catherine's work among the Giliaks, it would seem, was her missionary activity in bringing them the word of the holy Gospel.

She herself learns their barbaric language so that she can speak with them, in order to instill in them the concepts of God, of goodness, of love for that which is close to them, of concern about their future life, and – with sensible arguments and on the strength of her own conviction and her charm – in order to convert them to Christianity.

By her own description, this message proved irresistable to the tribespeople, who were so "captivated and filled with timid admiration for our pious Sunday rituals and other religious festivals" that they came on their own volition *en masse* and asked for sacred ablution. There was little wonder that the natives would begin to deify her and, by extension, to recognize all their Russian masters as benefactors and saviours. "Observing her, living close to her, these pitiful half-people, half-animals began to see her as a divine being, and to see the Russians as defenders and friends, honest, firm, and good-hearted."[45] This was the sort of comforting conclusion that made it possible for nationalists in European Russia to secure for themselves, out of the desolate wilderness of the far-eastern taiga half-a-globe away from home, precisely that role and corresponding self-image they so desperately sought.

By the mid-1850s, a stream of letters, reports, articles, and eventually books by those explorers who followed in the Nevel'skois' footsteps began to appear in European Russia. In these materials, and especially in those designated for the popular periodical press, the psychological process of "discovery" that we

[44] Nevel'skoi, *Podvigi*, pp. 418–419.
[45] Nevel'skoi, *Pogvigi*, p. 419; Vend, *L'Amiral*, pp. 164–165.

have observed on a private level with the Nevel′skois was formalized and reenacted publically, for a broad audience. Along with often highly technical descriptions of the physical geography of the region, its topography and its plant and animal life, the Russian advances on the Pacific were imbued with the eminently virtuous aura of salvation and enlightenment of a heathen population, and in this way rendered meaningful in terms of the larger nationalist vision. Indeed, there can be little doubt that for most European Russians, it was precisely this latter aspect that seemed to be of the greatest significance. "To whomever has lovingly followed the successes of Russia's movement into Asia over the past two decades," wrote one participant at the end of the 1860s, the names of the explorers of the Amur region are precious, for they "represent the dawn of Russian civilization over the dark shaman East."[46]

The author of these lines, Mikhail Veniukov, was one of the more outspoken and prolific of these explorers. Veniukov, who was subsequently to become a well known commentator on Russia's advance into Asia, began his career during this period in the Far East, where he was the first Russian to explore the headwaters of the Ussuri river in 1858.[47] The primary task of his mission was to prepare topographical surveys of the region, but in his published account, which appeared in 1859 in the journal of the Russian Geographical Society, he devoted considerable discussion to the native inhabitants that he encountered there. Veniukov echoed Nevel′skoi's view of the Russians as the saviours of these helpless peoples from the pernicious domination of the evil Manchurians. Conversations with the Gol′di people on the Ussuri, he recounted, "convinced even me that they bless their fate for the fact that the Russians have appeared on the Ussuri, [for we] are able to rule subject primitive peoples without ruining their life, and [moreover] have been long awaited there as saviours (izbaviteli) from the cruel yoke of the Manchurians."[48] At the same time, he left no doubt as to the abysmally low level of cultural development of Russia's new charges, and pointed to this fact in explaining why this splendid river valley, despite its abundant and rich resources, should be so sparsely populated.

> Here is manifested in all its force that never-changing law which determines that the successes of humanity even in the propagation of the race are in direct correspondence with the mass of blessings that are supplied by civilization. The hunters and gatherers who inhabit all of East Asia are limited in their demands by their ignorance, and they wander in the vast forests among the wild mountains exposed to all the destructive influences of Nature. Finally, unable to withstand the cruel contact with organized tribes, these peoples will forever be unable to grow and multiply . . . Entire Gol′di families die out under the influences of the more powerful Manchurians.[49]

[46] Veniukov, "Ob uspekhakh," p. 51.
[47] Semenov-Tian-Shanskii, *Istoriia*, I, pp. 205–206; Alekseev, *Russkie geograficheskie issledovaniia*, pp. 48–49. [48] Veniukov, "Obozrenie," pp. 190–191 (quote), 200, 204.
[49] Veniukov, "Obozrenie," pp. 232–233.

While that section of Veniukov's report containing a heavily detailed topographical description of the Ussuri valley might have presented no more than a mixed interest for readers in the nation's capital, no one could have remained unmoved by those paragraphs in which he described the pathetic Gol′di. Their implications were eminently clear. The Russians simply had no choice but to adopt these children of nature and assume full responsibility for their welfare. This would be accomplished through developing in this area "all types of activity with which European civil life is rich," in other words Russian settlement and Western forms of economic and social development.[50]

Other explorers' accounts elaborated upon how the activity to which Veniukov referred was already helping the indigenous population, and with what enthusiasm it was being received. Richard Maak's first book on the Amur was quickly followed with a second recounting an expedition up the Ussuri.[51] In both works, Maak had been of two minds about the significance of the presence of native peoples. When he was concerned to demonstrate the suitability of the region for Russian colonization, he claimed that it was "almost entirely empty" and that the indigenous population was "extraordinarily thin," but when his point alternatively concerned Russian interaction with these natives they somehow became a major presence.[52] Indeed, in his 1861 work he claimed to have gathered extensive ethnographic material, and although he did not present it (explaining that it would appear in a planned companion volume), he took the opportunity in the opening passages to assure the reader of his most important determination, namely the beneficent influence which the new Russian presence had already begun to exert. He confirmed that the "heavy yoke" of Manchurian domination over the indigenous population along the Ussuri had ended with the appearance of the Russians.[53] The natives – "extremely ignorant" as they were of "everything that does not concern the satisfaction of their material needs" – regarded the Russians as their "liberators" (osvoboditeli), and their attachment to these saviors was becoming ever-more apparent. The peoples of the Amur region had already grown used to various aspects of the Russian way of life – so much so, for example, that Russian bread had become one of their most essential staples. "It may be assumed," he continued, that these simple folk, "among whom Manchurian influence has left few traces of enlightenment, will graft the Russian element onto themselves and will discard many of their current crude morals and customs, if their frequent contact with the Russians . . . becomes more intense."[54]

Maak's conclusions were fully corroborated by the naturalist Gustav

[50] Veniukov, "Obozrenie," p. 228.

[51] Semenov-Tian-Shanskii, *Istoriia*, I, pp. 206–207; Alekseev, *Russkie geograficheskie issledovaniia*, pp. 50–51.

[52] Maak, *Puteshestvie na Amur*, p. v; Maak, *Puteshestvie po doline*, p. vii; Slezkine, *Arctic Mirrors*, p. 96.

[53] Maak, *Puteshestvie po doline*, vii.

[54] Maak, *Puteshestvie po doline*, pp. 9, 3–4.

Radde, a participant in the Great Siberian Expedition who spent 1857–1858 in the Khingan mountains.[55] "The savage Giliak is submitting to the strictness of [European] laws," he reported approvingly, "and the poor Gol′di and the Amur Tungus – who have done nothing up to this point but suffer oppression – are joyful at the protection offered them by the Russian cossacks."[56] Indeed, Dmitrii Romanov claimed that even ordinary Manchurians themselves "eagerly develop contact" with the Russians and would do so much more if their officials did not try to obstruct them.[57] The satisfaction that a Russian in the country's European capitals might have derived from such depictions can be easily imagined, for they represented the incontrovertible testimony of eye-witnesses to the easy and natural success Russia was enjoying in its endeavors to exercise its national mission in Asia. What, after all, could be more impeccably civilized than Russian bread?

A particularly vivid example of how explorers on the Amur rendered the region meaningful within the framework of nationalist concerns can be seen in an account written by Nikolai Przheval′skii. Przheval′skii was subsequently to gain world renown with his four expeditions into Central Asia, but his first scientific foray was to the Ussuri valley in 1866–1867.[58] His task was to survey how the Russian settlements that had been established in the region some years earlier were faring, but like Veniukov he devoted almost as much attention to the indigenous inhabitants.[59] Przheval′skii's report appeared in the widely-read *Vestnik Evropy* (The European Messenger). He obviously shared Veniukov's cultural bias and his dim view of the indigenous peoples of the region, and he wrote in the following graphic manner about an Orochei tribesman he happened to encounter.

[55] Semenov-Tian-Shanskii, *Istoriia*, I, pp. 188–192.

[56] [Radde], "Pri-Amurskaia oblast′," p. 102. Although judging by these comments Radde's support for Russia's national interests would not seem to be in doubt, his professional work in the Far East was attacked nonetheless for being too committed to "universal science" and too little concerned with Russia alone. A letter sent from the Amur region to the *Russkii Vestnik* in St. Petersburg criticized him (and other participants in the Great Siberian Expedition) for not gathering "practical information" which could "bring benefit to the region" and concentrating instead on "collecting spiders." R., "Iz Irkutska," pp. 145–146. The great irony, of course, is that this expedition was as we have seen conceived precisely as a practical exercise concerned exclusively with furthering Russia's national–political interests in the Far East. Despite this, the naturalists sent out from St. Petersburg remained too cosmopolitain for the taste of at least some locals. Still in Irkutsk, Radde responded to his critics with a rebuttal, but the defense of "pure science" that he offered there, together with his German origins (he came from Danzig) and the fact that he was not yet writing or lecturing in the Russian language, could hardly have helped his standing as a representative of Russian national science. Radde, Untitled article, pp. 9–10; "Gustav Raddes asiatische Reisen," p. 275. Murav′ev, however, appreciated his efforts sufficiently to name a postal station on the Amur after him. Veniukov, "Vospominaniia," p. 302.

[57] Romanov, "S ust'ia," p. 106.

[58] Semenov-Tian-Shanskii, *Istoriia*, I, pp. 214–215; Alekseev, *Russkie geograficheskie issledovaniia*, pp. 61–65; Dubrovin, *N. M. Przheval′skii*, p. 68.

[59] Przheval′skii, "Avtobiografiia," p. 537; Dubrovin, *N. M. Przheval′skii*, p. 50–51; Rayfield, *Dream*, pp. 20–21.

What a small difference there is between this person and his dog! Living like a beast in its lair . . . he forgets all human strivings and, like an animal, cares only about filling his stomach. He eats meat or fish, half-cooked on coals, and then goes hunting, or sleeps until hunger compells him to get up, start a fire in his stinking smokey hovel and once again feed himself.

This is how he spends his entire life. For him, today is no different from yesterday or from tomorrow. Not feelings, desires, joys, hopes – in a word, nothing spiritual or human exists for him.[60]

In spirit, Przheval′skii's reactions are identical to those of Catherine Nevel′skoi, and in both cases the same psychological process is taking place. The encounter with such utter degeneration and human decay disgusts, but at the same time the prospect of a population in such abject need immediately works to stir a sense of responsibility, indeed an imperative to do everything possible to help the savages better their wretched lot. In this way, Przheval′skii's account once again has the effect of securing a ready arena upon which an ambitous national mission of salvation could be exercised. This is precisely what we have observed, to repeat, with Madame Nevel′skoi; the difference is that on the pages of one of Russia's most popular journals Przheval′skii transforms what for the admiral's wife was a personal experience into one which the entire reading public can effectively share.

Przeval′skii did not leave this imperative up to the imagination of his readers. He described the desire of Koreans living adjacent to Russia's new southern border in the Far East to settle on Russian territory. His recommendation was that the members of this group – "crude and impenitent in their ignorance"[61] – be allowed to do so. They should not, however, be settled on the border region near their old homeland, but should be moved further to the north, to Lake Khanka or even to the Amur. The point was to relocate these Koreans to places where Russian settlements already existed, so that they could live

among our peasants, with whom they could become better acquainted and from whom they could assimilate something new. Then, gradually, the Russian language and Russian habits, together with the Orthodox religion, would begin to penetrate to them, and perhaps with time an heretofore unknown miracle would take place: the regeneration in a new life for these groups who came from tribes as stubborn and immobile as the other peoples of the Asiatic East.[62]

It was precisely because they could offer such a palpable prospect of this sort of *chudo* or miracle that Russian geographers and explorers were able to contribute to the preoccupation with a national mission. Przheval′skii made the fulfillment of this miracle into an uncomplicated affair indeed, involving no more than the occupation and settlement of the region. The superior Russian

[60] Przheval′skii, "Ussuriiskii krai," p. 569.
[61] Przheval′skii, "Ussuriiskii krai," p. 577.
[62] Przheval′skii, "Ussuriiskii krai," p. 576.

ways would then presumably dissipate among the native peoples as if by osmosis.[63]

In all fairness, we should note that the commentators we are considering were not uniformly negative in the judgments they pronounced on the indigenous populations of the Amur and Ussuri regions. Maak, for example, commented on the "good- and open-heartedness," the "honesty and helpfulness," and even the "charitable inclinations" demonstrated by some of the groups he encountered during his first voyage. After the second, he went so far as to venture a muted concern about the possible negative influence that the Russian settlers might exert if they were not able properly to "value" (otsenit′) as he delicately put it the attachment of the natives to them.[64] Przheval′skii also distinguished in a fundamental way between the different groups he encountered, and could speak with great sympathy about some. The Gol'di, for example, he described as having warm and loving family relations; indeed, they compared entirely favorably in this respect to the moral turpitude he encountered in some of the Russian cossack settlements, where wife-sharing among other things was apparently not unknown.[65]

Beyond these lurking uncertainties, the process we have called cultural construction – the attempt, in other words, to identify and describe the peoples of the Far East in a manner that rendered them relevant and useful to Russian nationalist priorities – had its outright critics at the time. A review of the literature about the Amur which appeared in a major St. Petersburg newspaper, for example, angrily denounced the "insatiable conceit" on the part of Radde, Veniukov, and others, who intentionally distorted their accounts of the indigenous peoples to make the latter look cruder and more primitive than they actually were.[66] The most compassionate defense of the Amur peoples, however, came from the pen of Grigorii Potanin, who in an article in *Russkoe Slovo* (The Russian Word) in 1861 denounced the notion hinted at by Veniukov that they were somehow destined to die out. "These groups were given their language, their poetry, their ways of living and thinking, in a word their nationality [like all other groups]," and the suggestion that contact with "civilized nations" will lead them to disappear from the face of the earth was nothing less than "murderous fatalism and a crass insult to human nature."[67] Yet for all of his genuine and, unfortunately, most unusual empathy with the indigenous inhabitants of the Far East, even Potanin shared the general belief in their relative backwardness, their need to develop "higher" forms of social and cultural life, and the notion that "the role of further developing civilization [in the Amur region] has fallen to Russia." And while he may have had the most sincere understanding for the concerns the natives expressed to him

63 Przheval′skii was not alone in expressing this idea: see Vasil'ev, "Otkrytie," p. 20.
64 Maak, *Puteshestvie na Amur*, p. 47; *idem*, *Puteshestvie po doline*, p. 10.
65 Przheval′skii, "Ussuriiskii krai," pp. 557–568.
66 Te-v, "Istoriia," pp. 108 (quote), 92.
67 Potanin, "Zametki," p. 204.

that Russification would lead eventually to their cultural dissolution, he di not entirely sympathize.[68]

Troubled voices such as Potanin's were few and far between, however, an we may feel safe in assuming that the educated Russian public was not overl disturbed by them. There was after all a much more voluminous literatur about the Russian Far East which offered an altogether more agreeable an encouraging picture. If authorities such as Przheval′skii and Maak were to scholarly and dry, then one could pick up popular descriptions and read abou the *vostorg* or "enthrallment" with which the natives first received the Russian after the generous cossacks had handed out glistening silvery two-kopel pieces.[69] And thanks to the new technology of lithography and the illustrate journals it made possible,[70] readers could now also admire pictoral depiction of the region. Here page upon page of detailed lithographs were reproduced accompanied by easily digested commentaries with tidbits on enticingly exoti subjects, such as how the natives periodically smeared the fish-skin window in their lodges with grease to make them more translucent, or how they trie to feed their wooden idols as a reward for a successful hunt. The pictures fea tured portraits of such heroes of the day as Murav′ev and Putiatin (alway dazzling and resplendent in their uniforms), scenes of new Russian settlement along the Amur (all neat, bustling, and prosperous), and of course a variet of likenesses of the indigenous inhabitants, ceremoniously outfitted in thei native garb and looking solemnly oriental and uncivilized.[71]

"A diploma with the title of a truly European nation"

In none of the considerations of Russia's mission in Asia which we have con sidered up to this point has a significant distinction been drawn betwee Russia and the West. Quite to the contrary, there seems to have been no ques tion that, as Russia penetrated ever more deeply into the East, it did so as representative of that "Europeanism" about which Romanov had spoken i such ardent tones. A vigorous sense of unity of purpose with all other impe rial powers was expressed again and again, as the Russians sought to articu late how they understood their own activities on the Pacific. "We will say i with the words of the Western powers," bluntly wrote a steamship captai from the Amur, "*we are spreading Christianity and civilization among wil tribes and peoples.*"[72] Even such a hostile and uncompromising critic of th West as Mikhail Pogodin shared this belief. Indeed, Pogodin appears to hav originally conceived of the salvation of the non-Western world as a pan European affair. "There is no other way to civilize (obrazovat′) Africa an

[68] Potanin, Review of R. Maak, *Puteshestvie na Amur*, p. 98 (quote); Potanin, "Zametki," pp. 208 211, 212. [69] *Puteshestvie po Amuru*, p. 54. [70] Brooks, *When Russia Learned*, pp. 65–66
[71] E.g. "Ekspeditsii russkikh"; Sibiriak, "Zametki."
[72] N. N. Kh-O., "O parakhodstve," p. 2 (emphasis in original).

Asia," he mused in his diary in 1826, than to send troops "from all of Europe" in a modern-day crusade against them. "Let Europeans sit on the thrones of the Ashanti, the Brahmans, the Chinese, and Japanese, and [let them] introduce European order there . . . The happiness of the human race depends upon this."[73] In later decades, to be sure, Pogodin came to appreciate Russia's special role in this process, but even at the height of anti-Western sentiment in the wake of the Crimean War, he was still capable of an eloquent declaration of Russia's fraternal kinship with its fellow colonizers. In 1857, he described the remarkable change in the public attitude toward Britain the moment that reports of the Sepoy revolt against the English population in India began to circulate in Moscow, with gruesome accounts of atrocities at Cawnpore and elsewhere. "We forgot at once that the English were our enemy, and we saw in them only Europeans, Christians, sufferers. We saw in them an advanced nation which barbarism was threatening with destruction, and a general compassion and sympathy was expressed from all corners."[74]

Something is very wrong with all of this. Our study began by pointing out that the nationalist ideology of which Russian messianism was a component part had emerged out of the confrontation with the West and was clearly directed against it. Nationalist doctrine aspired to nothing less than the elaboration of a Russian identity that was radically independent, with an autonomous and exclusively Russian structure of cultural and social values. The notion itself that Russia was in some way charged with a mission of salvation came from the attempt to fashion such an identity, and we have seen that, for some at least, this mission would begin with the salvation of Europe itself. Our consideration of the Russian mission in the Far East, however, betrays a fundamental ambiguity. Simply put, while in certain situations the the Russians was concerned to distinguish themselves emphatically and stand apart from the West, elsewhere they were inclined to associate themselves with Europe and emphasize their own Europeanness. This was far more than simple ambivalence, it was a veritable schizophrenia, and one that was embedded in the very core of Russia's nationalist sentiment. Driven by these contending dynamics, Russian nationalism was fraught with a dualism, a sort of geoideological split that suggested more than anything the two faces of Janus, aligned in opposing compass directions. The familiar nationalist affirmations about Russia's exclusivity and superiority were appropriate only as long as Janus faced West – that is, as long as the Russians confronted Europe directly. Yet as Pogodin's observations make clear, the moment the geographical focus swung round to the East, the Russians began to see themselves as the bearers of a civilization and enlightenment which they themselves recognized as being thoroughly European. Nicholas Riasanovsky has summarized this contorted situation very well, noting that the very Russians "who vehemently denounced Europe

[73] Barsukov, *Zhizn'*, II, p. 17. [74] Pogodin, "Vtoroe Pis'mo," p. 16.

and postulated a fundamental constrast and opposition between Russian and Western principles, nevertheless, as soon as they turned to consider Asia, identified themselves with Europe, with the West."[75]

However, even the ambivalence which Riasanovsky notes does not convey the full intricacy of the Russian attitude. While many Russians, to be sure, simply "identified themselves" with Europe in Asia and left the matter at that, there were others for whom, on a deeper psychological level, the prospect of Russia in Asia provided an opportunity not to put aside the old problem of their hostile relationship with the West but rather to scrutinize it anew. The novelty in this opportunity was precisely the element of a new geographical location. The theater in which the familiar issues of Russia *contra* Europe were now being articulated and thought through, in other words, was not Moscow or St. Petersburg or any other venue in "European" Russia, but rather far away in Irkutsk, Yakutsk, and on the Pacific. This condition of geographical dislocation involved two elements, both of which are profoundly important for our study. In the first place, it suggested that Russia's much-touted redirection of attention and energy to Asia in the wake of the Crimean War was in fact not a turn away from Europe at all. Indeed, it could be argued that Asia *per se* was of merely instrumental importance to the entire enterprise, useful essentially in its capacity as a new and unfamiliar arena upon which Russia could continue to rehearse the familiar confrontation with the West. Yet at the same time, the very fact that the geographical venue of the Russia–Europe juxtaposition had been relocated out to Asia and the Amur fundamentally altered, or rather promised to alter, the quality of the juxtaposition itself. It was this latter circumstance that the Russians were ultimately to find so compelling. Put most simply, through their progressive civilizing activities among the Siberian heathens, the Russians believed in effect that they could transform the quality of their own civilization. And this transformation, in turn, meant that the Russia which confronted the West on the shores of the Pacific was qualitatively different from Russia west of the Urals, and one far better equipped for precisely this confrontation.

This rather convoluted perspective emerged with particular clarity on those not-very-numerous occasions when the Russians tried to work through the meaning of their new presence in Asia with some precision and consistency. One example was a long and reflective essay published anonymously by Mikhail Petrashevskii in 1857 in the *Irkutskie Gubernskie Vedomosti*. As his title – "Some Thoughts about Siberia"[76] – indicates, Petrashevskii did not limit his discussion specifically to the Amur and Ussuri valleys, although it was the excitement and anticipation generated in Eastern Siberia by the occupation of

[75] Riasanovsky, "Asia," pp. 17 (quote), 6–9, 14–16; Sarkisyanz, "Russian Imperialism," pp. 66.

[76] The article was signed "Brd." On Petrashevskii's authorship, see Semevskii, "M. V. Butashevich-Petrashevskii," no. 2, p. 24; Leikina, *Petrashevtsy*, p. 139; Dulov, "Sibirskaia publitsistika," pp. 4–5.

the Amur valley that had focused his attention on these questions in the first place. Already in his opening statement, he subordinated the significance of Siberia and Asia overall to the larger juxtaposition of Russia and Europe, and his evaluation of this larger juxtaposition did not reflect favorably upon his homeland. "[T]here is still a great deal of material foreign to us which we must assimilate from the West in order to reach the same level as the Germanic or Anglo-Saxon peoples," he stated, and added that Russia would long be compelled to make use of the results of Western science. Russia's relationship with Asia, however, presented a very different picture. Here the roles were reversed, and it was Russia which emerged as the educated and advanced society. "In relation to [the peoples of the East]," he affirmed, "we have the superiority of social life and civilization on our side." More precisely, Petrashevskii saw Russia's level of development as somewhere between that of Europe and Asia, such that both the principle of "Europeanism" as well as that of Asiatic stagnation were characteristic to it. The important point, however, was that Europeanism dominated. Thus, "[i]f in Europe we Russians are the younger brothers in a moral sense, then in the circle of Asian peoples we are justified in claiming seniority." If there was some satisfaction or comfort to be derived from its intermediate position, however, Petrashevskii made very little effort to emphasize it as he drew out the implications, for Russia's inferior status *vis-à-vis* Europe remained uppermost in his mind. The "genuine and abolute moral value" of his homeland was "meager," and even the overall balance of Russian superiority over the Asians was "modest" and "limited." Yet however thin, the margin of superiority nonetheless remained in Russia's favor, and this was enough to make one vital point perfectly clear.

> [F]or us, it is necessary to be to these peoples [of Asia] that which the European peoples are for us. We must believe that our settlements in Siberia or in Siberian Asiatic Russia have been predestined . . . first to teach these peoples the results of science and civilization, and then to introduce them fully into the sphere of universal intercourse between all peoples. Here [in Asiatic Russia] is the environment in which the moral and industrial strengths of Russia can manifest themselves freely and independently, with the least constraint.[77]

If he believed that Russia and Europe's shared destiny as the bearers of civilization was of importance by virtue of the improvements that they could bring to the lives of millions in Asia, Petrashevskii certainly did not say so. What emerges out of his arguments instead is the point we have just made, namely that Asia represented a novel geographical arena upon which the original Russia–Europe juxtaposition – so unfavorable for the former – could be reconsidered and, ultimately, readjusted. For although Europe's superiority over Russia may well have been "unquestionable," it was so only as long as the geographical frame of reference for Russia remained west of the Urals. In

[77] Brd, "Neskol'ko myslei," pp. 3–4.

Asia, Petrashevskii insisted, Europe's natural predominance dissipated, and the two became much more like equals. Indeed, by virtue of its singular geographical position and historical experience, Russia possessed a series of distinct advantages, and Petrashevskii repeated the familiar point that Russia was far better equipped than Europe for the task of civilizing Asia.

> The influence of the Russians on these peoples ought to be incomparably more firm and reliable than the influence of all the other European peoples, for the Russians are not, like them, strangers from across the ocean who, sticking like polyps to the ocean coast, can be thrown back into the sea at the first movement of discontent among the native peoples. For the Asiatic peoples the Russians have already ceased to seem like newcomers from unknown lands, [and instead] are like old neighbors. This simple territorial fact has great political significance.[78]

It must not be forgotten, Petrashevskii went on, that for a distance of over seven thousand *versty* virtually any Russian border post, firmly rooted in native Russian soil, could serve as a source for the "spread of our moral and industrial influence" in the neighboring regions of Asia. The Europeans, by contrast, had no choice but to rely upon a few remote and scattered outposts, all of which were separated from the homeland by thousands of miles of ocean. Russia's superior geographical position, the effect of which was to "weaken the significance of that superiority and those advantages over us which their civilization gives [the Europeans] *in Europe*," was an indication that *in Asia* Russia was not merely the equal of Europe but indeed its superior.[79]

Having established in this way the novel sort of parity between Russia and the West that Asia made possible, Petrashevskii went on to consider what the future might hold. These concluding thoughts are in a sense the most interesting and revealing, for they betray the very essence of the ambiguity in Russian attitudes toward the West that we noted above. On the one hand, his tone became markedly hostile. The recent war, he observed, was essentially the beginning of the struggle between the Slavs, the Germans, and the Anglo-Saxons for political predominance, and although Russia lost the war, it learned a valuable lesson. The coming contest will be decided "not in Europe but in Asia, where the rivalry is already operating under conditions that are most favorable for us." Petrashevskii no doubt had the occupation of the Amur region very much in mind as he wrote these lines. Yet at the same time that the future Russian successes in the East would serve as a means to triumph over the West, they would paradoxically represent incontrovertible proof to the world that Russia was worthy of the designation "European" in the first place. Through its missionary and civilizing activities in Asia, Russia could definitively prove its moral worth and join fully in the European community without any qualification or hesitation.

[78] Brd, "Neskol'ko myslei," p. 4. [79] Brd, "Neskol'ko myslei," p. 5 (emphasis added).

Our present position in Asia, its strengthening or weakening may be considered an indication of the level of development of our society, a general conclusion about our social life, a touchstone for the evaluation of the degree to which we have assimilated the principles of Europeanism, which are general principles of humanity. By its position Siberia is destined to achieve for us a diploma with the title of a truly European nation!

"This, then," Petrashevskii concluded, "is the task for the Russian in Siberia." In order to accomplish it, the support of Russians in European Russia will certainly be forthcoming, for otherwise the latter will "be constrained to forfeit their *droit be cite* [sic] in Europe."[80]

A rather different expression of essentially the same logic can be seen in an essay by the geographer Peter Semenov. Semenov, a committed nationalist, was one of the staunchest patrons of the geographical study of Asia in the period under consideration, and himself conducted an important expedition to the Tian-Shan mountains of Central Asia in the 1850s (in recognition for which the distinction "Tian-Shanskii" was appended to his surname). Semenov believed that the exploration and study of the Russian East contributed to the nationalist project by embracing Russia's *samopoznanie* – that is to say, by helping Russia become aware of itself, as it were geographically. He had an intense conviction in the need for Russia to pursue its "historical work" in the East, and was particularly stimulated in this regard by the annexation of the Amur.[81] Although he did not visit the region himself, he was a major authority on it. The essay in question, which appeared in 1855, was ostensibly devoted to a physical–geographical description of the region, but he took the opportunity in it to reflect on the significance of Russia's advances in the East within a broader national–historical framework. Over the past 30 years, he wrote, Russia had made great strides in the study of Asia, conducting explorations around the Caspian sea, in Central Asia, Mongolia, and of course in the Far East.[82] "By all of these routes," he affirmed, "Russia moves forward, as Providence itself has ordained, in the general interest of humanity: the civilizing of Asia."

As Semenov developed his exposition, however, it emerged that Asia itself was not the only, and indeed not even the primary, object in this mission. As with Petrashevskii, the main point once again was the manner in which its civilizing activities would reflect upon Russia's relationship to Europe. Semenov repeated Petrashevskii's depiction of the salvation of Asia as a task for Russia alone, with the same claim that Russia occupied a special geographical and historical position between East and West and therefore was more naturally suited to bring civilization and enlightenment. He also alluded to divine intentions in this regard, suggesting that Russia was in effect God's chosen instrument to carry out this grand mission.

80 Brd, "Neskol'ko myslei," p. 5.
81 Semenov-Tian-Shanskii, *Memuary*, I, p. 261.
82 Semenov, "Obozrenie," p. 253.

Chosen by God as an intermediary between East and West, having received its Christianity from the capital of an Eastern Empire [Byzantium] and spent its adolescence as the European hostage of an Asiatic tribe . . ., Russia is equally related to Europe and Asia and belongs equally to both parts of the world. For this reason it is more capable than other nations of fulfilling the role which its geographical position and history have designated for it.

It was more than geographical location or historical preconditioning, however, that indicated Russia's enhanced suitability *vis-à-vis* the West for the task of civilizing Asia. Beyond this, Semenov felt that the actual record of accomplishment thus far demonstrated it as well. While the Russians had consistently acted in a manner that was beneficial to the local Asiatic population and put their interests first, the nations of Europe were interested in nothing more than callous subjugation and merciless exploitation. At every moment, he asserted, Europeans

who have attained the highest stage of contemporary civilization, oppress hundreds of millions of peoples of a different, less developed race [and] exploit these Asiatics for their own mercenary ends like inert matter. [The Europeans] convert them like stupid animals into the blind instrument of their material interests, and do not give them even a single ray of their own enlightenment.

Thus European activity in Asia was of a purely predatory nature, uninspired by the paternal desire to help the ignorant Asiatic savages or half-savages that animated Russian activities there. Indeed, the nationalist Semenov strove to disassociate his own country from the brutal and bloody legacy left by the Europeans in the non-European world.

The Russians do not annihilate – either directly, like the Spanish at the time of the discovery of America, or indirectly, like the British in North America and Australia – the half-wild tribes of Central Asia [and the Far East]. Rather, they gradually assimilate them to their civilization, to their social life and their nationality.

"For this reason," he continued, "each new step of Russia into Asia is another peaceful and sure victory of human genius over the wild, still unbridled forces of Nature, of civilization over barbarism."[83]

If the desperate atmosphere of the Crimean War – still raging as Semenov wrote his essay – explains why his tone was rather more belligerent than than of Petrashevskii, then the fact that his perspective was characterized by precisely the same ambivalence is all the more striking. He was attempting nothing less than to maintain the standard Russian nationalist stance against Europe in a geographical context that transformed Russia itself not merely

[83] This contrast between "good" and "bad" civilizers was a popular one for Russian nationalists, both at the time and subsequently. Grigor′ev, "Ob otnoshenii," p. 10; Tiutchev, "Rossiia i Germaniia," p. 286; Kohn, "Dostoevsky," p. 512. Indeed, it appears to have retained its appeal to the present day: for an example from the 1980s, see Kuniaev, "Velikii put′," pp. 6–7.

into a representative of Europeanism but into a *superior* representative of Europeanism. The claim for Russian exclusivity really lost its meaning at this point, and had to be replaced by a vaguer claim, namely that Russia was simply better at carrying out its mission that other European powers. This, however, amounted in turn to nothing more than the assertion that Russia could beat the West at its own game. And this is precisely what Semenov, like Petrashevskii, wanted to do. Although he did not use the word, it was clear that the quality of Russia's Europeanism was his most fundamental concern. Russia's civilizing activities on the Amur and elsewhere in Asia were important to him ultimately because he believed that they were a uniquely effective means to endorse and enhance his country's status in this regard, and he ended his essay on the following acerbic note.

> Just let the childern of this West say now that we still stand on a low level of civilization! If this low level is already producing such marvelous fruits for the interests of humanity in general, then we are fully justified in expecting even more from a higher [level of civilization], which will quickly develop in view of the rapid pace characterizing the history of our development.[84]

The unconcealed bitterness with which Semenov made this point might be taken as a hint that he was himself not entirely convinced of the point he was making.

In both of these essays, the ultimate standard of value was European civilization, and it was a standard of which Russia clearly fell short. The project of bringing enlightenment and civilization to the East was therefore not a goal in itself, and certainly not an altruistic exercise to serve the "interests of humanity in general," as Semenov would have it. Asia represented instead nothing more than a means or medium, which the Russians could use to make a very big and important point. In Asia, they thought that they possessed a uniquely favorable environment upon which they could demonstrate their worth and even their superiority in the framework of European values, and indeed in a manner which would have to be recognized and acknowledged by the "children of the West" themselves. In a word, by civilizing Asia the Russians obviously believed that they could and would civilize themselves.

[84] Semenov, "Obozrenie," p. 254.

7

Poised on the Manchurian frontier

"What cannot be held had better be ceded"

The belief that Russian navigation on the Amur river was vital for the logistic support of settlements in the North Pacific was an old one, going back well into the eighteenth century. It was a perspective that Murav′ev at first strongly endorsed, and upon assuming office he used this rationale to press the importance of the river. In a memorandum to the tsar on the eve of the Crimean War, however, we have seen how the governor-general gave the first indication of a reorientation in his own thinking about Russia's position in the Far East. In its North American territories, he suggested, Russia was faced with what was essentially a hopeless situation. Despite all efforts to maintain Russian authority and control in these remote regions, it would ultimately prove impossible to resist the expansive pressures of a dynamic American empire which sought almost intuitively to occupy and incorporate them. American dominion over the entirety of North America was simply inevitable, and thus Russia was left with no choice but to relinquish its Alaskan territories as gracefully and as profitably for its own interests as possible. Murav′ev sought to mitigate the shock of this point, however, with the argument that the forfeiture of Russian North American did not in any any way signal an imperial retreat from the Far East. Quite to the contrary, Russia would sacrifice its position in Alaska in order to realize its more natural and more valuable position to the south in East Asia, over which, he insisted, Russia was the natural lord.

Murav′ev made these points in 1853. In the years that followed, the argument that Russia's New-World possessions had lost their value and that Russia would be well served by getting rid of them became increasingly popular in governmental circles. The possibility that the United States might be a suitable cash customer had been raised already during the Crimean War, when a fictitious sale was clandestinely considered as a ploy to secure Alaska against possible allied attack. Discussion along these lines persisted after the war, and while it was still conducted in strict secrecy, the issue was now approached with greater seriousness. Partisans of the movement to transfer the territories grew increasingly more insistent, and they offered a variety of compelling reasons to justify Russia's taking take the unprecedented step of peacefully relinquishing such a sizeable portion of its imperial domains. Economic considerations were stressed for example by Alexander II's reform-minded uncle Konstantin Nikolaevich, who assured Foreign Minister Gorchakov in 1857 that because Russia's American colonies "bring us a very small profit," their loss "would not be greatly felt."[1] The point he was making, of course, was that by mid-century the fur trade – the original rationale for the Russian presence in the region – had withered into insignificance. While the most recent research on the Russian–American Company indicates that its attempts at the time to diversify its economic pursuits by supplying timber, fish, and ice to California and the Hawaiian Islands had in fact met with some fiscal success, and that the discouraging picture of economic uselessness was consequently rather overdrawn,[2] this was not the perception at the time. Another factor strongly influencing opinion in favor of relinquishing the territories related to the problem of their administration, and arose out of the new political climate in the late 1850s. The Russian–American Company, established under Paul I in 1799, not only possessed an imperial monopoly on the Pacific fur trade but was empowered with near-autocratic privileges in the territories under its jurisdiction, where it dominated virtually every aspect of life. In the half-century of its existence, it had managed to acquire a notorious reputation, not only for its questionable judgment in business matters but for its striking brutality in dealing with the indigenous population. In spirit, its mode of operations was not fundamentally at variance with Nicholas I's system of government, but it was inevitable that it would clash with the new mood of social and political reform which animated the late 1850s. After the death of the old tsar, the Russian–American Company remained in effect as a living reminder of the heavy-handed autocratic arbitrariness and caprice of an earlier period, which had been cast off with such satisfaction and to such apparent good advantage in other parts of the Empire. Consequently, there was a great deal of opposition to the Company on the part of liberals and reformers, and a desire somehow to dismantle its power. The Naval Ministry in particular was a reservoir of such sentiment, and in 1861 it dispatched two inspectors to the North Pacific to conduct a review of affairs there. The predisposition of these agents can be gauged from their comment after the accidental death of a sailor on their boat before even arriving in Alaska: "the poor seaman . . . passed from this world to the next, *where he will undoubtedly be better off than in the service of the Russian–American Company*."[3] Predictably, the report of the review commission was highly critical, and although it did

[1] Quoted from Miller, *Alaska Treaty*, p. 38; also see Okun' *Russian–American Company*, pp. 220, 226. [2] Gibson, "Sale," pp. 20–21; Bolkhovitinov, "How it was decided," p. 124.

[3] Quoted in Gibson, "Sale," p. 19n (emphasis in original).

not recommend selling the territories, it nonetheless contributed in no small measure to the shift of interest away from Russia's northern Pacific colonies.[4]

Overshadowing all of these concerns, however, were logistic and strategic considerations, and these would appear to have been decisive in Russia's final decision to divest itself of its American territories. The degree to which the economic viability of the territories was exhausted might have been open to some question, but the inability of the Russian government to protect them or prevent foreign military or economic encroachment was not. This lack of control was demonstrated among other things by the illegal activities of American whaling ships in Russian waters – in the 1850s it was estimated that over 100 such vessels visited the Sea of Okhotsk yearly[5] – and it was commonly agreed that the Crimean War had demonstrated Russia's helplessness in these maritime regions beyond any question.[6] As if to underscore St. Petersburg's complete lack of authority there, the Russian ambassador to the United States de Stoeckl reported with some alarm in 1857 on rumors circulating in Washington that Brigham Young and the Mormons were planning to establish themselves in "our American possessions." The tsar himself was apparently swayed by his emissary's insistence that Russia would be powerless in the face of this sect's "most warlike resolution," and on the margins of the communiqué he noted the need for "settling henceforth" the issue of Alaska.[7] Indeed, even such a committed supporter of Russia's North Pacific presence as Baron Ferdinand Vrangel′, a noted Arctic explorer who served from 1830 to 1835 as the governor of Alaska and from 1840 to 1849 as the head of the Russian–American Company, was constrained to admit in the late 1850s that "*fears of the future*" and "*anticipatory prudence*" might compel the government to rethink its possession of these otherwise valuable lands.[8] The prospect of an irrepressible wave of American invaders grew yet more ominous in the early 1860s, as increasing evidence pointed to the existence of gold in Alaska and the experience of California in 1848 threatened to repeat itself.

Murav′ev's point from the early 1850s on the inevitability of American intrusion into Alaska became the dominant theme of those supporting the cession of the territory to the United States. In describing America's expansionist urges the governor-general had even conceded a sort of natural legitimacy to them, and this indulgence was echoed by others in the government. "Following the natural order of things," wrote Konstantin Nikolaevich, "the

[4] *Doklad komiteta.* Another highly critical account of the conditions and administration of Russian Alaska was published in installments in the organ of the Ministry *Morskoi Sbornik*, and then issued as a separate volume: *Materialy dlia istorii.* Also see Okun′, *Russian–American Company*, pp. 253–254, 256; Jensen, *Alaska Purchase*, pp. 47–48.

[5] Cole, *Yankee Surveyors*, pp. 153–154; Kuznetsov, "Sibirskaia programma," p. 10; Sgibnev, "Vidy," p. 691; Romanov, "Prisoedinenie," p. 372.

[6] Gibson, "Sale," pp. 23–25; Bolkhovitinov, "How it was decided," p. 116.

[7] Quoted in Miller, *Alaska Treaty*, p. 45. On Alexander II's early support of the sale, see Kushner, *Conflict*, p. 136.

[8] Quoted in Miller, *Alaska Treaty*, p. 42 (emphasis in original).

United States of America is bound to aim at the possession of the whole of North America. . . . No doubt they will even take possession of our colonies without much effort and we shall never be in a position to regain them."[9] And if these comments left the impression that the director of the Naval Ministry did not find the prospect overly disturbing, then one of his subordinates, the commander of the Pacific fleet Rear Admiral Andrei Popov, was positively enthusiastic. Popov's thoughts are particularly revealing, and we will quote from them at length. Noting the cynicism which Europeans entertain about "the dogma known in the political encyclopedia as the 'Monroe Doctrine' or the doctrine of 'manifest destiny,'" he made the following points in a memorandum to foreign minister Gorchakov in 1860.

> [A]nyone who has lived the North American life cannot fail to understand instinctively that this principle is entering more and more into the veins of the people, and that new generations are sucking it in with their mothers' milk and inhaling it with every breath of air. Even one who has not lived in America, if he can free himself for the time being from the conceptions of a Europe long since bound up by artificial conditions, will understand that a people which has developed so rapidly and so successfully was bound to appreciate that the main reason for this development was the absence of the restricting influence of neighbors. This people tries to maintain this invaluable advantage by all the means at their disposal and the question of the destruction of the influence of neighbors leads in practice to not having any. The geographic situation of the North American continent facilitates this healthy endeavor, and the Americans, as a people, would be criminally blind or careless if they did not for their part apply every means to assist, thus helping in the unhindered development of a firm political independence.
>
> Once the natural tendency of the Union toward the realization of the idea of *manifest destiny* is understood, it is not necessary to live in America in order to agree that twenty millions of people, . . . all of whom are inspired by the same thought, will sooner or later carry this thought into execution. They are already putting it into practice rapidly by this absorption of adjacent nationalities, and a similar fate awaits our colonies. It is manifestly impossible to defend them, and what cannot be held had better be ceded in good time and voluntarily.[10]

The eagerness and apparent empathy with which Popov greeted the "healthy endeavor" of American continental expansion, even at the acknowledged expense of Russia's own territory and position, is a remarkable tribute to the fraternity that Russians felt at the time toward the United States.

There should be no mistake, of course, that these vigilant guardians of Russia's imperial welfare in the Far East could endorse with such equanimity and composure what was for all practical purposes to be a forced retreat from Alaska only because it was not in fact a retreat at all. The move away from

[9] Quoted in Miller, *Alaska Treaty*, p. 46; Kushner, *Conflict*, pp. 135–136.
[10] Quoted in Miller, *Alaska Treaty*, pp. 54, 38 (emphasis in original); Jensen, *Alaska Purchase*, pp. 20–21.

Alaska, that is to say, was to be made good by Russia's advances on other fronts in the Far East. This had been the final and most momentous point of Murav′ev's 1853 memorandum, and the connection between the sale of Alaska and the annexation of the Amur region was invoked repeatedly as the former issue moved toward the center of the government's attention. "Russia must endeavor as far as possible to become stronger in her center," wrote Konstantin Nikolaevich in 1857, "in those fundamentally Russian regions which constitute her main power in population and in faith, and Russia must develop the strength of this center *in order to be able to hold those extremities which bring her real benefit*."[11] He almost certainly had the Amur region in mind here, but perhaps because the Russian occupation of these territories had not yet been offically recognized, he did not openly indicate it. De Stoeckl, however, did make the connection explicit two years later.

Our interests lie on the Asiatic coast, and we should direct our energies thither. In that area we are in our own territory and in a position to exploit the production of a vast wealthy region. We shall take part in the extraordinary activity that is being developed on the Pacific, our establishments will vie with similar establishments of other nations, and, in view of the solicitude which our august monarch has given to the coastal region of the Amur, we must not miss the opportunity to attain in this vast ocean the high position of which Russia is deserving.[12]

Russia's new acquisitions in the Far East, in other words, promised to offer precisely what Russian America did not. This theme echoed again in the mid-1860s, as discussion around the problem of Alaska entered its final phase. The Minister of Finance Mikhail Reutern, noting in 1866 that "we have now firmly established ourselves in the Amur territory, where the climatic conditions are incomparably more favorable," felt comfortable endorsing the sale,[13] and finally, on the very eve of the decision, Konstantin Nikolaevich took the trouble to make explicit the connection to which he had only alluded ten years earlier. "[I]t is urgent to abandon [Alaska] by ceding it to the United States and to render all of the government's solicitude to our Amurian possessions, which form an integral part of the empire and which by all accounts offer more resources than the northerly shores of our American possession."[14]

This point concerning the vital connection between the occupation of the Amur region and the sale of Alaska a decade later has been stressed in several recent studies.[15] An important elaboration remains to be made, however,

[11] Quoted in Miller, *Alaska Treaty*, p. 46 (emphasis in original).

[12] Quoted in Okun′ *Russian–American Company*, pp. 239–242; Miller, *Alaska Treaty*, p. 50; McPherson, "Projected Purchase," p. 86. De Stoeckl had conferred with Collins in Washington in 1856, prior to the American's departure for Russia, was impressed by his plans, and promised him assistance in dealing with the Imperial government. Collins, *Siberian Journey*, pp. 47–48; Jensen, *Alaska Purchase*, p. 20.

[13] Quoted in Bolkhovitinov, "How it was decided," pp. 117–118.

[14] Quoted in Gibson, "Sale," p. 31.

[15] Bolkhovitinov, "How it was decided," p. 124; Gibson, "Sale," p. 29.

regarding the question as to precisely what Russia's vision of its future position in the Far East actually involved. It is usually assumed that the Amur valley itself was seen in the 1850s and 1860s as the axis for this future – in other words, as the area that would serve as the main base for the subsequent development of the Russian presence in the Far East. This is an entirely logical assumption, for it suggests that the annexation of the Amur river and the accompanying relocation of Russia's international boundary supplied an appropriate historical closure to the process of transition, and it even lends this transition a neat and seemingly convincing geographical symmetry – a beginning in the northeastern-most extremity of Russia's imperial domains and a termination as it were at their new southeastern-most limit. It was moreover an assumption sincerely shared by many of the main actors at the time, and perhaps more importantly by virtually all of those who subsequently chronicled Murav′ev's glorious achievement. And yet this assumption was, and remains, erroneous. The annexation of the Amur river supplied the initial impetus for this transition, to be sure, and the region itself was the geographical starting point, but once the shift away from Alaska had been initiated, a process was set in motion that led almost immediately to the transgression of the Amur. In stark contradiction to the view that the acquisition of the river valley was a geographical *dénouement* to this process, Russian attention and activity were instead consistently directed beyond it, ever further to the south, and indeed from the very moment that the annexation of the Amur took place.

"The principal and final goal . . . had to be the Ussuri"

If the resolution among the advocates of a forward policy in the Far East concerning Russia's need to divest itself of its North American colonies was universal, there were appreciable divergences in the evaluation by these same individuals of the geopolitics of the Amur annexation itself. It is highly significant in this regard that Murav′ev did not decide upon the need to abandon Alaska until the onset of the Crimean War. For over half-a-decade prior to this, all of his efforts as governor-general in Irkutsk had been directed toward the retrenchment and fortification of Russian fur-trading settlements, in particular those on Kamchatka, which in the event of hostilities would be exposed to the danger of siege and even occupation by enemy forces. These concerns were expressed in a copious stream of communications back to the capital. The need for Russia to annex the Amur – the conclusion to which all of these communications pointed with monotonous regularity – was motivated almost entirely in terms of strategic considerations for the defense and provisionment of Kamchatka and the Okhotsk coast. That Murav′ev had not yet decided on the need to give up Russian America is clearly demonstrated in an observation made in 1849: "I dare to say . . . that to abandon our possessions in North America, in the event that the Russian–American Company should not be in a position to administer them, would not be in line with the aims of the government."[16] Yet even after he changed his mind about Alaska, and despite his grandiose assertion of Russia's future destiny as the lord of the East Asian coast, he still did not entirely reject the view that the North Pacific was an important arena for Russian activities in the Far East. This perspective indicated that for him, as Stanton notes, "the real value of the Amur was to serve as a shorter and more convenient route of communication from the Transbaikal region to Kamchatka."[17]

Murav′ev's lingering committment to the preservation and even the enhancement of Russia's North Pacific territories was apparent in his position on the important question regarding the best location of Russia's principal naval port in the Far East. It will be remembered that the lack of an adequate port facility was one of the most serious problems afflicting these territories, and that beginning in the 1820s considerable exploratory effort had been expended on the search for an alternate location which could replace the ill-fated settlement at Okhotsk. Murav′ev was convinced that the best solution would be to abandon Okhotsk in favor of Petropavlovsk, a port on Avachinsk Bay on Kamchatka's southeastern coast. This, he felt, would not only provide an excellent natural harbor, but also appeared to supply the best defense possible against feared English advances on the peninsula.[18] To underscore his committment to this proposal, Murav′ev became the first Siberian governor-general to take the trouble of actually visiting remote Kamchatka, in 1849. Nicholas I was convinced, and he ordered the relocation of the port in the same year.[19] In pressing subsequently for permission for a Russian descent down the Amur, Murav′ev argued mainly on the basis of the need for supplying Petropavlovsk.[20] It was only in 1855, after an English landing and attack on Petropavlovsk demonstrated that the enhanced security of a port on Kamchatka was illusory that Murav′ev requested and received permission to move Russia's main naval fortification of the Pacific south to Nikolaevsk on the mouth of the Amur.[21]

While all this was happening, however, a rather different perspective on Russia's new position in the Far East and on the Pacific was being articulated by a group of individuals who up to this point had been among Murav′ev's strongest sources of support. These were the so-called *konstantinovtsy*: a circle

[16] Quoted in Okun′, *Russian–American Company*, p. 221. [17] Stanton, "Foundations," p. 158.

[18] "Pis′mo N. N. Murav′eva-Amurskogo," pp. 413–416; Barsukov, *Graf . . . Amurskii*, I, pp. 218–220; Nevel′skoi, *Podvigi*, pp. 95–97.

[19] Struve, *Vospominaniia*, p. 46; Barsukov, *Graf . . . Amurskii*, I, p. 213; Kabanov, *Amurskii vopros*, pp. 95–96.

[20] Nevel′skoi, *Podvigi*, pp. 317–319. This is not to deny that Murav′ev also intended to use the descent in his struggle for the annexation of the Amur as a further affirmation of Russian domination over the river, a point which Nevel′skoi missed. On the other hand, it is certainly incorrect to suggest, as did Veniukov, that the mission to Kamchatka was merely a pretext. [Veniukov], "Primechanie," p. 2.

[21] Stankov, "Pochemu i kak," pp. 24–25; Nevel′skoi, *Podvigi*, p. 359.

of individuals closely grouped around the head of the Naval Ministry Konstantin Nikolaevich. In its upper echelons this group included such notables as Lütke, A.V. Golovnin, Reutern, Putiatin, and Popov, and it was represented at lower ranks as well, as we will see.[22] The events of the Crimean War, they felt, had taught two very important lessons. The first of these was the critical importance of naval power in the intensifying rivalry with the other European countries. Russia must at least be able to hold its own again the naval forces of its European antagonists, which meant being in a position to strike directly at their enemies the way the latter had struck at Sebastopol and Petropavlovsk. The second lesson of Crimea was the danger inherent in Russia's navy being restricted to narrow and "closed" seas, where squadrons could effectively be bottled in and thus severely limited in times of war. The answer of the *konstantinovtsy* to both of these concerns was a new policy of developing a network of year-round naval operations in international oceanic waters across the globe.[23] Existing facilities on the Baltic and Black seas were important but by no means sufficient in order to support this endeavor, for they were not only situated on "closed" seas but moreover were all rendered inoperable for part of the year due to freezing harbors. "No one denies the usefulness for us of Peter's window into Europe [in the Gulf of Finland], but it is precisely no more than a window in its dimensions and – importantly – the fact is that it is closed in by the Sound and the Belt. Here we are always potentially, and on occasion actually in the position about which Krylov wrote, namely that 'you see an opportunity but can't take advantage of it' (chto vidit oko, da i zub neimet)."[24] Russia needed naval bases which could insure unobstructed year-round access to open oceanic waters, and in the late 1850s Konstantin Nikolaevich resolved to secure them. His first step toward this end – an ambitious if rather far-fetched attempt in 1859–1860 to establish a Russian naval base at Villafranca on the Mediterranian island of Sardinia – was unsuccessful, but served as an indication of his determination to realize this program in practice.[25]

It can be readily appreciated that Russia's advances in the Far East were of the greatest moment to this new naval policy. Not only was the Pacific itself one of the important new oceanic arenas upon which Russia was now to maintain a significant naval presence,[26] but the prospect of new territorial gains held out the hope that here at least Russia would be able to find a maritime bastion that would satisfy the navy's new needs. As they studied the geographical configuration of the areas in question, however, the *konstantinovtsy* were at once disturbed by the geostrategic implications of the particular policy Murav'ev was pursuing. The annexation of the lands to the north of the Amur river and the river itself, which was the governor-general's long-stated objec-

[22] Kipp, "Grand Duke," pp. 19–20n. [23] Kipp, "Grand Duke," pp. 231–232, 245.
[24] Kpt, "Gde budet," p. 395. [25] Kipp, "Grand Duke," pp. 246–248.
[26] Kipp, "Grand Duke," p. 139.

tive, would leave the right bank and all the territory to the south and east in Chinese hands. Consequently, there would be no real alternative to Nikolaevsk as Russia's main naval base on the Pacific. It was, however, precisely this eventuality that they wanted at all costs to avoid, for reasons which they insisted were clear from a simple glance at the map. Situated on the very northern end of the narrow Tatar straits, the only satisfactory maritime access that Nikolaevsk offered was northwards into the Sea of Okhotsk, which was not only far removed from the most populous regions and scenes of commercial activity in East Asia, but was in any event effectively blocked off from the open waters of the Pacific by Japan's northern islands and the Kurile island chain.

If Russia was going to be in a position to maintain a significant naval presence on the Pacific, they concluded, it would need secure and open access to the south, into the Sea of Japan. Even this would be rather less than ideal, but it was certainly more desirable than anything Nikolaevsk and the mouth of the Amur could offer. And in order to achieve this goal, it seemed clear to them that the acquisition of the Amur river alone was not going to be enough. Russia's main port, in other words, would have to be located somewhere along the continental coast to the south of the Amur and Nikolaevsk, which meant in turn that Russia's territorial aims in the Far East as sketched out by Murav′ev – which stopped at the river – would have to be reformulated as well. This point had not escaped the American Collins:

> But Russia having possession of the upper waters of the Amoor, she now also needs the Ousuree, because this river runs off to the south, towards Corea, and approaches the Sea of Japan at a point where there is a good harbor (Pahseeat Gulf) . . . which remains unfrozen during the winter – a desideratum "devoutly to be wished" by Russia on the Pacific, and which she will find necessary in her growing affairs in this region . . . The possession of this point would be one of the best steps Russia could take, perfectly proper and justifiable in view of her future position on the coast of Tartary and the necessities of her commerce. The occupation of this wild uninhabited coast by her will be alike beneficial to civilization, to commerce, and to Christianity.[27]

The views from the early 1850s of Captain Gennadii Nevel′skoi , who as we have seen was one of the original forces behind the resurrection of the Amur question, indicate that some Russians at least had arrived at these conclusions long before Collins. Nevel′skoi 's explorations of the mouth of the Amur and the western coast of the Tatar straits provided critical support for Murav′ev's cause in the capital, and the governor-general's ultimate success would by no means have been assured without it. Although he and Murav′ev thus worked together toward what seemed to be a common goal, however, Nevel′skoi's evaluation of the situation confronting Russia on the Amur and in the Far East in general reflected the concerns of the *konstantinovtsy*, and thus diverged

[27] Collins, *Siberian Journey*, pp. 251–252.

from that of the governor-general. Indeed, the divergence was apparent already in 1849, the year of Nevel′skoi's first expedition, when he objected to Murav′ev's plan to make Petropavlovsk Russia's main port on the Pacific. Kamchatka, he pointed out, was geographically detached from the rest of Siberia, and thus a Russian naval base there would be exposed and easy prey to an enemy fleet. The only way to insure the security of a Pacific port, he countered, would be to establish it on the continental mainland, where it could be reliably supplied and supported from the interior. Nevel′skoi thus proposed at this time that the future port be relocated away from Okhotsk to the mouth of the Amur itself. In contrast to Kamchatka, the Amur would secure a connection from the interior of Siberia not only to the Sea of Okhotsk to the north, but – rather more tenuously – through the Tatar straits to the Sea of Japan to the south as well. He underscored moreover the value of the Amur valley not only as a link to the ocean but as an "convenient route" to the heavily populated regions of Manchuria, via its tributaries to the south, especially the Sungari. This was important, he concluded, "in view of our commerce with and political influence on neighboring China."[28]

Despite his endorsement of Nikolaevsk over Petropavlovsk as a site for Russia's Pacific port, however, Nevel′skoi did not believe this to be an entirely satisfactory alternative either. In the years following his initial exploration of the Amur estuary, and once again acting on his own initiative, he pressed further south through the Tatar straits, charting the Manchurian coastline in the quest for a port facility more fully suitable to the priorities of the *konstantinovtsy*. This goal proved elusive, but one thing at least appeared to be entirely clear. Russia's territorial expansion in the Far East must not stop with the Amur, but should rather be extended to include the valley of the Ussuri river, a southern tributary which flowed from headwaters in the Sikhote-Alin mountain range northwest into the lower Amur. This proposition had no precedent, for throughout Russia's long history of involvement in the Far East the Ussuri had never been mentioned in connection with the Amur and certainly was at no point ever recognized to have any particular significance to Russian interests in the region. Nevertheless, Nevel′skoi now insisted vociferously upon its central importance to the entire venture.[29] In 1855, as he was preparing to leave his post in Eastern Siberia and return permanently to St. Petersburg, he presented his thoughts formally in a report to the governor-general.

Nevel′skoi had two principal points to make, the first of which related to the geographical scope of the territories that Russia should be striving to acquire in the Far East. "[T]he Amur and the Ussuri regions," he stated, "represent an indivisible whole."

> [T]he mouth of the Sungari and the entire Ussuri basin with its coastline down to the Korean border must constitute the inalienable property of Russia, the more so as the

[28] Nevel′skoi, *Podvigi*, pp. 95–97. [29] Nevel′skoi, *Podvigi*, pp. 144–145, 183–184.

Amur river by itself represents here only a base (bazis) for our activities and does not at all [by itself] represent the full significance of these regions for Russia. Indeed, the entire strength of this region and its *political importance* for Russia, as the current war clearly shows us, is comprised by the southern coast of the Ussuri basin.[30]

Nevel′skoi's second point was more general, and related to the prospects for the future development of the far-eastern region as a whole. This area was quite rich in its natural endowments, he readily concurred, and therefore without question constituted an important acquisition for Russia, and one which was destined some day to make a significant contribution to the country's national development. However, owing to the peculiar geographical conditions of the Amur valley and the overall logistics of the region, prospects for its settlement and development could be realistically entertained only for the distant future. On this point he was emphatic, insisting that "in this region we will have to remain for a long time as if in an army camp (nadolgo eshche v etom krae nam dolzhno ostavat′sia kak v lagere)." In particular, he rejected any suggestion of a parallel between the Amur valley and the United States in regard to the prospects for settlement and development in the near future. "[O]wing to its enormous deserted expanses, its geographical situation and climatic conditions, commerce and industry can never develop as rapidly here as we observe in the North American states, and in particular California." It would be "more than foolish," he continued, "to be seduced by the example of America, and await here [the same development] that is taking place there in this respect. The Amur region is in all respects entirely different from America, and for this reason the government should . . . not be seduced by illusions and by the example of the United States and California."[31]

Nevel′skoi certainly must have realized that the views expressed in this memorandum were quite out of harmony with the image of the Amur region that Murav′ev himself was assiduously encouraging. It is not altogether easy, however, to imagine that he appreciated the extent to which his sober points amounted to a refutation of the most fundamental and indeed the most cherished ideas about the significance of Russia's acquisition in the Far East. His basic message was that the Amur – Murav′ev's pride and joy, the salvation of Siberia, and Russia's link to the great Pacific civilization of the future – was utterly worthless in and of itself. The access it provided to the sea, rather than being convenient and satisfactory, was inadequate to such an extent that the region's real outlet to the Pacific (to the extent the region was to have one at all) could only be along the coast well to the south of Nikolaevsk. In a word, while for Murav′ev the annexation of the Amur valley had always been and remained a culmination and an end in itself, for Nevel′skoi it was, and could be only a first step. On this final point he wanted to leave no doubt whatsoever:

[30] Nevel′skoi, *Podvigi*, pp. 364–365 (emphasis in original).
[31] Nevel′skoi, *Podvigi*, pp. 365–366.

"*the principal and final goal*" of Russian territorial expansion in the Far East "*had to be the Ussuri basin and its coastal region.*"[32]

Murav′ev in 1853 may well have made the extravagant claim for the ears of the tsar that Russia was the natural master of "the entire Asiatic coast of the Pacific Ocean," but his dispute with Nevel′skoi indicated that he had no practical intention of attempting to realize this vision. The implicit differences between the two perspectives came to the surface in regard to the issue of how the boundary with China should be drawn. As we have seen, the Russian government had resolved in the late 1840s to initiate negotiations with China in order to settle once and for all the location of the border. From the very beginning of his tenure as governor-general, it was Murav′ev's fervent desire to be empowered as Russia's plenipotentiary in these talks, although his view of what an appropriate border would look like differed from that of St. Petersburg. This aspiration was finally realized by an order of Nicholas I in 1853. It was to take five further years of preparation and waiting, however, before circumstances in the Russian Far East and within China itself made it possible to convene negotiations on the border, in the town of Aigun on the right bank of the Amur. Murav′ev's unchanging position from the very beginning had been that the river itself should form the boundary between the two countries. This position was consistent with his earlier interpretation of the river's significance as a communications and supply route within Russia's northern Pacific complex overall. He was convinced of the truth of his claims that the Amur was easily navigable, and that with it Russia's secure link to the Pacific would be guaranteed.[33]

Nevel′skoi, on the other hand, argued that to run the border along the Amur would violate the geographical unity of the region and in fact render the Amur useless. He had made this point to Murav′ev while he was still stationed in the Far East, as we have just seen, and he continued to repeat it after returning to the capital in the mid-1850s. He apparently entertained more than one alternative as to how precisely this geographical unity might be maintained. According to Veniukov, at one point he advocated drawing the boundary in a straight line from the Amazara river, a tributary of the Amur just below the confluence of the Shilka, southeast to the northern boundary of Korea and the sea, thereby extending Russian domination over most of northern Manuchuria.[34] An alternative, and distinctly more modest possibility was for the border to run along the Amur river down to the confluence of the Ussuri, and then up the latter to Lake Khanka and across the Sikhote-Alin range to the ocean. Murav′ev finally agreed upon this second option at the border conference at Aigun in 1858, but still only partially and at the very last moment, after realizing that the Chinese had been so weakened by the

[32] Nevel′skoi, *Podvigi*, p. 331 (emphasis in original); Butsinskii, *Graf . . . Amurskii*, pp. 31–34.
[33] Nevel′skoi, *Podvigi*, p. 375. [34] Veniukov, "Vospominaniia," pp. 96, 287.

T'ai-P'ing rebellion and the threats of further European intervention that they would be unable to resist.[35] He took advantage of this situation by obtaining a joint Russo-Chinese condominium over the territory stretching from the Ussuri east to the ocean.[36] With the Treaty of Peking, signed in November, 1860, this jointly controlled territory became exclusively Russian property. In an unmistakable acknowledgment of the very point that Nevel′skoi had endeavored to make about the superior Pacific access of the region south of the Amur, St. Petersburg gave this land the official name of the "Maritime region" (Primorskaia oblast′), in contrast to the "Amur region" to the north.[37]

Other *konstantinovtsy* as well gave voice to Nevel′skoi's insistence that the expansion of Russia's territorial dominion in the Far East had to go beyond the geographical limits of the Amur river itself. The same Rear Admiral Popov who spoke with such empathy about American Manifest Destiny speculated in 1860 that the Russian population of Alaska could be relocated to the sea-coast south of the Amur estuary, from where "it may be presumed, not satisfying ourselves with a pointless battle against the natural hinderances [in the Amur region] we will move farther south into Korea." In contrast to Russia's "detached colonies" in the North Pacific, he asserted, expansion southwards along the coastline of the continent would be "founded on the inner forces of Russia."[38] Konstantin Nikolaevich's adjutant Ivan Likhachev complained in 1859 that boats sailing from European Russian ports to the Far East should not be limited to the "road to the Amur" and that "there is no need to halt after [only] the first step."[39] This position was vigorously endorsed by many of those exiled Decembrists in Eastern Siberia who had a naval background, such as Baron V.I. Shteingel′ and Mikhail Bestuzhev. Shteingel′ had noted the need

[35] Veniukov, *Iz vospominanii*, I, pp. 251–252, 237, 243; Mancall, "Major-General Ignatiev's Mission," p. 58. Worried that the border issue might be resolved by one of the other Russians who were negotiating with the Chinese government at the time, Murav′ev was desperately anxious to conclude a successful agreement at Aigun. This inclined him resist advancing demands for territorial concessions beyond the Amur river itself, insofar as these could complicate and possible scuttle the delicate negotiations. He did so only when it became clear that the Chinese, in their enormously pressured position, would agree without difficulty. Zavalishin, "Dokumenty," p. 412; Zavalishin, *Zapiski*, II, p. 389; Veniukov, "Vospominaniia," p. 82.

[36] Veniukov, *Iz vospominanii*, I, pp. 58–59; Nevel′skoi, *Podvigi*, pp. 400–401.

[37] The rivalry between Murav′ev and Nevel′skoi was apparent even as they were both serving in the Far East. Barsukov, *Graf . . . Amurskii*, I, pp. 346–347; 395. Most subsequent accounts have tended to accord Murav′ev major credit as the mastermind of the Amur annexation and depict Nevel′skoi as his obedient assistant: Nevel′skoi was "Murav′ev's hand and instrument" (Sullivan, "Count N. N. Muraviev-Amurskii," p. 168); Murav′ev was the "chief architect" of Russia's new far-eastern policy (Mancall, "Russia," p. 322); Murav′ev sent Nevel′skoi to explore the Amur estuary in 1849 (Lin, "Amur Frontier Question," p. 7). Also see Efimov, *Gr. N. N. Murav′ev-Amurskii*, pp. 37–46. In his memoirs, Nevel′skoi supported his counterclaim by insisting that the resurrection of the Amur question in the 1840s and its resolution the following decade was the work of naval officers serving in the Far East at the time. Aside from his daughter and the historian Butsinskii, however, his claim – which has a good deal of validity – unfortunately seems to have been lost. Nevel′skoi, *Podvigi*, p. 411; Vend, *L'Amiral Nevelskoy*, pp. 11–12; Butsinskii, *Graf . . . Amurskii*, pp. 25–27, 47.

[38] Quoted in Miller, *Alaska Treaty*, p. 55.

[39] Quoted in Zhitkov, "Vitse-Admiral," pp. 6, 7.

for more southern ports in 1854,[40] and Bestuzhev brought the matter up rather irately with Murav′ev himself in 1857 (or so he claimed). "Is it really the case," he demanded in a conversation with the governor-general,

> that the Amur drew us to the ocean in order that we would fold our arms and gaze admiringly from behind the ice floes and sandbars of its mouth at the commercial activities of other nations? If you want to take part in this activity and maintain fleets [on the Pacific], then it is necessary to prepare suitable ports for them, and not glaciers and harbors closed [by ice], which are all that Russia can boast of [on the Amur]. Yes! it will be a sin if we Russians, out of our customary apathy, do not now take advantage of favorable circumstances and seek out for ourselves an open port to the south.

Bestuzhev reported that Murav′ev interrupted him at this point. "You sailors are so capricious," he sighed, "perhaps you want [to extend Russia's domination all the way] to the Gulf of Pechili [Bo Hai]?"[41] Murav′ev apparently intended his objection to be sarcastic, but this was in fact precisely what at least some of his critics had in mind. Anticipating what was to become the preoccupation of Russian far-eastern policy four decades later, voices were raised already at this time calling for the extension of Russian influence into Korea. Korea should be "opened" as quickly as possible, demanded one commentator, for at a time when the ports of Japan and China had already become accessible for Western trade, its hermetic isolation was "completely abnormal."[42]

An important but little-remembered incident some years after Murav′ev's conversation with Bestuzhev, which has been curiously overlooked in virtually all subsequent accounts of the Amur annexation, serves as an indication of the determination of Konstantin Nikolaevich and his like-minded associates to shape events in the Far East in line with their perspective. By the end of the 1850s, even Murav′ev had come to agree that as a port facility Nikolaevsk was highly problematic. Yet while he fixed his own attention on what is today the Peter the Great Bay at the very southern tip of Russia's newly acquired coastline as a potential alternative, insisting somewhat illogically that this relocation would give the Amur "enormous significance in regard to Japan, Korea, and the coast of even China itself," voices within the Naval Ministry were advocating a very different alternative.[43] Ivan Likhachev, an assistant and

[40] Shteingel', "Pis′ma," p. 317; Azadovskii, "Putevye pis′ma," p. 210.

[41] Bestuzhev described this discussion to Dmitrii Zavalishin in a letter of August, 1862, in Zavalishin, "Dokumenty," p. 412. As a part of the bitter discord of the 1880s and 1890s between Zavalishin on one hand and the other Decembrists exiled to Eastern Siberia on the other, involving charges and countercharges of distorting what had actually transpired during their exile, Frolov maintained that the documents which Zavalishin reproduced here cannot be used to judge Murav′ev. [Frolov], "Vospominaniia," p. 472; Azadovskii, "Neosushchestvlennyi zamysel," pp. 218–219. Nevertheless, I see no reason to doubt that this is an accurate indication of Bestuzhev's views. This judgment is supported, among other things, by Mark Azadovksii, who points out that Bestuzhev demonstratively sided with Nevel′skoi in espousing a perspective on the Amur issue that was "far broader" than that of the governor-general. Azadovskii, "Putevye pis′ma," p. 211.

[42] Kpt, "Gde budet," p. 395.

[43] Romanov, "Prisoedinenie," no. 7, p. 114 (quote); Zhitkov, "Vitse-Admiral," pp. 21–22.

trusted confidant of Konstantin Nikolaevich, had spent several years in naval service in the Far East in the early 1850s, and through the experience became a strong supporter of expanding Russia's naval presence on the Pacific.[44] In May 1860, he submitted a detailed memorandum to the Grand Prince setting forth his own evaluation of the overall situation confronting Russia in the Far East. His note emphasized just how much the stakes in these far-flung territories had been raised since Nevel′skoi first spoke out a half-decade earlier. Conceding that the Treaty of Aigun had gained for Russia the advantage of the coastline south of the Amur estuary, Likhachev pointed out that the country's overall naval position in the Far East remained highly problematic nonetheless, for this coastline provided access only into the Sea of Japan, which was itself cut off from the Pacific by the Japanese islands and was thus in its own way an "internal" sea like the Black Sea and the Baltic. This disadvantageous situation would inhibit the future development of Russia's naval power on the Pacific in the same way it had been inhibited elsewhere.[45] To overcome this obstacle, Russia had to secure free and unrestricted access in and out of the high seas of the Pacific, and the only way to do this would be to establish a Russian naval base on what Likhachev called the "main gate" leading out of the Sea of Japan, namely the Japanese island of Tsu Shima. Tsu Shima was located between Japan and Korea in the middle of the Straits of Korea, about 500 miles south of the southernmost point of Russia's new Pacific boundary. Likhachev characterized the importance of this option in the most urgent terms. "I take the liberty of expressing my opinion," he wrote, "that although this new acquisition would not come cheaply, it would not be in vain, for without it, all of our efforts toward the development of our naval significance on the Pacific Ocean would prove to be a futile and pointless waste of capital, time, and labor."[46]

Likhachev's proposal had the strong support of Konstantin Nikolaevich, who was able to obtain the consent of the tsar to proceed with a provisional landing on the island. Likhachev himself was sent to survey the situation in the Far East firsthand, and in April 1861, after inspecting Russia's newly acquired coastline, he wrote back to Konstantin Nikolaevich. He expressed disappointment at having confirmed that the Peter the Great Bay did indeed freeze each winter for 4–5 months, and restated the need for Russian access to an "open, unfettered (neskovannoe) ocean." He now declared unequivocally that Russia's "future main military port on the Pacific should be on Tsu Shima," ordered the landing on the island for that spring, and proceeded to fortify the Russian position there.[47] The project, however, was aborted almost immediately. The Russian build-up elicited strong objections from the Japanese as well as the British – the latter making a demonstrative landing of their own on the island in June – and in St. Petersburg, where the Foreign

[44] Kipp, "Grand Duke," pp. 82–83. [45] Quoted in Belomor, "Tsu-Shimskii epizod," p. 248.
[46] Quoted in Belomor, "Tsu-Shimskii epizod," p. 249.
[47] In Belomor, "Tsu-Shimskii epizod," p. 75.

Ministry had been critical from the outset, support began to waver. By December 1861, Likhachev had been ordered to abandon the island and return to St. Petersburg.[48] The choice for relocation of Russia's Pacific port away from Nikolaevsk now fell by default to Vladivostok, the new settlement founded by Murav′ev , and although the wisdom of this decision continued to be questioned in naval circles, there it was to stay.[49]

Russia's short-lived episode on Tsu Shima had no practical consequences. The ideas and activities of Likhachev, Nevel′skoi, Bestuzhev, and other *konstantinovtsy* are significant, however, for they provide an indication of an important transition in perceptions of the Amur that was taking place quite literally as the region was being annexed. On a basic level, they shared the same conviction in Russia's future destiny on the Pacific that we have examined in earlier chapters. They perceived the emergence of a new and progressive universe of commercial and political activity in the Far East, and were certainly as determined as anyone that Russia had to take its rightful part in it. Yet in their pursuit of this vision, a critical geographical disengagement had been set in train. Up to this point, the prospect of Russian activity on the "Mediterranean of the future" had depended entirely upon the Amur river as the medium which would make participation on this arena possible. It was the Amur which was to be Russia's reliable and convenient link to the Pacific, it was the Amur which would open up Siberia to the free exchange of goods and ideas, and finally it was the Amur valley which would nurture the growth of a mighty nation and a vast city. This bold vision, however, was conspicuously absent from Bestuzhev's as well as from Nevel′skoi's arguments, and both indeed made quite the opposite points instead. The Amur itself was cut off from the ocean by sand bars, ice floes, and indeed glaciers, and was thus entirely incapable of serving as a base for Russia's activities on the Pacific. The very argument, in other words, which up to this point had been invoked to support the Russian annexation of the Amur river was now used to condemn the river and abandon it for a port which, unlike Nikolaevsk, would be truly maritime and "open." This was an extraordinary conceptual inversion, which demonstrated better than anything else the enormously volatile quality of the geographical images of the far-eastern *terra incognita* with which the Russians were operating. In effect, no sooner had the Amur valley been successfully occupied by the Russians than it began to be conceptually blocked out and surpassed in the quest for a more southern location.

"Nothing short of the Wall of China"

The eminent historian of Russian–American relations Nikolai Bolkhovitinov has recently observed that the sale of Alaska signaled the recognition of a

[48] Belomor, "Tsu-Shimskii epizod," pp. 73–74, 84; Kipp, "Grand Duke," pp. 253–256.
[49] Kpt, "Gde budet"; Afonas′ev, "Nikolaevsk"; Burachek, "Gde dolzhen byt′."

"continental and not a maritime future" for Russia in the Far East.[50] Our consideration of the *konstantinovtsy*, however, suggests this was not necessarily a conclusion shared by all parties involved. Insofar as their views of Russia's position on the Pacific were formulated in terms of the larger project of enhancing the country's status as a global naval power, the perspective of Konstantin Nikolaevich and his co-thinkers in the Naval Ministry retained a predominantly maritime focus. For them, the principal benefits of Russia's territorial advances were geostrategic, and were ultimately to be realized on the open waters of the Pacific ocean. At the same time, however, an alternative vision of Russia's future in Asia was being articulated which, as Bolkhovitinov suggests, was indeed explicitly continental. It fully accepted the general geographical relocation of the main significance of Russia's new acquisitions to the south, away from the Amur valley, but it had a very different object in mind. Its focus was territorial, and the principal concerns were not with Russia's presence on the ocean but rather with its economic and political penetration into Manchuria and northern China. As this perspective was elaborated, the redefinition of what the Amur river actually represented in a strict geographical sense, a process begun by Nevel′skoi , was pressed ever further. From earlier visions of the river forming the empire's proper and "natural" boundary in southeastern Siberia, the image of the Amur was now transposed into one of a waterway which occupied an entirely intermediate position with regard to the regions to the south. It was intrinsically joined with the Ussuri valley, but beyond this formed the main axis of an extensive and tightly unified network of rivers spreading out to cover the vast interior-continental region of Manchuria. In these terms, the primary significance of the Amur valley was the access it afforded not to the ocean but rather into the vast northern provinces of the Chinese empire. From here it was only a small step to see the annexation of the Amur as a move not so much east toward the ocean but rather overland to the south.

To be sure, there was nothing really novel about the identification of the Amur valley as a part of Manchuria. The Manchu dynasty in China had always considered the region to be the natural northern frontier of its native homeland, and many of the Western sources from which the Russians got their information concurred in this physiographic classification. A series of letters about a visit of Christian missionaries to Manchuria published in the late 1840s in *Syn Otechestva* (Son of the Fatherland), for example, included extensive geographical descriptions of the country and depicted the Amur and Ussuri valleys as natural–geographical components of the larger region. These churchmen from the West, preoccupied as they were with the salvation of pagan Oriental souls "still devoted to crude superstition," obviously had nothing to say about the potential significance of this geographical

[50] Bolkhovitinov, "How it was decided," p. 124.

configuration for the Russians, but the implications were signaled as soon as Russian commentators began speaking to the same point.[51] V.P. Korsh's 1851 essay "The Amur River Basin and its Significance," which as we have seen was one of the very first attempts to explain the importance of the Amur region for a broader public audience,[52] indicated by its very title that there was more than just the river at stake. "The Amur basin waters the northeastern part of China and covers an enormous territory between Lake Baikal, the Gobi desert, China proper, the Korean peninsula, and the Stanovoi Range." To the northeast, it extended to the coast of the Sea of Okhotsk, and to the south as far as the city of Mukden (Shenyang) and further to the Yellow Sea.[53] Attention was drawn yet more explicitly to the direct connection between Manchuria and the Amur in a number of essays and translations by the historian Vasilii Vasil′ev published in 1857. The documents had been prepared at the request of Murav′ev, who was seeking to generate as much publicity about the region as possible,[54] and they appeared prominently in the journal of the Russian Geographical Society. In the most extensive of these, entitled simply "A Description of Manchuria," Vasil′ev made it clear that the Amur was an important part of the region which extended southwards from the "remote limits of our Siberia," beginning at Nerchinsk and continuing to the ocean, stretching northeast as far as the Shantar islands.[55] In his geographical descriptions of this massive area, Vasil′ev focused practically all his attention on the lands lying south of the Amur, that is to say on a region which lay beyond the Russian empire but to which the annexation of the river the following year would provide access.[56]

The suggestion that the Amur constituted an important physical–geographical part of Manchuria was, as just noted, a prelude to the broader perspective on the acquisition of the river as the first step of Russia's penetration into northern China itself. The appeal of such a vision lay not so much in strategic considerations as in a simple factor shared by all nations of the industrialized and imperialist West, namely the attraction of rich and untapped markets. A veritable universe of commercial potential seemed to present itself in Manchuria, for its densely packed population had been entirely shielded up to that point from all outside influence. Once again, the American Collins provided the most florid imagery for the manner in which this Manchurian market would inevitably draw Russia down from the Amur southwards, and once again he elaborated the point with a comparison drawn from American experience. Writing to the American Secretary of State in 1857, he declared:

[51] "Pis′ma o Manzhurii," pp. 27 (quote), 48ff. [52] See chapter 3.
[53] Korsh, "Bassein," pp. 1327–1328. [54] Vasil′ev, "Opisanie Man′chzhurii," p. 10n.
[55] Vasil′ev, "Opisanie Man′chzhurii," p. 2.
[56] Also see Vasil′ev's "Zapiska o Nindute" and "Opisanie bol′shikh rek," both of which were translations of eighteenth-century Chinese sources.

The probability is that Russia will find it necessary, in order to give peace and security to the trade on this important river, from her Siberian possessions into the ocean, to follow our example in the acquisition of Louisiana; for the whole of Manchooria is as necessary to the undisturbed commerce of the Amoor as Louisiana was to our use of the Mississippi; consequently, in my opinion, nothing short of the Wall of China will be a sufficient boundary on the south, and that is not so remote to Russia here at this day as the Rocky Mountains in Jefferson's day were to us, much less the mouth of the Columbia river, which we acquired then . . . Twenty thousand cossacks would overrun and hold the country as easily as our little army moved on to Santa Fe and conquered New Mexico.[57]

Virtually all foreign observers of Russia's advance in the Far East agreed with Collins that Russian domination was destined to spread south – *Petermanns Mittheilungen* predicted that "all of Manchuria" would "soon be a Russian province"[58] – and these prognostications were loudly echoed by Russian commentators as well.[59]

If in the final analysis Murav′ev proved unwilling or unable to grasp the logic by which Russia should seek to extend its domination of the Asiatic coastline far to the south of the Amur itself, he had from the very beginning been keenly interested in expanding Russian influence overland into China, and consequently needed no encouragement from Collins. The prospect of other European powers progressively broadening the scope of their operations and influence within the Middle Kingdom was at once powerfully tempting and powerfully aggravating for him, and it provided a constant stimulus for his own fervent desire that Russia should not fail to take its rightful part in this process. Unlike the other powers, however, Murav′ev understood that Russia had somewhat more flexibility in how it chose to define its role in China. Involvement there could take one of two forms, either hostile aggression as part of the Western incursion or alternatively benign support and aid to the Chinese government against both this incursion as well as domestic unrest. Thus, while Murav′ev remarked in 1849 on the need to maintain the necessary beneficent support in order to aid Russia's weak neighbor, he was apparently thinking even at this early date of a military campaign against Manchuria.[60] He displayed the same ambivalence some years later, when China was being rocked by the T'ai-P'ing rebellion. In 1854 he wrote that if Russia were to assume control over Mongolia as a protectorate, it could then consider offering its patronage and protection to the beleaguered dynasty in Beijing.[61]

[57] Collins, *Siberian Journey*, pp. 95–96. Collins perceived great potential for American interests in Russia's penetration of Manchuria. As a result of this advance, the "immense countries" of Manchuria "would then be open to our [American] commerce and enterprise, . . . unlocking to us a country of vast extent and untold wealth." Also see Atkinson, *Travels*, pp. 449–450.

[58] "Die neuste russische Erwerbungen," p. 175 (quote); Ravenstein, *Russians*, p. 142; Swan, "Explorations," p. 178; [Liudorf], "Amurskii krai," pp. 216, 221.

[59] "Donesenie Kollinsa," p. 7; [Lamanskii], "Mnenie," pp. 55–56.

[60] Barsukov, *Graf . . . Amurskii*, I, p. 211; Matveev, "Dekabrist Zavalishin," p. 18.

[61] Barsukov, *Graf . . . Amurskii*, II, pp. 112–113.

In a very different tone, however, he affirmed the following year that Russia "must be ready for all eventualities, and at the very least to demonstrate to the Chinese our readiness to advance into Mongolia and Manchuria if they take it into their minds to obstruct next year's descent [down the Amur]."[62]

Murav'ev returned to his project to expand Russian influence to the south two years later, when in a letter to his assistant Korsakov he spoke of the desirability of separating Manchuria and Mongolia from China and establishing them as "separate kingdoms, under the patronage of Russia."[63] He ruminated on this prospect in a message to Konstantin Nikolaevich, which was filled with chagrin at the thought that Russia might fail to act as triumphally as the other European powers.

> In the final analysis, it might even be better if we have to compel [the Chinese] with force to accept our suggestions [regarding the Amur], and why should we not exercise the same right which the English and French have appropriated for themselves, by which they compel the Chinese with arms to accept their suggestions?[64]

Murav'ev persisted in this tone to the very end of his tenure in Eastern Siberia, and in 1860 was still insisting on the need to prepare for an armed advance into Manchuria and Mongolia.[65] The government in St. Petersburg, for its part, cautiously maintained a neutral stance toward China until the late 1850s, resisting not only the agitated entreaties from the governor-general of Eastern Siberia but those of the Western powers as well, who extended repeated invitations for Russia to join in crushing the T'ai-P'ing revolt.[66] Indeed, Murav'ev's seemingly irrepressible eagerness for a military campaign into China even occasioned jokes in St. Petersburg to the effect that, in addition to the "Count of the Amur" he wanted to become "Count of Mongolia" (alternatively the "Mongolian Count": Murav'ev -Mongol'skii) as well![67]

Murav'ev's desire to expand Russia's sphere of domination in the Far East through direct military action may have been frustrated, but there were other means by which the same essential goal could be realized. The primacy of the issue of commercial access to Manchuria was clearly reflected in the three treaties which Russia concluded with China in 1858 and 1860. This is an important point to note, for these treaties have tended to be read in retrospect essentially as agreements on the transfer of territory and the establishment of

62 Zaborinskii, "Graf . . . Amurskii," pp. 644–645.

63 Barsukov, *Graf . . . Amurskii*, I, p. 485.

64 Barsukov, *Graf . . . Amurskii*, I, pp. 144–145.

65 Barsukov, *Graf . . . Amurskii*, I, pp. 485–486, 571, 602, II, pp. 283–284.

66 Popov, "Tsarskaia diplomatiia," pp. 184–189, 191; Costin, *Great Britain*, p. 247. Russia finally did decide to extend an offer of assistance to the Chinese court only in 1858, and the resulting military mission, led by N. P. Ignat'ev, was ultimately transformed into the diplomatic mission which concluded the Treaty of Peking. Mancall, "Major-General Ignatiev's Mission," p. 60; Quested, *Expansion*, p. 174.

67 Kropotkin, *Dnevnik*, p. 155. Nevertheless, some 40 years later Kropotkin claimed that "no one of us could think at that time the Russia would ever try to establish her rule of Manchuria." Kropotkin, "Russians," p. 270.

new international boundaries. In fact, they included much more than this. In the Treaty of Aigun the governor-general was careful to obtain for Russia exclusive rights of unrestricted navigation and commerce along the entire extent of the Amur's two principal tributaries to the south, the Ussuri and Sungari rivers.[68] In a communication to the tsar, Murav′ev bragged about the great value for Russian trade of the access he had just secured into a Manchuria "up to this time inaccessible."[69] Immediately upon signing the treaty, he demonstratively put the issue to the test by sending an expedition up the Ussuri and by himself sailing some 20 *versty* up the Sungari.[70] Expeditions to other parts of Manchuria followed, including two led by Peter Kropotkin.[71] Chinese concessions at Aigun were further confirmed in the Treaty of Peking, which additionally restored the rights of Russian merchants to travel by caravan from Kiakhta to Beijing, a privilege surrendered in the Treaty of Kiakhta of 1727. The Russians were now entitled to conduct all of their commerce with China on the Amur and Ussuri rivers free of tariffs, and Russia's exclusive right to navigate and trade on the Sungari was reaffirmed.[72]

So significant indeed was the function of the Amur as a continental gateway into northern China that, at least for some of those on the scene, it took precedence over the factor of access to the ocean. It was clearly to the south and not to the east that the eyes of the merchants at Kiakhta were expectantly cast in 1854, when they composed the following verse to celebrate the governor-general and his first descent down the Amur:

> Ura! Nash mudryi Nikolai!
> Tvoi orly pariat vysoko . . .
> Molchi Mongol! Ne spor Kitai!
> Dlia russkogo i Pekin nedaleko.
>
> Hurrah! Our wise Nikolai!
> Your eagles soar on high . . .
> Be silent, Mongol! Don't argue, China!
> For the Russian even Beijing isn't far away.[73]

The explorer Richard Maak commented that while the acquisition of the Amur region had undoubtedly brought the Russians closer to the Pacific and

[68] Sladkovskii, *History*, pp. 82–83; Mancall, "Major-General Ignatiev's Mission," p. 58; Lin, "Amur Frontier Question," p. 27.

[69] Aleskeev, "Amurskaia ekspeditsiia," p. 490.

[70] "Maximowicz's Reise," *passim*; Veniukov, "Vospominaniia," p. 285; Semenov-Tian-Shanskii, *Istoriia*, I, pp. 205–206; Nevel′skoi, *Podvigi*, p. 405; Alekseev, *Russkie geograficheskie issledovaniia*, pp. 48–49.

[71] Kropotkin, "Poezdka," pp. 663ff; Kropotkin, "Dve poezdki," pp. 65ff; Kropotkin, *Memoirs*, pp. 199–203; Anisimov, *Puteshestviia*, pp. 28ff; Usol′tsev, "Svedeniia," p. 173. Ironically, Kropotkin returned from his trips highly skeptical about the possibilities for Russian commerce in the region; see below, p. 254.

[72] Romanov, *Poslednie sobytiia*, p. 24.

[73] Quoted in Barsukov, *Graf . . . Amurskii*, I, pp. 364, 522.

thereby to America and Japan, "even more important" were the commercial and political implications for Russian relations with the Chinese. The southern tributaries of the Amur "traverse densely populated provinces of China," insuring that these regions were destined to come into the "closest relations" with Russia. The ultimate possibilities, however, he saw extending well beyond the limits of Manchuria proper. "Both the density of population and the convenience of movement across all of central and southern China compells us to assume that the southern tributaries of the Amur will become commercial routes upon which, in exchange for Russian goods, the riches of the entire Celestial Empire and, perhaps, even of India will be delivered to us."[74] Suggestions were even made to relocate Russia's main administrative center in the Far East away from Nikolaevsk to an inland position – specifically to the hub of the river network at Khabarovsk – precisely in order to facilitate Russian development of the continental Manchurian trade.[75] Indeed, one commentator affirmed in a St. Petersburg newspaper that the entire significance of the Amur river had to be seen in terms of adjacent territories in northern China. "The acquisition of Manchuria is so important that without it the Amur would lose half of its value, if not more. It is more correct to say that the Amur should be only a first step (vvedenie), and the most essential part of [our] acquisition is Manchuria."[76]

The attraction of regions to the south of the Amur for the Russians was considerably enhanced by the belief that they were particularly promising for agricultural development. This was especially true of the Ussuri valley. Mikhail Veniukov, who had led the first expedition up the river in 1858, returned gushing about the Ussuri as "one of the best provinces of East Asia for all forms of agriculture," and after his own expedition up the river in the following year Maak assigned it "the first place without doubt among all of the parts of the Amur region" in terms of its potential for agricultural settlement.[77] Kropotkin signaled the agricultural abundance of the "incredibly rich" Ussuri valley with a piquant reference to the fact that the farmers there "have been eating [fresh] pickles for a long time already, while Nikolaevsk is still being supplied by American canned goods." Indeed, the reputation of the Ussuri grew so quickly – fired, no doubt, by the failures of agricultural colonization along the Amur itself – that by the early 1860s it had actually begun to replace the Amur in the minds of the local population as the real promised land. "[W]ith its southern vegetation, its warm, non-Siberian climate, with the beauty of its forests, [and] the fertility of its soil," Kropotkin continued, "one constantly hears remarks about the valley of this river as a Siberian El

[74] Maak, *Puteshestvie na Amur*, p. v.
[75] Afonas'ev, "Nikolaevsk," p. 130.
[76] Kpt, "Gde budet," p. 395.
[77] Veniukov, "Obozrenie," p. 228; Maak, *Puteshestvie po doline*, p. 67. On Maak's expedition, also see "Die neusten Expeditionen," p. 315.

Dorado."[78] Decades later, in an essay on the renewed Russian activity in Northern China at the turn of the century, he recalled the origins of this interest in the 1850s: "no sooner was [the Amur region] taken than the eyes of the new settlers were already turned farther southward."[79] "It is a good thing that we acquired the Ussuri region and the right bank of the Amur," concluded the author of a detailed survey of the Amur region in 1865, "for without it the settlement of the Amur's left bank . . . would be entirely hopeless," and the river valley would "remain a desert."[80]

The Russian preoccupation with the prospect of the untapped markets of northern China bespoke a certain unity of purpose that was felt with the other Western powers, all of which were united by a common determination to overcome the resistance of the Chinese government to allowing them into markets in the internal regions of the country. It was precisely because of this resistance, proclaimed a brief item in the journal of the Russian Geographical Society, that

> the English, Americans, French, *and Russians*, hand in hand, forced their way (probivalis′) into the Celestial Empire. They wanted to gain access to its internal regions, to sail with steamships up the major rivers into the very heart of this enormous country. They wanted to conduct trade with China not only along the ocean coast.[81]

Yet despite such heartfelt affirmations of solidarity, the Russians nevertheless defined and defended their interests to a significant degree in clear distinction from, if not indeed in opposition, to the other European powers. The same sort of ambivalence in Russia's attitude toward China that we noted above in the case of Murav′ev – the alternative postures, that is to say, of friend or foe, ally or predator – was exploited to the fullest as the Russians pursued their diplomatic and political designs in the Far East throughout the late 1850s. They made full use of it for their agreement with the Chinese at Aigun,[82] but the spectacular culmination was in the negotiations leading up to the Treaty of Peking in 1860. In a display of remarkable diplomatic agility, the youthful plenipotentiary Nikolai Ignat'ev[83] used his country's indeterminate position to manipulate among all of the contending parties and secure what was effectively the most decisive and influential position for Russia. As he explained subsequently in his account of the negotiations, his winning strategy was on the one hand to assure the Chinese that Russia "had observed the strictest neutrality" in China's conflicts with the European powers and that it "sincerely desired to save the Manchu dynasty," while on the other to convince

[78] Kropotkin, "Iz Vostochnoi Sibiri" (December 1864), pp. 14–15. Kropotkin also reported on rumors that the Sungari valley was similiarly "extraordinarily fertile," but of course Russian settlement here was precluded.

[79] Kropotkin, "Russians," p. 270. Also see Kropotkin, *Memoirs*, pp. 268–269.

[80] Noskov, "Amurskii krai," no. 4, pp. 18 (quote), 26, 4–8.

[81] "O torgovle," p. 53 (emphasis added).

[82] Veniukov, *Iz vospominanii*, I, pp. 251–252.

[83] In 1860 Ignat′ev was only 28 years old. Koot, "Asiatic Department," p. 98n.

the Western belligerents of the "value of joint moral activity in China on the part of all representatives of European powers," of which Russia, of course, was one.[84]

There was, however, something more fundamental than this sort of opportunistic jockeying for advantage that served to set the Russian position in China apart from that of all other Europeans. This was the geographical fact that Russian economic relations with China were conducted exclusively by means of overland commerce, while for all other parties the China trade depended upon extended trans-oceanic links. The venture noted earlier in this study, in which a handful of Siberian merchants in the late 1850s sailed down the Amur and southwards along the Chinese coast to investigate the potential of maritime trade with China's ports on the Pacific, was ultimately nothing more than a testament to the power of the vision of the Pacific as the Mediterranean of the future, for it produced no appreciable results. The practical irrelevance of maritime links between China and the Russian Far East was well demonstrated by the fate of the Treaty of Tientsin which Russia and three other Western powers jointly concluded with China in the same year as the Treaty of Aigun. This treaty, negotiated by Putiatin, substantially expanded the privileges that other nations had already obtained from the Chinese. For the Russians, it was significant in that they now received these privileges as well, including access to treaty ports, rights of extraterritoriality, and so on. Yet the agreement remained a dead letter for Russia, precisely because it did not deal with the most important issues of continental boundary delimitation and commercial access overland. Murav′ev vehemently opposed the negotiations at Tientsin;[85] indeed, the agreement he secured at the same time at Aigun should be seen more than anything as a deliberate alternative to it.[86] As Sladkovskii has pointed out, the Russians conducted virtually no maritime trade with China and there were consequently no Russian enclaves in any of the Chinese ports. The Treaty of Tientsin acquired some practical significance only in 1862, at which time a supplemental "Regulations for Overland Trade" was appended to it, further stipulating conditions for Russian commercial activity in Mongolia and Manchuria.[87]

It was no coincidence that Russia's negotiator at Tientsin was a naval figure and an important member of Konstantin Nikolaevich's circle. The differences between the two treaties can be read as an indication of a deeper distinction between the emerging maritime and continental perspectives on Russia's position in the Far East, and they were explicitly noted as such by at least some

[84] Ignat′ev, *Otchetnaia zapiska*, p. 4; Mancall, "Major-General Ignatiev's Mission," p. 56 and *passim*; Ritchie, "Asiatic Department," pp. 309–310; Hudson, "Far East," p. 704.

[85] Barsukov, *Graf . . . Amurskii*, I, pp. 144–145; Zavalishin, "Dokumenty," p. 412; Zavalishin, *Zapiski*, II, p. 389; Veniukov, "Vospominaniia," p. 82; Veniukov, *Iz vospominanie*, I, pp. 237, 243.

[86] Mancall, "Major-General Ignatiev's Mission," p. 61.

[87] Sladkovskii, *History*, pp. 72, 83–85; Hudson, "Far East," p. 703; Cameron, *China*, p. 177.

commentators at the time. In his eulogy to the signing of the Treaty of Peking, Romanov for example made this point with considerable emphasis. With surprising ease, he dismissed the importance of both the commercial treaty Russia had concluded with Japan as well as the Treaty of Tientsin. The Treaty of Peking, however, served to broaden significantly the scope of Russia's overland commercial activity in China and thereby "corresponds more [than the other treaties] to the views and interests of Russia's merchantry." Romanov offered an interesting justification for this conclusion, suggesting that Russian merchants felt it somehow more natural to trade over land than over water. "[W]e notice everywhere the general fact that Russian merchants engage skillfully and actively only in domestic trade or in foreign overland trade." A repulsion, he maintained, was felt toward maritime commerce, and it would be impossible to foster artificially a more open view in this regard. For this reason, Russia "has no choice but to facilitate that route which Russian commerce has chosen voluntarily for itself, and to broaden the circle of its activities." The Treaty of Peking, he observed, "entirely satisfies these needs."[88] To fully exploit Russia's new and favorable commercial position in China, he suggested, the transport of goods across Siberia itself would now have to be expedited, first and foremost through the introduction of steamship routes on the rivers of Siberia.[89]

The intrinsic importance of northern China to Russia's newly acquired territories along the Amur, and the implications of the continental–maritime contrast for Russia's position in the Far East and Asia in general, were explored rather more thoughtfully by the same Vasil′ev whose geographical depictions of Manchuria we have noted above. The acquisition of the navigational rights on the Amur, he wrote in 1859, has opened up a region equal in size and position to France exclusively for Russian commerce. "This is central and northern Manchuria, watered by the Sungari, which is properly considered to be the main tributary of the Amur."[90] With the development of activity on the Amur, the population of these territories would inevitably be drawn into Russia's sphere of influence, and the region would present great potential for Russian commerce, which could operate here without foreign competition. Moreover, Russia possessed an important advantage over America and the other European powers in their common drive to expand their commercial activities in China. This was the fact that Russia was geographically contiguous with these territories, and thus was the only power which had access there overland.

This circumstance was especially significant, Vasil′ev suggested, when

[88] Romanov, *Poslednie sobytiia*, p. 31. This observation is particularly interesting if compared with Korsh's assertion that it was precisely the Chinese fear of the ocean and their inability to navigate on it which was irrefutable evidence of their backwardness and inability to advance independently their civilization; see above, pp. 183–184. [89] Romanov, *Poslednie sobytiia*, p. 32.
[90] Vasil′ev, "Otkrytie," p. 22.

viewed in the light of the major technological developments of the day, which allowed science to "conquer fantasy" and which brought the modern world to the threshold of a momentous transformation. Transport and communication overland, he pronounced, were about to triumph over movement across the oceans. The particular "fantasy" which made this development possible, of course, was the proliferation of the railway. Although at this early point railroad construction in Russia was still in its infancy, the far-reaching implications of its future development seemed quite apparent to him nonetheless. "The more Russia will develop," he stated confidently, "the greater the significance that overland communications will acquire."[91] And while he did not entirely dismiss the need for Russia to develop its maritime connections with China's Pacific ports, he saw no practical role of any sort for the Amur in this endeavor, insofar as commercial navigation on the river would render the import of Chinese goods into Russia neither more convenient nor cheaper than existing overland routes. To the extent that maritime links between the two countries were necessary, they should continue to be maintained by Russian ships sailing around the world from ports on the Baltic and Black seas, as had been the case before the annexation of the Amur. Indeed, if anything would enhance Russia's maritime intercourse with China, it was not the acquisition of the Amur but rather the contruction of the Suez Canal, which was begun in the year that he was writing and which promised to shorten substantially the trip to the Pacific from Russia's European ports.[92]

Thus Vasil′ev understood the significance of the Amur region not in terms of the access to the ocean that it provided, but rather as a threshold and gateway into the territories of Manchuria. From this standpoint, in turn, he saw the advance in the Far East as but one element of the broader wave of Russia's penetration across continental Asia, a process that also included the absorption of the Kazakh steppe and the territories of what were later to be called Russian Central Asia. He was one of the first to make an explicit association between all of these far-flung arenas of Russian expansion in the second half of the nineteenth century. The imperial colossus that was taking shape out of this expansion had affinities with European empires, to be sure, but at the same time he believed that the essential geographical contiguity of the Russian empire served both to set it apart and give it a decisive advantage. Pondering the prospect of the spread of the railroad network, a vision of a vast Russian-dominated network of inner-continental commerce began to take shape in his imagination. "From the depths of Manchuria, from the banks of the Amur, from the foothills of the Altai, merchants of all regions, with every sort of diverse product, will bring their wares here." The time was not far off "when railroads will extend from Russia to the border of China, and eventually they will penetrate into the interior of that country. When this

[91] Vasil′ev, "Otkrytie," pp. 17–18. [92] Vasil′ev, "Otkrytie," pp. 18–19, 26.

happens, the relations between the West and the East will be completely transformed."[93] Collins had also looked forward to the day when "the whole of Tartary, Bucharia, Thibet, and Northern China" would combine with the Amur region to form one immense zone dominated by Russia.[94] The critical difference between them was that the American had envisioned the Amur as a vital axis for this trade, while Vasil′ev most emphatically did not. Indeed, the latter even argued that it was far more urgent to construct a rail line southeast across Kazakhstan into Mongolia than across Siberia to the Russian Far East.[95]

With Vasil′ev, I would suggest that we have moved very far indeed from the position of the *konstantinovtsy*, to a point where there is not simply a contrast but a positive incompatibility between the maritime and continental perspectives on Russian development in the Far East. The most significant aspect of Vasil′ev's argument was his emphatic turn away from the ocean and his minimizing of its significance for Russia. In its place, he offered a vision of Russia as a land-based power, the attention of which would be focused upon the continental interior of Asia. There was in his essay an unmistakable echo of the point made by Mikhail Pogodin some five years earlier, when the historian fervidly demanded that Russia should "lay new roads in Asia or search out old ones."[96] Vasil′ev may have lacked Pogodin's stylistic flourish, but his proclamations were sober and well calculated nonetheless. They were based on the revolution in transportation heralded by the railroad, which in his estimation was to redefine Russia's presence in and interaction with the inner Asian realm adjacent to it. The Amur, to repeat, was important only by virtue of the overland access it afforded into this interior. Indeed, it might be noted that with his insistence on the primacy of the railroad, much of even this significance of the river as a gateway into Manchuria would be eroded.

[93] Vasil′ev, "Otkrytie," pp. 26, 18.
[94] Collins, *Siberian Journey*, p. 322.
[95] Vasil′ev, "Otkrytie," pp. 29–30.
[96] See above, p. 68.

8

The Amur and its discontents

"A malignant ulcer"

At the end of the 1850s, amidst the excitement and optimism about the annexation of the Amur which we have been considering, a very different tone began to be heard in the discussion of the region and river. This was a tone not only of reservation, but even of incredulity in regard to the the dizzying vistas which were being depicted for the reading public of the country. It was not so much the acquisition as such of the far-eastern territories as such which was called into question – although this was something of an issue as well, as we will see – but rather the way in which the entire process was being handled, and in particular the careless and simple-minded exuberance with which the affair was being presented to the rest of the country. An officer who had taken part in establishing the first settlements on the Amur complained bitterly in a letter from Moscow back to Eastern Siberia about the "lies and boasting" and the "fallacious information" which newspapers and magazines in the nation's capital were disseminating about affairs in the Far East.[1] Other critics expressed their consternation more publicly. In one of his earliest essays, the Siberian historian and ethnographer Nikolai Iadrintsev – a committed regionalist and sensitive in any event to the way Russians west of the Urals viewed his homeland – spoke mockingly about the promoters or "panegyrists" of the Amur. For them, he ridiculed, "there is no problem in constructing a telegraph to San Francisco, building a railway across all of Siberia, filling Russia's ports on the Pacific with ships from the entire world, and finally, conquering Asia." It was this final pursuit that they loved above all else. "They foresee the clash of Siberians and the English in India, and they fix their gaze so intently on China that one can imagine they dream about becoming khans themselves." "We created a thousand broad plans for our trade, wealth, and power in Asia," Iadrintsev observed elsewhere, "and our patriotic conceit knew no bounds."[2]

[1] Imberkh, "Pis′mo," p. 218.
[2] Iadrintsev, "Sibir′ pered sudom," p. 24; *idem*, *Sibir′ kak koloniia*, p. 332.

In a lengthy essay in the widely read *Otechestvennye Zapiski* (Notes of the Fatherland), another commentator offered an extended parody on the buoyant and confident mood of expectation which the advocates of expansion had tried to generate:

Peacefully, without a struggle or a drop of blood, the Russian state has increased its territory in the east by an entire country – and what a country it is! In size alone it exceeds any of the most important European states. And what is not to be found in this blessed, luxuriant country! It is watered by a river that is another Mississippi in its size and depth. This river, the most beneficent of all rivers in Asia, is for its entire course . . . convenient for navigation on the largest scale.

Soon, very soon, in a year or at the most two, steamers will scurry up and down it, parting its virgin waters. On the Amur we will establish extensive commercial relations with the half-wild Manchurians, with the Chinese and Japanese. The Amur will bring us close to America, the Americans will come to us, we will meet them here face to face . . . and we will even open the Amur to them for colonization . . . On the Amur we will succeed in a short while in matching the active Yankees in every respect. We only need an example and a model, which we will be able to adopt very quickly.

The most wonderful climate and the most fertile soil in the world will make the Amur region a rich, inexhaustible breadbasket for the entire eastern region . . . We will cultivate grapes on the Amur – for what cannot be cultivated under this blessed sky! – and soon wine from the banks of the Amur will overshadow the glory of the famous Rhine wines. We will start up the production of silk, and nothing will stop us from turning the entire region into the most plentiful industrial market in a short time.

Although our commentator used this hyperbole deliberately, in order to satirize the contemporary positive appraisals of the Amur region, he was not in fact exaggerating their tone a great deal, and we can easily recognize all of his points – with the possible exception of silkworms – from our earlier discussion. "From complete ignorance about the Amur region," he concluded, "we quickly adopted the most complete and unlimited enthusiasm. In our fervent (pylkaia) fantasy, unrestrained by any positive information, miracles resounded."[3] The lack of accurate and reliable information indeed became the most general criticism of the excitement surrounding the Amur and one that was repeated again and again. How can we imagine that we have found an El Dorado here, it was asked, as long as next to nothing is known in European

[3] I. A., Review of R. Maak, *Puteshestvie na Amur*, pp. 10–11. This article was a review of Richard Maak's chronicle of his historic first voyage down the Amur river four years earlier. Maak's work, ostensibly presented as a scientific–geographical narrative, was at the same time a sort of grand advertisement for Russia's recent advance in the Far East. Its positive message was communicated not only by the text, which sang the praises of Russia's new lands on the Pacific on practically every page, but by its physical appearance as well. The oversized, lavishly embossed, and most handsomely leatherbound volume – complete with a full-page portrait of the hero Murav'ev resplendent in his general's attire – was obviously designed to make a statement about the innate importance and weightiness of its subject. Other reviewers of the work, including Nikolai Dobroliubov, took a similarly sceptical tone: "The euphoria (vostorgi) excited by the Amur, is premature and exaggerated." [Dobroliubov], "Puteshestvie," pp. 405 (quote), 402–403, 418.

Russia about this region? Shouldn't we rather be careful, not accept at face value the wonderous accounts that are offered, and demand facts and figures instead of giving free reign to our "fervent fantasy?"[4] Out of these concerns, and the suspicions they stirred that something was amiss in the events unfolding in Russia's new far-eastern territories, a perspective critical of the Amur euphoria began to develop. Along with it, a descriptive scenario and interpretation of events took shape which diverged dramatically from the proclamations of Murav′ev and his spokesmen.

The originator, inspiration, and chief source of information for this tendency was a remarkable individual, Dmitrii Irinarkhovich Zavalishin (1804–1892), whose youthful views on Russia's destiny in North America have already been noted.[5] Zavalishin had contacts with the circles which rose in mutiny against the accession of Nicholas I in 1825, and although he was not present in St. Petersburg at the moment of the actual December uprising, he was nonetheless arrested and exiled along with dozens of others to the remote reaches of Eastern Siberia. There he spent the following four decades.[6] Zavalishin was a colorful individual, possessed of enormous energy and enormous egocentricity in equal measure, who missed no opportunity to exaggerate – usually extravagantly – his own role in the historical events to which he had been witness.[7] The latter included most prominently the annexation of the Amur. He first met Murav′ev in the town of Chita east of Lake Baikal in the late 1840s, as the governor-general was assuming his post. As with other exiles, relations between the two were at first quite friendly, and Murav′ev sought out his advice on various issues,[8] but they soured quickly and Zavalishin was to become the governor-general's nemesis and single most passionate critic. Indeed, the antipathy between the two was obsessive enough for Bakunin to comment that if one were to "take away [Zavalishin's] hatred of Murav′ev, he [the Decembrist] would die tomorrow." "He intentionally lies, slanders, distorts, and dreams up his 'facts.' "[9] Yet while there was no denying an element

[4] I. A., Review of R. Maak, *Puteshestvie na Amur*, p. 11; Te-v, "Istoriia," pp. 73, 81, 85. Skepticism about how "fantasy has been given free rein" in regard to the Amur was expressed in the foreign press as well. Gerstfeldt, "Über die Zukunft," pp. 93 (quote), 106. [5] See above, p. 27.

[6] Matveev, "Dekabrist Zavalishin," p. 15; Pasetskii, *Geograficheskie issledovaniia*, pp. 24, 74, 79.

[7] E.g. Zavalishin, *Zapiski*, II, pp. 276, 289–290, 309, 311, 313. Collins met Zavalishin when the latter was 52 and described him as "hale, hearty, sprightly, and active" – qualities apparently still intact some 15 years later when Zavalishin married and fathered six children. Collins, *Siberian Journey*, p. 161; Shatrova, *Dekabrist D. I. Zavalishin*, p. 137; Matveev, "Dekabrist Zavalishin," p. 16. Zavalishin, incidentally, did not return Collins' high estimation, reckoning that behind the American's plans for a trans-Pacific telegraph lurked "ulterior political motives." Svatikov, *Rossiia*, p. 31.

[8] According to Zavalishin, at the conclusion of their first meeting Murav′ev said to him: "Dmitrii Irinarkhovich, I recognize that I am fully ignorant [of conditions in this region], and I give myself over entirely to your direction." Zavalishin, *Zapiski*, II, p. 313.

[9] [Bakunin], *Pis′ma*, pp. 120, 143 (quotes), 144–148. For more positive appraisals by others who knew Zavalishin in Eastern Siberia, see Kropotkin, *Dnevnik*, p. 61; Butsinskii, *Graf . . . Amurskii*, pp. 60–63.

of personal vendetta on both sides, there was nevertheless much more to Zavalishin's campaign against Murav′ev and the "Amur enthusiasts" than Bakunin was willing to recognize.[10] The Decembrist's criticism of Murav′ev and his successor M.S. Korsakov did so much damage that he was officially deported from Eastern Siberia to Kazan′ in 1863, acquiring thereby the distinction of being the first and apparently only Siberian exile ever to be sent back involuntarily to European Russia.

In 1858 and 1859, taking advantage of an opportunity created by Murav′ev's absence on a trip to Japan, Zavalishin published a series of sensational exposés of affairs in the Far East. Most of these appeared in *Morskoi Sbornik* (Naval Journal), the journal of Konstantin Nikolaevich's naval ministry,[11] to which Zavalishin had special access through its editor Ivan Zelenyi, an old friend of his.[12] These exposés became the major source in European Russia for information on this region which differed from more official accounts. In them, Zavalishin disputed virtually all the claims which were made about developments in the Amur region, including the development of commerce, the progress of settlement, and even the navigability of the river itself. The picture that he offered was a bleak one indeed, and is well conveyed in an observation from 1859. "The Amur at present brings no benefit to Russia, but represents rather a malignant ulcer, . . . which only exhausts the energies of the region."[13] In addition to Zavalishin, two other individuals might be mentioned whose writings served in a similar way to cast a critical light on affairs in the Far East. These were Peter Kropotkin, who served in a variety of capacities in a local cossack regiment, and Sergei Maksimov, a young journalist sent out to the Amur from the Naval Ministry in St. Petersburg. Their arguments found a receptive audience, and received ever-growing support, both in Eastern Siberia as well as European Russia.[14]

The defenders of the situation in the Far East were almost all spokesmen for Murav′ev in one way or another. The most notable among them was his relative Bakunin, whose exhortations in favor of the governor-general and his cause we have already examined, but the most prolific by far was Dmitrii Ivanovich Romanov, a military engineer in Murav′ev's service. Romanov had apparently been designated at an early point as a sort of combined propagan-

[10] Svatikov, *Rossiia*, p. 27; Shatrova, *Dekabristy i Sibir′*, p. 161; Shatrova, *Dekabrist D. I. Zavalishin*, p. 135. [11] Zavalishin, *Zapiski*, II, p. 397. [12] Svatikov, *Rossiia*, p. 36n.

[13] Zavalishin, "Zamechaniia," p. 124. Also see *idem*, "Po povodu statei"; *idem*, "Amur"; *idem*, "Otvet g. Nazimovu"; *idem*, "Otvet g. Romanovu."

[14] Butsinskii, *Graf . . . Amurskii*, pp. 60–63; Azadovskii, "Putevye pis′ma," pp. 212–213; Shatrova, *Dekabristy i Sibir′*, pp. 160–161. While denouncing the Decembrist's "lies, pernicious gossip, distortions, and inventions of facts," Bakunin nonetheless conceded that they had a strong effect.

> In Moscow and St. Petersburg, people insist most sincerely that the Amur is a bluff (puf), . . . that the Amur has destroyed Russia, that millions of rubles and thousands of lives have perished there – that the Amur, in a word, has become an ulcer for the country.

[Bakunin], *Pis′ma*, p. 120.

dist and what today might be called a "spin doctor" for the events unfolding in the Russian Far East, and in this capacity his exertions were prodigious. In addition to a book-length essay on the historical background of the Russian occupation of the Amur valley, he produced over a dozen articles appearing in Eastern Siberia and St. Petersburg, in which he energetically, and often enough rather desperately sought to rebut the specific criticisms of Zavalishin and others.[15] He attempted thereby to counteract the growing image of the Amur as Russia's malignant ulcer, an effort in which he was not ultimately destined to succeed. Others in Murav′ev's retinue to contribute to his campaign against the detractors of his cause in the Far East were A. Sgibnev, N. Nazimov, F. Gubanov, and A. Gvozdev.[16] We have already considered some of this literature in our examination of the various images associated with the Amur, but the critique raised by Zavalishin and others served to cast it all in a rather different light.

The confrontation of these two opposing groups or camps, both anxious to propagandize their version of the truth concerning the *amurskoe delo* in the Far East, resulted in a flurry of articles and at least one book, by Maksimov.[17] As might be expected, the contending scenarios that were offered were the source of not inconsiderable confusion for those attempting to follow events in the Far East.[18] The critical attention that increasingly came to be focused upon the handling of affairs in Russia's newly acquired territories prepared the way for what was to be the highest irony of the Amur epoch, namely the fact that the Russian presence in the Far East came to be condemned in the name of the very principles that had inspired the annexation of the Amur in the first place.

"This unappealing desert inspires unbearable grief"

One category of problems with Russia's new acquisitions in the Far East related to what may be called "natural" conditions – that is to say, factors inherent in the physical environment. Almost nothing had been known about this environment prior to annexation, and this circumstance had made it conveniently possible for Russians to entertain images of the region as an El Dorado or a blessed land. The actual experience of occupying the region put these images to the test, and it became apparent almost immediately that at the very least they were unrealistically positive. The most immediate of these environmental problems centered on the physical properties of the Amur river

[15] Romanov, "Po povodu statei"; *idem*, "Prisoedinenie"; *idem*, "Na stat′i g. Zavalishina"; *idem*, "Amur – zlokachestvennaia iazva Rossii"; *idem*, "S ust′ia Amura"; *idem*, "Amur"; *idem*, "Poslanie"; *idem*, "Izvestiia."

[16] E.g. Sgibnev, "Vidy"; Nazimov, "Otvet"; Gubanov, "Pis′mo"; Gvozdev, "Otvet."

[17] Maksimov, *Na Vostoke*.

[18] Semenov, "Amur," pp. 188–189; Shatrova, *Dekabrist D. I. Zavalishin*, pp. 123–124.

itself. Russian navigation on the Amur had begun triumphally with Murav′ev's military descent in 1854, but by the end of the decade, despite the settlement of the river banks and considerable efforts on the part of the government and independent merchants, regular communication along the entire course of the river had still not been established.

In assessing this situation, critics were quick to point to difficulties presented by the river. One major problem was its excessive shallowness, especially in its upper course. This was a concern about which Russians in the area had been sensitive even as Collins was proclaiming the brightest prospects for the development of steamship traffic on the river, and his light-minded assurances that the largest oceanic steamers would be able to ply the entire extent of the river's waters without any hindrance were discretely qualified in the Russian press as they were reported.[19] As it turned out, the use of deep-sitting vessels on the Amur did in fact prove to be highly impractical, if not impossible; indeed, according to some observers the river's shallowness precluded passage of any boats other than flat-bottomed barges or rafts. In all events, it was commonly agreed that two different types of ship at least would be necessary for navigation along the entire course of the river.[20] Nor was it only in its upper reaches that the lack of sufficient depth created problems for navigation. The Decembrist Mikhail Bestuzhev characterized a trip down the Amur in the summer of 1857 as a "difficult, delayed, agonizing river voyage or, more correctly, a dry-land descent along the bed of the Amur,"[21] and Peter Kropotkin was sufficiently impressed by the difficulties for navigation in the lower Amur to observe that "the lack of water [in the river channel] would almost make it more advangateous to supply the Ussuri valley and our ports south of the Amur from around the world [by ocean], rather than down the Amur" itself.[22] All in all, Zavalishin concluded ironically, movement up the Amur was so very difficult that one could reach Irkutsk from Nikolaevsk just as quickly by sailing up the Okhotsk coast to Ayan and proceeding on from there overland![23] And if the simple physical inconvenience of the trip were not enough, steamship operators added insult to injury by charging extortionist rates for tickets.[24]

The attempt to make the Amur river the principal axis of transport in the Russian Far East was further complicated by climatic conditions. The most obvious problem, which was at the same time the most obvious difference with the Mississippi, was the fact that from some time in October to some time in

[19] "Donesenie Kollinsa," p. 7.

[20] Kropotkin, "Iz Irkutska" (no. 14: April 1864), p. 12; Afonas′ev, "Nikolaevsk," p. 113; Zavalishin, *Zapiski*, II, pp. 354–355.

[21] Cited in Azadovskii, "Putevye pis′ma," pp. 230 (quote), 231–232.

[22] Kropotkin, *Dnevnik*, p. 92. [23] Zavalishin, "Amur," pp. 44–45.

[24] In 1860, passage from Nikolaevsk to Khabarovsk cost 75 rubles, to Blagoveshchensk 125 rubles, and all the way to the confluence of the Argun and Shilka rivers – 200! Maksimov, *Na Vostoke*, p. 267n.

April – in other words, for over half the year – the entire length of the river became in the words of one observer "one vast sheet of ice."[25] The Amur was not of course the only Russian river to freeze over during the winter, but due to the dense forests crowding its banks and the fact that its ice cover was highly uneven, it was more difficult than elsewhere to replace water traffic with communication either by horse overland or by sleigh across the river itself.[26] Nor did the summer months necessarily bring relief, for the entire lower course of the river, from Khabarovsk to the ocean, was subject to the great cyclonic storms characteristic of the region's summer monsoon climate, which could be of such an intensity as simply to destroy all wooden vessels in their path.[27] The combination of these various obstacles to normal navigation on the Amur led to the conclusion that "all of these rich locations [along the river] lose a significant part of their value, for they are cut off for a significant part of the year" from other populated areas.[28]

The Amur "enthusiasts," of course, denied the negative conclusions which those who enumerated these complications drew from them. In their writings, however, one finds a substantial degree of concurrence that problems did indeed exist, and consequently that the vision of the Amur as Russia's Mississippi of the East was in need of at least some qualification. Maak, for example, spoke about the problem presented by the shallowness of the Amur estuary and its numerous sandbars and reefs, although he still chose to affirm "resolutely" that the river was navigable over its entire length.[29] And even Dmitrii Romanov, who waxed ecstatically about the unbroken line of communication from the Baltic to the Pacific that succesful navigation upon the Amur ostensibly brought into being – an accomplishment that "no other government in the world" could claim – was hard-pressed to deny many of Zavalishin's assertions, such as the need for two or even three different types of boats to navigate its entire course. Indeed, in regard even to the damning claim that the trip from Nikolaevsk to Irkutsk was quicker via Ayan than up the Amur, Romanov could say nothing more than it was "to a certain extent unjustified."[30] In fact, he inadvertently did the critics one better, by sketching a variety of major projects on the Amur that he reckoned would be necessary to improve navigation, including the construction of locks, dams, and reservoirs. Ultimately, Romanov sought to mitigate the problem of the river's shallowness not by denying it, but by arguing that these sorts of complications were encountered on rivers in European Russia such as the Volga or Neva,

25 Atkinson, *Travels*, p. 495 (quote); Zavalishin, "Amur," pp. 48–49; I. A., Review of Maak, *Puteshestvie na Amur*, p. 15; Malozemoff, *Russian Far Eastern Policy*, p. 2.
26 Kropotkin, "Iz Vostochnoi Sibiri" (no. 43: December 1863), p. 8.
27 Kropotkin, "Iz Irkutska" (no. 14: April 1864), pp. 12–13.
28 Kropotkin, "Iz Vostochnoi Sibiri" (no. 43: December 1863), p. 8.
29 Maak, *Puteshestvie na Amur*, p. 255.
30 D. R., "Pis'ma," pp. 368 (quote), 15, 28; Romanov, "Po povodu statei," pp. 1556–1557.

where they in no way prevented steamship companies from operating upon them "with a great profit for the pocket."[31]

Another problem with the Amur river was the question of the access to the sea offered by the port of Nikolaevsk at its mouth. The strongly negative attitudes on the part of the Naval Ministry toward this facility which we have already examined had been articulated in confidential government documents, but similar evaluations eventally began to appear more publically. The loudly touted notion that the miniscule settlement of Nikolaevsk-na-Amure would quickly grow to rival San Francisco as a Pacific entrepôt was quickly put to rest, and within a decade of its founding it had instead earned a reputation as an unfit port situated in a dreadful locale, representing in general something not much less than a hell on earth. "I don't know where this antipathy to Nikolaevsk comes from," wondered Kropotkin:

> I only know that opinions about this "city of all possible scandals and abominations" – these are the actual words – are to be heard constantly, not only on the Amur itself but from practically everyone returning [to Eastern Siberia] from the Amur. If you ask about the reason for this hostility, you are answered with dozens of tales about the procrastination of the administration, about the unjust treatment of [Russian] merchants, about . . . the dreadful inflation, about boredom, scandals, and so on.[32]

Sentiments about this "cursed city" were voiced among others by Bestuzhev, who it will be remembered was an enthusiastic advocate of the acquisition the Amur itself. His eager anticipation wilted, however, as soon as he actually ventured to the region itself, and to judge from a rather poignant passage in a letter back to his sisters in European Russia in 1857, the conditions he encountered at Nikolaevsk played a not inconsiderable role in his disillusionment.

> I am writing this letter to you under the influence of the most sorrowful feelings. In don't want to pretend, as I have in earlier letters, and paint the laughing colors of the rainbow with my pen, in order not to cause you anguish. You will understand my situation from the following seven words: I have begun to winter in Nikolaevsk![33]

Foreigners as well could be disdainful, as for example Ravenstein, who remarked contemptuously that its main tavern could "scarcely compare to a low German beer-house." And although the German merchant Otto Esche was altogether more positive, noting that thanks to the presence of numerous Germans and German speakers from the Baltic provinces "something of a German spirit dominates" in Nikolaevsk, this circumstance could hardly have helped endear the town to patriotic Russians in the Far East.[34]

As a port, Nikolaevsk did indeed possess a variety of negative physical fea-

[31] D. R., "Pis′ma," pp. 16–17 (quote), 21–23.

[32] Kropotkin, "Iz Vostochnoi Sibiri" (no. 45: December 1863), p. 7 (quote); Kropotkin, *Dnevnik*, p. 261. For a description of the intense provincialism of life in Nikolaevsk, see Maksimov, *Na Vostoke*, pp. 318ff.

[33] Quoted in Azadovskii, "Putevye pisma," p. 230.

[34] Ravenstein, *Russians*, p. 198; "Ekspeditsiia g. Ottona Eshe," p. 40.

tures. Like the river itself, the mouth of the Amur was shallow, too shallow to allow for free and easy access by ocean-going vessels.[35] It was plagued by countless sandbars, which made navigation not only difficult but positively dangerous, as Ivan Goncharov indicated in his description of the frequent groundings that his ship experienced as it attempted to pass through the Amur estuary in 1854.[36] It appeared, moreover, that the estuary of the Amur remained ice-bound even longer than the river itself, and was inaccessible for some seven months of the year. This circumstance alone led a foreign observer to the conclusion that it could "never become a first-rate commercial port," and many Russians eventually came to agree.[37] Due to all these negative factors it was impossible to obtain insurance for boats sailing in the mouth of the Amur, which served as yet another disincentive to Nikolaevsk's development.[38] Compounding the problems for navigation posed by the mouth of the Amur were the difficult physical–geographical conditions of the surrounding region. "The nature that surrounds Nikolaevsk," offered an on-the-spot observer, "is extremely wild, gloomy, and lifeless, and the prospect of a grey wooden city in the midst of this unappealing desert inspires unbearable grief."[39] Particularly problematic, once again, was the climate. As was the case with the entire Okhotsk sea-coast, the maritime location did not ameliorate the arctic–continental winter regime, and with ferocious winter blizzards the climate at Nikolaevsk became the most notorious of all the settlements on the Amur. "Not much good can be accomplished," our commentator continued,

> where the mercury freezes in the thermometer and doesn't thaw sometimes for a month, where the snow lies deeper than two meters, where 10 degrees below zero Celsius is considered a thaw, and where a blizzard may rage for days on end, so fiercely that . . . if some necessity forces you outdoors you risk freezing on the street, especially at night.

And if the horrors of winter were not enough, at other times of the year the climate was "conducive to scurvy."[40]

In the late 1850s, as we have already seen, all of these assorted problems had stimulated an active search for a more satisfactory location for a Russian commercial and military port. The decision to relocate Russia's main Pacific bastion to Vladivostok did not entirely solve the problem, however, for the Amur could not simply be abandoned, and indeed steamship navigation on it remained in principle highly desirable. Thus, the idea was put forward that De Castries Bay, located some 100 miles south of the Amur estuary, could

[35] Griffin, *Clippers*, p. 343. [36] Goncharov, *Fregat "Pallada"* (1986), pp. 487–490.

[37] Atkinson, *Travels*, p. 495 (quote); Noskov, "Amurskii krai," no. 5, p. 33; Afonas′ev, "Nikolaevsk," p. 98.

[38] Afonas′ev, "Nikolaevsk," pp. 104, 108–109; Burachek, "Gde dolzhen byt′," p. 13; [Liudorf], "Amurskii krai," pp. 208–210. [39] Afonas′ev, "Nikolaevsk," p. 93.

[40] Afonas′ev, "Nikolaevsk," pp. 95–96. On the extraordinary blizzards in Nikolaevsk, also see Kropotkin, *Dnevnik*, p. 155.

effectively replace Nikolaevsk as the coastal terminus for the river. For many, this seemed the prefect solution: it was sufficiently deep, reportedly closed by ice for a shorter duration, and – because it was situated below the difficult De Castries (today Nevel′skoi) straits, was more easily accessible to ocean traffic approaching from the south.[41] Nature had endowed this bay with "remarkable advantages," creating it "as if intentionally to serve as the receiving point for goods entering and leaving the Amur."[42] All that was necessary would be to connect the bay with the Amur river, which could be done by constructing a short railway line due west. Such a line would have the advantage not only of making it possible to avoid the mouth of the Amur entirely, but would also considerably shorten the trip down the Amur to the ocean. This project was much discussed, and Murav′ev, citing water levels in the Amur estuary so shallow that they would not even suffice for a rowboat, was enthusiastically promoting it in St. Petersburg as early as 1858.[43]

The construction of this short railway line may have seemed like an easy panacea to the advocates of development of the Amur, but it in fact betrayed that something was very wrong in the picture they had been depicting of the Russian Far East. By admitting the desirability and indeed even necessity of a supplementary railway, even one as modest as the route in question, they were in effect admitting the imperfection of the Amur route to the ocean. The excellence of this river route, however, and the unconditional suitability of the Amur mouth for ocean-going vessels had been one of the major premises behind its annexation in the first place. The real natural conditions in the region simply did not correspond to this envisioned excellence and suitability, and Murav′ev, in trying to ameliorate the situation, was forced to begin to undermine the very arguments he had made at the beginning of the decade. Moreover, the gap between fancy and reality in regard to the Amur was so great, the unfitness of the river for navigation under the conditions of the time so incontrovertible, that the notion of a railroad, intended originally as a small-scale supplement, was in the decades that followed to overshadow and eventually replace the vision of the Amur river as Russia's vital far-eastern link to the Pacific. With admirable prescience, this development was foreseen by Zavalishin in 1859 in his reactions to the proposals we have just discussed:

> But wait just a minute. If things . . . have come to the point of building railway lines, then would it not be better to forget about the Amur entirely and make this railway our main concern, instead of having it fill a supplementary role? . . . Would it not be more natural to search out a route for it . . . which would lead, for example, from an ice-free

[41] Griffin, *Clippers*, pp. 336, 336n.

[42] Afonas′ev, "Nikolaevsk," pp. 103–104 (quote); Burachek, "Gde dolzhen byt′" p. 8; D. R., "Pis′ma," pp. 24–27; Maak, *Puteshestvie na Amur*, pp. 215–216.

[43] Barsukov, *Graf . . . Amurskii*, II, pp. 173–174; Romanov, "Proekt," *passim*; D. R., "Pis′ma," No. 16, pp. 367–370, No. 17, pp. 24–25; Romanov, "S ust′ia Amura," p. 109; "Russkie na Amure," p. 271; *Istoriia reki Amura*, p. 20.

harbor or major trading point to the nearest point in Russia, rather than to some sort of place on the lower Amur?[44]

In fact, this was precisely what was to happen when railroad construction in the Far East was finally undertaken by the Russians in the 1890s, with the result that the line avoided the Amur region entirely and crossed Manchuria instead, to meet the Pacific at Vladivostok.

"Everywhere you find filth, hunger, and poverty"

Another important source of disillusionment with the Amur related not to the natural features of the area but to the way in which the settlement of the region had been carried out and its economic life administered. These problems were not environmental but human, and logically pointed to the individual who wielded absolute authority in the area and ultimately bore full responsibility for the state of affairs there. Indeed, the fact that these complications originated from willful human action made them in a sense even more incriminating and inexcusable for public opinion than an inhospitable climate or too-shallow river depths, and it was the former which did more than anything to fade the brilliant colors originally associated with the Amur annexation.

Among the most pressing practical tasks with which the annexation of the Amur confronted Murav′ev was the need to establish some sort of permanent Russian settlement in these vast and desolate territories. This imperative was motivated by military–strategic considerations as well as those of economic development. Attempts to deal with it were frustrated, however, by the paucity of population across all of Eastern Siberia. The optimal solution would have been to rely on voluntary peasant migration from European Russia, and indeed the very first resettlement legislation in Russia, adopted in the month following the emancipation of the serfs in 1861, opened up to the Amur region to free colonization and even provided significant incentives, including exemptions from taxes and conscription. The results, however, were anything but encouraging. Despite some strong initial interest on the part of religious sects such as the Molokane, Dukhobory, or Mennonites, and even indeed from ordinary peasants in European Russia,[45] the enormous distances and lack of familiarity with local conditions worked decisively to limit an

[44] Zavalishin, "Amur," p. 50.

[45] There seems to be no doubt that the curiousity of at least some peasants west of the Urals was indeed piqued by tales of the Amur. Maksimov recounted the following interrogation he received from a peasant woman near Viatka in the early 1860s, after she learned he was on his way to the Far East: "Now just what is this ′Mur all about?" she queried, with interest and a little suspicion. "Folks have just now begun to talk about it, as if it only appeared on earth yesterday! . . . Are things really so good out there, on this ′Mur of yours?" However, other peasants cautioned him "not to be seduced" (ne soblazniai) by the glowing reports of developments on the "′Mur." Maksimov, *Na Vostoke*, pp. 112–114.

optimistically projected "stream of voluntary migrants" to something less than a trickle.[46] By 1861, only about 12,000 souls had settled in a vast region the size of France.[47] Of the roughly 600,000 peasants who migrated across the Urals from European Russia in the three decades following 1861, it is estimated that no more than 60,000 made it as far as Eastern Siberia. The number who actually pushed on to settle in Russia's far-eastern territories could not have amounted to more than a fraction of the latter figure.[48]

In the absence of free serfs for the purposes of colonizing the Amur, Murav′ev resorted to measures that were to earn him a good deal of infamy in the public opinion of the time. Beginning in 1857, he began to establish settlements in the Amur valley with colonists drawn from the Transbaikal region, simply by ordering them to resettle. The first group to be transferred in this manner was the Transbaikal infantry, a unit he had formed in the early 1850s out of the poorest agriculturalists.[49] It soon became clear that many more people would be needed, and Murav′ev accordingly resorted to an unprecedented measure. He ordered the freeing of large numbers of hard-labor convicts and formed them into another cossack unit, which he then ordered to settle on the Amur. To objections that proper settlement required not only men but families he made a very simple response. He gathered up the women from the public houses in Irkutsk and other towns, sent them on barges down the Amur to these men, bade them choose partners quickly, and had a priest marry them all in one common ceremony on the banks of the river![50]

A variety of problems were bound to plague efforts at colonization carried out in this manner. The fact that a significant proportion of the settlers were not agriculturalists at all but rather taken from the criminal population undermined the effectiveness of the entire enterprise, and indeed in a manner that

[46] "Neuste Nachrichten," pp. 474–475 (quote); Barsukov, *Graf . . . Amurskii*, I, pp. 560, 601, II, pp. 296–297; Putilov, "Po povodu rasskaza," p. 253.

[47] Of these, the number of voluntary peasant families could have been as little as 250. Stephan, *Russian Far East*, p. 64.

[48] Vevier, "Introduction," p. 35; Malozemoff, *Russian Far Eastern Policy*, pp. 9, 255n; Treadgold, *Great Siberian Migration*, pp. 69–71; Butsinskii, *Graf . . . Amurskii*, pp. 49–53. One indication of the rapid evaporation of whatever appeal the Amur region may have offered as an arena for agricultural resettlement can be gleaned from a promotional brochure published in St. Petersburg in 1865 and directed at those in Germany who were thinking of emigrating. While strongly advocating the advantages of a "new" (i.e. reformed) Russia over the United States, the author specifically recommended *against* Siberia in its entirety, "da die Verhältnisse der Landwirtschaft und Viehzucht dort zu abweichend von dem sind, was sie bisher betrieben haben, und beide Industrien da selbst am Ende auch zu wenig erfolgreich sind." The Far East was not even mentioned. Brandt, *Botschaft*, p. 35.

[49] [Veniukov], "Primechanie," p. 2; Ravenstein, *Russians*, p. 145; Romanov, "Prisoedinenie," p. 139; Malozemoff, *Russian Far Eastern Policy*, pp. 1ff; Forsyth, *History*, p. 215.

[50] Kropotkin reports that the women were convict laborers, like the men. "Iz Irkutska" (no. 19), p. 11; Kropotkin, *Memoirs*, p. 185; Anisimov, *Puteshestviia*, p. 17. Potanin ascribes these summary actions to orders from St. Petersburg, but there is little doubt that Murav′ev was responsible for them. "K kharakteristike," p. 605; also see [Veniukov], "Primechanie," p. 3.

was immediately apparent to observers in the region. It was the prim judgment of a British visitor that the colonists on the Amur were for the most part "extremely indolent" and "carry on their agricultural operations in the most primitive manner," but there were Russians who were prepared to agree with him. "The most characteristic traits of the majority," observed Peter Kropotkin, "is their inclination to drunkenness and sometimes to barbarism," and even a devoted supporter of Murav′ev's efforts remarked that with their very presence these military colonists were "a burden for the honest settlers."[51] All these problems were exacerbated by the manner in which the settlements along the Amur were located and set up. The prospective colonists were not allowed to choose for themselves the areas on which to build homes and start farming, but were instead told where to settle by the government. As may be expected, the government chose the locations hastily and carelessly, and with more concern for establishing a chain of posts directly on the river than with identifying optimal areas for habitation and agriculture. Thus it was that settlements were established, houses built, dense forest growth cleared and fields plowed, only to discover that the land was infertile, that the area flooded periodically each year, that it was subject to infestation by rodents or insects, or some other catastrophic complication. This led to the abandonment of many settlements just a few years after they were built for more favorable locations, where all the initial labor had to be entirely repeated.[52] The effects of all this on the development of agriculture, to say nothing of the overall morale of the early settlers, need not be spelled out.

There were scattered accounts in European Russia testifying to the "complete satisfaction with their existence" on the part of the Amur colonists,[53] and in a particularly impassioned passage Bakunin insisted to Herzen that their fate contrasted gloriously to the deprivations endured by "North-American colonists on the banks of the Mississippi and Arkansas rivers," for the colonists on the Amur were settled "in areas that are free and open, healthy, and amazingly fertile . . . The people have a wonderful life there."[54] By and large, however, the colonization of the Russian Far East increasingly came to be seen in a very critical light. The fact that resettlement had been effected for the most part by administrative fiat and not by voluntary migration was highly proble-

[51] Ravenstein, *Russians*, p. 150; Kropotkin, "Iz Irkutska" (no. 19: May 1864), p. 11 (quote); Kropotkin, "Iz Vostochnoi Sibiri" (no. 34: October 1864), p. 6; Veniukov, "Postupatel′noe dvizhenie," p. 79.

[52] "Amurskii krai," p. 2; Kropotkin, "Iz Vostochnoi Sibiri" (no. 44: December 1863), p. 14; *idem*, "Iz Irkutska" (no. 19: May 1864), p. 10; *idem*, *Dnevnik*, pp. 90–91; *idem*, *Memoirs*, p. 186; Veniukov, "Postupatel′noe dvizhenie," pp. 78–79; Maksimov, *Na Vostoke*, p. 290. For general accounts of the settlement process, see Zavalishin, "Amurskoe delo," pp. 89ff; Butsinskii, *Graf . . . Amurskii*, pp. 36–47.

[53] "Ekspeditsiia . . . Ottona Eshe," 1858, p. 41.

[54] [Bakunin], *Pis′ma*, p. 132. Also see To l . . . zin, "Pis′ma," pp. 377–380. For an example from the foreign press, see "Kolonisation im Amur-Land," p. 33.

matic, for it suggested that it lacked the all-important ingredient in Russia's era of national reform, namely civic freedom. This was an objection, it might be noted, that was voiced not only by those like Zavalishin or Petrashevskii who were in any event critical of affairs in the region,[55] but even by those who otherwise strongly supported the governor-general's endeavors.[56] Another source of notoriety were the disturbing and frequently quite harrowing accounts of the brutalities that accompanied the colonization process. Rumors circulated, for example, of an incident of cannibalism that was said to have occurred in the mid-1850s among a battalion of soldiers who, dutifully following absurd orders from above, were stranded by the ice on the lower Amur and forced to winter there without provisions.[57] Such accounts, however, remained unsubstantiated at the time, and more compelling overall were the verifiable reports delivered by eye-witnesses on the scene to the press in European Russia.

Among the most poignant of such accounts were the numerous brief articles which Peter Kropotkin sent back from 1863 to 1865 to *Sovremennaia Letopis'* (Contemporary Chronicle), a supplement to a major St. Petersburg newspaper. Kropotkin served on a flotilla which carried provisions down the Amur to the new settlements,[58] and thus was reliably informed about the situation in the region. His articles depicted a sorry scene indeed, in which his sense of decency and humanity was outraged by the carelessness and insensitivity of officials and the suffering of the settlers. The following episode will serve to illustrate this. Invited in by some cossack families at a settlement near Blagoveshchensk in 1863, he asked casually how they were getting along. His questions stirred memories of the first difficult period on the Amur, and a heavy-set elderly woman stood up and began to wail in a broken voice, her body trembling:

> Kto na Amure ne byval
> Tot goria ne znaval,
> Kto na Amure pobyvaet
> Tot gore vsiakoe uznaet.
>
> Whoever hasn't been to the Amur
> Has not yet seen grief,

[55] Zavalishin, "Amur," pp. 68–70; *idem*, "Amurskoe delo," p. 79; [Petrashevskii], "Mestnoe obozrenie," pp. 33–35; Dulov, "Sibirskaia publitsistika," p. 12; I. A., Review of Maak *Puteshestvie na Amur*, pp. 12–13.

[56] Zenzinov, "Iz Nerchinska," p. 357; A. K., "Amurskii krai," p. 142. "Ekspeditsii russkikh," p. 26; Azadovskii, "Putevye pis'ma," p. 212.

[57] [Gertsen], "Neschastnye soldaty," pp. 527–528; [Veniukov], "Primechanie," p. 2. According to Kropotkin, when Murav'ev's second-in-command Korsakov, who had issued the orders, was asked about trying to get some provisions to these unfortunates, he replied: "And where the hell am I supposed to get it from? Let them eat rocks!" Kropotkin, *Dnevnik*, pp. 75 (quote), 83.

[58] Kropotkin, *Memoirs*, p. 187; Karpov, *Issledovatel'*, pp. 16–17.

> Whoever comes to the Amur
> Will see grief of all kinds.
> . . . .
> And the cossacks dug
> They dug pits,
> They starved there
> They cursed the Amur.[59]

Their present state was overshadowed by desperation as well. "[I]n all of these villages," our correspondent concluded, "the peasants have become convinced that they will never achieve anything here, and that their labor will be wasted in vain."[60] This was certainly a far cry from the heaven on earth and El Dorado which many had considered the Amur to be. In his diary Kropotkin shed the reserve which he exercised in his articles for publication, and gave full vent to his anger at the abomination he was witnessing. For the Chinese and the Manchurians, he wrote, the Amur is suitable, for their life cannot be changed and they are going to die out anyway. "But why should these poor [Russian] peasants suffer a miserable life and die out? For what reason, after all? For the sake of the settlement of the Amur. And what the devil did they need the Amur for?"[61] In a series of public lectures on the Amur region delivered in 1860 in St. Petersburg, Gustav Radde echoed Kropotkin's criticism of the manner in which colonization had been carried out. "Der Amur kränkelt in seiner Jugend," he concluded: the Amur was Russia's "sickly child."[62]

It was a common perception that the lands of the Ussuri valley were particularly promising for agricultural settlement, and thus peasants discouraged with their failure on the Amur often tried to relocate there.[63] Here as well, however, the actual situation was anything but ideal, as can be seen from Nikolai Przheval'skii's survey of cossack settlements along the river in the mid-1860s. Przeval'skii's observations, published in the popular journal *Vestnik Evropy* (European Courier), took on very much the tone of an exposé of the wretched conditions in the area. The settlements he investigated were formed of cossacks either ordered in from the Transbaikal region or transfered from the Amur. The better-off among them had been able to escape this draft by purchasing replacements, and those who could not hated their new home and considered themselves exiles. The poverty among them was quite striking, there was much hunger during the winter, and their general state of health was poor. Their morale was at an extremely low level, and to underscore this point Przheval'skii indicated that, for the appropriate price,

[59] Kropotkin, "Iz Vostochnoi Sibiri. (Selo Khabarovka)" (no. 44: December 1863), p. 13; Kropotkin, *Dnevnik*, p. 133. The song was apparently a popular one on the Amur; for a slightly different version, see Maksimov, *Na Vostoke*, p. 178.

[60] Kropotkin, "Iz Vostochnoi Sibiri" (no. 45: December 1863), pp. 5–6.

[61] Kropotkin, *Dnevnik*, p. 181.

[62] "Gustav Radde's Vorlesungen" (1860), p. 267.

[63] Kropotkin, "Iz Vostochnoi Sibiri" (no. 45: December 1863), pp. 5–6; Kropotkin, *Dnevnik*, p. 138; Maksimov, *Na Vostoke*, pp. 293–294; Kaufman, "Pereselenie," p. 275.

they were even prepared to grant strangers the favors of their wives and daughters. "In general, everything which you see on the Ussuri, both the cossacks and their way of life, has an extremely unpleasant effect, especially on a newcomer. Everywhere you find filth, hunger, and poverty, such that your heart involuntarily aches at the sight of it all."[64] As noted earlier, in Przheval′skii's account even the savage existence of Asiatic tribes compared favorably to the utterly miserable life of the cossacks. His essay created enough of a sensation in European Russia to bring about his expulsion from the Siberian branch of the Russian Geographical Society in Irkutsk, which did not appreciate this sort of negative publicity about the region.[65]

In view of problematic natural conditions on the one hand, but even more the arbitrariness, brutality, and lack of intelligent planning, it was not surprising that the efforts at developing food production in the Amur region in the 1860s were by and large unsuccessful. The dimensions of the failure, however, were striking. The region not only failed to emerge as an agricultural supply base for the other Russian settlements in the Far East and the North Pacific – all of which precisely as before continued to depend upon provisioning from European Russia and the American west coast – but quickly proved to be incapable of supporting even its own population. Supplies of grain and meat consequently had to be brought in from European Russia, the Transbaikal region, and even from adjacent territories in Manchuria.[66] In Maksimov's estimation, "the Amur cannot exist at the present time without the Transbaikal region," and would hardly be able to do so "for long into the future."[67] "Things would be really hard for us without the Manchurians," admitted a settler on the Ussuri, and need for foodstuffs at Nikolaevsk was so desperate that local traders reported a handsome profit to be turned by importing cattle from Yakutsk![68] In a word, from the very moment of their acquisition, Russia's new territories on the Pacific constituted what one report termed a "consumer" (potrebitel′) and not a "producer" region, which required a constant infusion of resources from the outside simply in order to exist.[69]

The irony of this situation was remarkable. In precisely the same way that Murav′ev's talk of a railroad to circumvent the Amur mouth was a betrayal of the vision of the river as a great natural route to the Pacific, so the fact that the Amur region had to be supplied with food from without was a betrayal of another vision equally important to the Amur epoch. Endorsing without

[64] Przheval′skii, "Ussuriiskii krai," p. 246.
[65] Dubrovin, *N. M. Przheval′skii*, p. 100; Rayfield, *Dream*, p. 45.
[66] Skal′kovskii, *Russkaia torgovlia*, p. 82; Malozemoff, *Russian Far Eastern Policy*, p. 3; Kpt, "Gde budet" p. 365; Sychevskii, "Russko-Kitaiskaia torgovliia," pp. 158–159; [Petrashevskii], "Po povodu," no. 40, p. 12. On Petrashevskii's authorship of the latter essay, see Dulov, "Sibirskaia publitsistika," p. 6. [67] Maksimov, *Na Vostoke*, p.136.
[68] Maksimov, *Na Vostoke*, pp. 199 (quote), 185; "Popytka Iakutskikh kuptsov," p. 709; [Liudorf], "Amurskii krai," pp. 216, 221.
[69] Burachek, "Gde dolzhen byt′ " p. 13.

qualification the centuries-old belief in the boundless agricultural potential of the region, the proponents of annexation had insisted that it could be Russia's granary on the Pacific. Practical experience, however, was quickly proving the exact opposite. Indeed, the difficulties besetting the agricultural development of the region led Zavalishin to question whether this project had ever really made any sense. "Is the main significance of the Amur region really in its agricultural lands?" he asked.

> For God's sake, we already have too much of this sort of land [in the empire], from which we have not yet begun to receive a benefit . . . Moreover, in regard to what has to be prepared for the future, *Russia should not scatter its energies but rather concentrate them and pull them together!*[70]

It was not Murav′ev's prophecy for the future of this region which was beginning to be realized, and even less Collins', but rather the bold and grim words of Nevel′skoi, who on leaving the Amur for good in 1855 foresaw that the region was "long fated to remain like an army camp." The overwhelmingly military character of the settlements along the Amur was repeatedly noted,[71] and at least one perceptive observer pointed out that even as an army camp things were not in very good order. "If these colonies remain significant only as a military post and do not become a [self-sufficient] Russian national colony in the full sense . . . then no matter how strong this military post may be, the possession of it will remain uncertain, and it may be easily lost at the first military defeat."[72] As we will see, the perceived tenousness of Russian authority in the region which this observer perceived was to endure far beyond the period of the present study.

The peculiar fate of commercial development in the Russian Far East contributed yet another dimension to the tarnishing of the region's image in Russia. We have seen that the expectation of the development of international trading activity on the Amur, stimulated by the expansion of European commerce on the Pacific since the 1840s, had represented one of the most precious visions associated with Russia's new territorial acquisitions. As the 1850s came to an end, however, it was becoming clear that these initial expectations had also been distinctly overblown. The Amur was not experiencing the development that the American Collins and his Russian co-thinkers had envisioned, and indeed was not likely to experience it in the foreseeable future. The problem, as Zavalishin and others explained, was not so much that the initial accounts which European Russia received were incorrect as that they gave a very misleading sense of the significance of the developments they described, and of what was to come.

[70] Zavalishin, "Amur," p. 76 (emphasis in original).
[71] Even a strong partisan of the development of the Amur noted that Nikolaevsk "in fact is nothing more than a fort." "Ekspeditsiia g. Ottona Eshe," p. 40.
[72] I. A., Review of R. Maak, *Puteshestvie na Amur*, p. 16.

It was quite true, for example, that foreign ships were appearing yearly in the mouth of the Amur in the late 1850s. Their numbers, however, were not great, the commerce that resulted was entirely insignificant, and by the mid-1860s they had ceased coming altogether.[73] And while it may have also been true that there was some trading of imported items such as porter, Manila cigars, and pineapples, there had nonetheless been no exchange of significant capital goods, and in view of the pressing needs of the new settlements this was nothing less than a travesty. "Instead of sugar, they ship in pineapple jam, instead of canvas and linen, they ship in petticoats and fancy hats, and instead of rye and wheat, they ship in American crackers (sukhari) and canned goods, . . . rum and sherry," remarked one observer dryly, while another complained about how difficult it was to sell the champagne, Rhine wines, and fine curtains imported up the Amur in Chita.[74] To top it all off, very few concurred with Romanov that goods on the Amur were now "blessedly (blagoslovno) cheap," and charges were leveled instead that merchants on the Amur conspired to fix prices for their items at extortionist rates.[75] In short, there had been no real development of significant international trade along the Amur at all, but rather a "sad case of commercial illusions," a fact which even the official local press was constrained to recognize.[76] And whatever the realities of the movement of goods up and down the Amur might be, Zavalishin concluded, there was no question that it would still be cheaper to have it all shipped out from European Russia, as before.[77]

To many observers, it was clear that the fault for the problems of developing commerce in the Amur region lay squarely on the shoulders of the local administration, whose policies from the outset had been designed to attract and favor foreign merchants. Thus, in 1856 Murav'ev had given foreign merchants the right to conduct trade on the lower parts of the river tax-free, and in 1858 he requested St. Petersburg to allow them to trade all the way to Irkutsk. The Russian government built a lighthouse to guide foreign ships into De Castries Bay, and once there they provided Russian pilots to assist with the difficult passage into and through the Amur estuary. No harbor fees were collected, and in 1861 an official edict proclaimed Nikolaevsk to be a duty-free port for foreign trade for a period of 20 years.[78] All these benefits to foreign traders, of course, came at the expense of their Russian counterparts, who

[73] In 1856, two foreign merchants ships arrived on the Amur; in 1857 six; in 1858 four; in 1859 eight; and in 1860 seven. Griffin, *Clippers*, pp. 343–344; Ravenstein, *Russians*, pp. 149, 158.

[74] Maksimov, *Na Vostoke*, pp. 318–319 (quote); Kropotkin, "Iz Vostochnoi Sibiri" (no. 42: December 1863), p. 10; Zavalishin, "Zamechaniia," p. 74. Indeed, Zavalishin's point that these luxury goods were in fact nothing new to the Russian Far East, and had long been imported to Okhotsk and Kamchatka, was well taken. Zavalishin, "Amur," p. 64.

[75] Kropotkin, *Dnevnik*, pp. 56–57; Maksimov, *Na Vostoke*, p. 318n.

[76] Afonas'ev, "Nikolaevsk," p. 112 (quote); untitled articles in *Irkutskie Gubernskie Vedomosti*, no. 11 (1857), p. 13, and *Amur*, no. 20 (17 May 1860), p. 258; Ravenstein, *Russians*, p. 920; Sychevskii, "Russko-Kitaiskaia torgovliia," p. 163.

[77] Zavalishin, "Amur," p. 62.

[78] Barsukov, *Graf . . . Amurskii*, II, pp. 209–210; Ravenstein, *Russians*, p. 415; Griffin, *Clippers*, pp. 338–340, 343.

were not given comparable incentives, and the government placed too many restrictions on the manner in which the resources of the region could be exploited for Russian entrepreneurs to be tempted to take any significant advantage of them. "With policies such as these," complained one critic, "we will never develop maritime commerce, and the Amur region will remain forever an area of poverty and insignificance."[79] And precisely as Petrashevskii had castigated the authorities for their handling of colonization, so another *petrashevets* criticized the heavy-handedness of governmental control and the absence of true *svoboda torgovli* or free trade, not only for its immediate negative effects in the region but more broadly as a betrayal of the progressive spirit of the times.[80]

The virtual still-birth of a bustling trade along the Amur was something that even the supporters of the developments in the Far East did not entirely deny. Rather than admit, however, that the problem was structural and derived from a fundamental misappraisal of the general prospects for development, they sought scapegoats whom they could blame without ultimately implicating their cause itself. One convenient explanation was that the local Siberian merchants were at fault for being too conservative and unwilling to engage in the new and untried commerce along the river. This notion was particularly appealing to Bakunin, who maintained that, although the Amur was a "perfect trade route," "the Siberian merchants, just like the Russians, are incorrigible routinists and Old Believers, who do not believe in finding new ways."[81] The failures of commerce on the Amur were also blamed on the traders themselves, notably on the Amur Company. This Company, founded in 1857 and intended to promote the development of extensive international commerce in the Far East, acquired instead the reputation of an avaricious and unscrupulous monopoly. Although Collins had initially greeted the activities of the Amur Company with the characteristically American optimism that the transfer of business "from the State into the hands of private parties will augment the quality and cheapen the prices" of goods in the Far East,[82] he ultimately had nothing but contempt for the "miserable stupidity" of its management.[83] Even the Jews – hardly a major presence in the Far East – did not escape

[79] Afonas'ev, "Nikolaevsk," pp. 117 (quote), 118, 121, 146; Kropotkin, *Dnevnik*, p. 78.

[80] Chernosvitov, "Neskol'ko slov," pp. 1–2; Afonas'ev, "Nikolaevsk na Amure," *passim*.

[81] [Bakunin], *Pis'ma*, p. 123. Another writer castigated the local merchantry for being too "spoiled (izbalovanyi) by the Kiakhta trade" to venture onto the Amur. N. N. Kh.-O., "O parakhodstve," p. 4. Also see [Speshnev], Untitled lead article, p. 4; [Belogolovyi], Speech in honor of N. N. Murav'ev, p. 3.

[82] Quoted in Griffin, *Clippers*, p. 335.

[83] Quoted in Quested *Expansion*, pp. 239–240; Kropotkin, "Iz Vostochnoi Sibiri" (no. 43: December 1863), p. 7; Kropotkin, *Dnevnik*, p. 57; Noskov, "Amurskii krai," no. 5, pp. 27–29; Chernosvitov, "Neskol'ko slov," p. 1; [Speshnev], Untitled lead article, p. 13. Even Bakunin branded these traders as "adventurists" and swindlers (despite the fact that he was a comfortably paid employee of this company). [Bakunin], *Pis'ma*, p. 187; Semevskii, "M. V. Butashevich-Petrashevskii," no. 3, p. 46n. And Veniukov recounted that the company sold spoiled goods, such that the population began to wish for the return of the petty and dishonest speculators whom the company was supposed to replace – at least their goods had been fresh! *Iz vospominanii*, I, p. 253–254.

implication, as it was claimed that commerce on the Amur had begun to decline at the moment that the Yankee traders who had established the original concerns at Nikolaevsk were replaced by San Francisco Jews. The former "fully deserved trust and respect . . . for their honesty and correct dealings," while the latter brought their customary "habits of deception and making a mess of things" with them to the Far East. "This transfer of the trade . . . from proper operations to swindling, from American traders to the Jews, very clearly indicates the condition and potential of trade at Nikolaevsk."[84]

While some of the problems which these critics identified were real enough, especially those regarding the Amur Company, the obstacles to the commercial development of the Amur region went much deeper than the corrupt practices of the merchants operating there. Commercial relations are based on an exchange of some sort of goods or services, and the most obvious fact in this regard was that in the Far East the Russians had nothing to offer. The original hope inspiring the annexation, namely that the Amur would facilitate a flow of goods produced in European Russia into China and onto the Pacific, proved to be entirely fanciful, for even had there been an available surplus of such goods west of the Urals (and there was not) there was no feasible means of transporting them across Siberia and beyond Irkutsk to the headwaters of the Amur. At the same time, the region produced nothing of its own with which it could barter. The traditional resources of the Russian Far East – fur and ivory – were not attractive commodities for foreign merchants in Nikolaevsk, and these traders were even less inclined to exchange their own goods for the inferior supplies of beef, tallow, hides, and wool that the Russians tried to sell.[85] Moreover, attempts to develop alternative local resources such as lumber or coal, for which there would indeed have been a ready export market, foundered for lack of support and interference from the government.[86] Estimating imports into Nikolaevsk in 1859 at £150,000 sterling against £3,000 of exports, Ravenstein's conclusion that "[a]n export trade . . . scarcely exists at all" was hardly an overstatement, and he was not the only one to note that as a consequence foreign skippers, for lack of any other merchandise, were frequently constrained to load their holds down with worthless ballast simply so that their boat would be heavy enough to sail away.[87] Local traders would have been loathe to agree with the conclusion of Ravenstein's compatriot Atkinson that in general, the Russians "cannot compete with the Saxon races" in supplying merchandise to the Far East and that they would find their greatest profit by serving as the middlemen and distributors for "European wares brought seaward," but the unpleasant truth was

[84] Afonas'ev, "Nikolaevsk," pp. 111–112.
[85] Griffin, *Clippers*, p. 345; "Otto Esches Expedition," p. 162.
[86] "Popytka Iakutskikh kuptsov," p. 708; Noskov, "Amurskii krai," no. 5, pp. 32; N. N., "Proekt ustava," pp. 7–9; "Ekspeditsiia g. Ottona Eshe," p. 41.
[87] Ravenstein, *Russians*, pp. 428 (quote), 149, 158; Griffin, *Clippers*, p. 345.

that they really did have nothing of their own to offer.[88] "In its present condition," observed Radde, "Eastern Siberia needs too little, and itself produces even less, for an import–export trade to be of much significance." This was if anything an understatement, and to at least some Russians at the time the point was perfectly clear.[89]

In the final analysis, it is difficult not to agree with Malozemoff that none of the expectations stimulated by the acquisition of the Amur river "came as near to being a total failure" as the hope that would it facilitate the development of foreign trade in the region. Indeed, as Mancall notes, the annexation actually heralded the *decline* of the Russian Far East as a commercial arena. Repeated attempts were made in the 1860s to use the river to replace the overland route through Kiakhta for the import of Chinese tea, only to be given up each time due to excessive costs and difficulties in transportation. Russia's dearly won access to seaports along the Chinese coast, codified by treaties in 1858 and 1860 that put Russian commercial privileges on the same level as those of the other European powers, remained a dead letter. Not a single Russian merchant vessel entered any Chinese port from the period 1860 to 1871,[90] and it was not the acquisition of the Amur river at all but rather the completion of the Suez canal in 1869 that finally was to enable Russia by the end of the century to maintain an appreciable presence on the Pacific.[91] It was, moreover, not merely in terms of oceanic trade via that the Amur proved to be so disappointing, for the confident expectations that its southern tributaries could be used to facilitate commercial links overland with northern Manchuria proved to be no less ephemeral. To be sure, several expeditions were sent to explore the commercial potential of Russia's new rights of navigation on Manchurian

[88] Atkinson, *Travels*, pp. 449–450. An American explorer in the Amur estuary in 1855 had had much the same thought as Atkinson, commenting that the river

> will be one of our links in our Japanese and Chinese trade, for whatever the fertility of the soil on the banks of the Amour or its tributaries, it will avail little to the Russians through want of laborers for sometime to come, and every thing may be more cheaply obtained from our country than from the interior of Russia."

Quoted in Cole, *Yankee Surveyors*, p. 152.

[89] [Radde], "Gustav Radde's Vorlesungen" (1861), p. 268. On Russian views see "Mestnoe Obozrenie," pp. 265–266; or the observations of V. V. Grigor′ev in Veselovskii, *Vasilii Vasil′evich Grigor′ev*, pp. 237–239. This point was unfortunately less clear to those Soviet historians who awkwardly tried to construe Russia's annexation of the Amur in 1860 in terms of Lenin's theory of imperialism – that is, as the attempt to secure external markets for the surplus of an overproductive national economy. E.g. Kabanov, *Amurskii vopros*, pp. 55, 83; also see Pokrovskii, *Diplomatiia*, p. 380; Quested, *Expansion*, pp. 22–23. Such an explanation is entirely untenable in the case of the Amur. For a far more realistic view from another Soviet historian, who treats the failure of commerce in the Far East in terms of the industrial and economic *backwardness* of imperial Russia, see Sychevskii, "Russko-Kitaiskaia torgovliia," p. 163.

[90] Malozemoff, *Russian Far Eastern Policy*, pp. 6 (quote), 7. At least one Soviet historian corroborated this emphatically negative appraisal of the Amur annexation, although he still asserted the "progressive" character of the Amur annexation. Sychevskii, "Russko-Kitaiskaia torgovliia," pp. 162–163. On the Amur as no cheaper than the Kiakhta route, see Ravenstein, *Russians*, p. 411.

[91] Skal′kovskii, "Suezskii kanal," *passim*; Mancall, "Russia," pp. 325–326.

rivers, but these returned with findings that were decidedly ambivalent about entire enterprise. "Whether or not navigation on the Sungari will be useful for commerce," concluded Kropotkin, who headed two such expeditions, "is impossible to say."[92] Indeed, even the long-established Russian–Chinese trade through Kiakhta declined definitively during this period, never to revive.[93] In 1862 the two governments agreed on an annexe to the Treaty of Peking which stipulated reciprocal rights to trade in each other's territory along the entire extent of the boundary, but this was taken advantage of only by the Chinese, who sold cattle, food, and Chinese liquor.[94] The Russians were not in a position to reciprocate, for once again they simply had nothing to offer.

The cumulative effect of the manifold problems associated with the occupation of the Amur valley was to invalidate one of the most cherished images of this period. We have seen that a perceived affinity with the United States had been of fundamental importance to the Russians in the period under consideration, in regard both to their general attempts to define themselves as a "New World" in distinction from Europe, and in their specific efforts to exploit their "Asian Mississippi" in the Far East. The picture of enlightened and progressive development through settlement and growth which they saw everywhere in America had great meaning for them, and in the early days of the occupation of the Amur they tried very hard to believe that they were actually duplicating it. It could hardly have been a pleasant task to confront reality and conclude, as some now did, that the experience in the Far East demonstrated not how much the Russians were like the Americans, but rather how elementally they differed. "What the Americans would have done with this blessed land, had it fallen into their hands," sighed Bakunin wistfully, in what was perhaps his only admission that all was not well in the Amur region. "But the Russian, and still more the Siberian, despite all the praise showered on him by jingoistic patriots, is as helpless as a child."[95] Maksimov expressed himself even more categorically on this point. While the Americans were the models of practical, well reasoned activity, to which the settlement and exploitation of an entire continent was witness,

> we see nothing at all of this sort on our new and virgin soil of the Amur. For this reason, the very comparison [with the United States] cannot even be made, despite all the contrived desire of those zealots among us who want it so badly . . . We should forget about America this very minute, and not come back to it . . . [We should not] seek out similarities where there are none, nor be seduced by foreign examples and models that do not apply to a single aspect of our own affairs.[96]

[92] Kropotkin, "Dve poezdki," p. 109 (quote); Kropotkin, "Poezdka," pp. 674–675.
[93] Sladkovskii, *History*, p. 80; Sladkovskii, *Istoriia*, p. 210; Mancall, "Kiakhta Trade," pp. 19–22, 26. [94] Malozemoff, *Russian Far Eastern Policy*, pp. 18–19.
[95] [Bakunin], *Pis'ma*, p. 125. Also see I. A., Review of R. Maak, *Puteshestvie na Amur*, p. 12; Zavalishin, "Otvet g. Nazimovu," pp. 116–117. [96] Maksimov, *Na Vostoke*, pp. 314–315.

Indeed, the Russians even began to reproach the Americans for having suggested that there might be a parallel between the two countries in the first place. Paraphrasing Fedor Tiutchev's famous poem, Veniukov complained that "the Yankees measured the situation using a yardstick that was too American, assuming that the Amur – that 'Asian Mississippi' – would develop just as quickly."[97] As the quotation marks indicates, the reference to the Mississippi was now entirely sarcastic. Through all of this, Nevel′skoi's original warning to Murav′ev had come back to haunt the Russians as a fact, with the effect of undermining the most important vision of the period. The sense of utter despair and failure was well conveyed by the Decembrist Raevskii in the mid-1860s.

> The Amur, about which we . . . used to dream, has now become a bottomless pit, into which over 30 million [rubles] have been poured, never to be seen again. Whether or not there will be any useful result is most unclear. The enterprise was ruined from the very beginning. In a word, from the nineteenth century we have moved back to the first half of the eighteenth, and if things continue on in this manner, we will progress yet further to the seventeenth.[98]

"Such lack of cultivation, such insolence!"

It had been governor-general Murav′ev's decision to stake his career on the fate of the Amur region, and thus it was in a sense fitting that his own image in Russia should have experienced the same vicissitudes as the endeavor to which he devoted so much of his energy. We have seen in an earlier chapter that during the initial period of his tenure in Eastern Siberia he had established contacts with the political exiles in the region and gave every appearance of sympathy with oppositionist sentiment against Nicholas I's despised order. In 1856, he made a most substantial gesture by granting permission for Mikhail Petrashevskii, banished to the remote Eastern Siberian countryside, to resettle in Irkutsk, and their relations at first were close enough for Petrashevskii to manage the governor-general's household while the latter's wife was sojourning in Paris.[99] Beyond such gestures, moreover, it should be noted that Murav′ev's reputation of as a liberal and reformer was supported by the tangible efforts at significant social reform that he made in the early years of his administration. Appreciation of the governor-general's reformist and even democratic inclinations – his "uninterrupted guardianship of the interests of the common people," in the words of one sycophant – was widespread.[100] As one of *Kolokol*'s anonymous correspondents in Irkutsk wrote

[97] Veniukov, "Vospominaniia," p. 110 (quote); [*idem*], "Primechanie," p. 4.
[98] Quoted in Shatrova, *Dekabrist D. I. Zavalishin*, p. 126.
[99] Semevskii, "M. V. Butashevich-Petrashevskii," no. 2, p. 21; Veniukov, *Iz vospominanii*, I, p. 273. [100] Miliutin, "General-gubernatorstvo," pp. 619–620.

quite movingly to Herzen in 1860: "It is only since the arrival of Murav′ev that the Siberian peasant, the political exile, and the hard-labor convict with no rights have all begun to learn about justice, about access to representatives of governmental authority, and about the existence of an official who steals nothing from them." A British visitor praised his enlightened policy in trying to help Eastern Siberia's indigenous population as well, which was being ruthlessly exploited by Russian fur traders.[101]

Writing some 40 years later, Kropotkin still warmly recalled this aspect of Murav′ev's administration. He recounted how he had personally been involved in the reform of the police and prisons and in the creation of a citizens' committee in Chita, chosen "by all the population, as freely as they might have been elected in the United States."[102] Perhaps the most significant of the projects initiated during Murav′ev's tenure were the efforts at enhancing *glasnost′* through the establishment of a local press. In 1858, the *Irkutskie Gubernskie Vedomosti* (Irkutsk Provincial News) began publication, and although it was nominally an "official" organ its editorial board nonetheless included political exiles such as Petrashevskii, L′vov, and Speshnev.[103] Just as this was happening, moreover, preparations were under way for yet another newspaper, one which would be entirely free of governmental sponsorship and thus a truly independent tribune for public opinion. The administration was willing, and material support was forthcoming from a group of Irkutsk merchants. Consequently, on 1 January 1860, the weekly newspaper *Amur* made its first appearance under the editorship of M.V. Zagoskin, gaining for Eastern Siberia the little-appreciated distinction of sponsoring the first independent newspaper in Russian history to be published within the boundaries of the empire.[104] In addition to reports about affairs in the Amur region itself – not all of which were entirely positive – early issues of the newspaper dealt frankly with local problems such as the condition of the serfs, judicial reform, and even official corruption.[105]

Yet despite the unprecedented openness in the public life of Eastern Siberia that Murar'ev encouraged, his position remained fundamentally ambiguous and contradictory, and it was not his progressive side which was ultimately to win out. As soon as the first issue of the *Amur* was distributed, he became furious with the newspaper's broad profile and its muck-raking tone. His chagrin was only heightened by what he read in the issues that followed.[106] Within a few months of its initiation, he had taken away its financial indepen-

[101] "Po dele Irkutskoi dueli," p. 616; Atkinson, *Travels*, p. 460.

[102] Kropotkin, *Memoirs*, p. 169; Karpov, *Issledovatel′*, p. 14.

[103] Koz′min, "M. B. Zagoskin," pp. 194–195; Kubalov, "Pervenets," pp. 56–57.

[104] Demen′tev, *Russkaia periodicheskaia pechat′*, pp. 398–399; Esin, *Russkaia dorevoliutsionnaia gazeta*, pp. 30–31; Aref′ev, "M. V. Butashevich-Petrashevskii," p. 178; Dulov, "K. Marks," p. 20; Kubalov, "Pervenets," p. 57. [105] Dulov, "Sibirskaia publitsistika," pp. 10–13.

[106] See his correspondence with Korsakov, in Barsukov, *Graf . . . Amurskii*, I, p. 586

dence and put such pressure on Zagoskin and other co-workers that the newspaper became a *de facto* mouthpiece of his administration. The troublemakers Petrashevskii and L′vov were summarily dismissed from the editorial board – for their refusal to print "lies about the Amur," Zavalishin claimed – and replaced by Romanov and Sgibnev, assistants of Murav′ev who could be trusted.[107] With his treatment of Petrashevskii in particular – who anonymously contributed the most sharply critical articles – Murav′ev lived up to all of his true instincts as a tsarist official. In March of 1860, some four years after having summoned Petrashevskii to Irkutsk and only two months after the newspaper began to appear, orders were issued banishing him back to what was effectively an early death in Yeniseisk.[108]

From this point on, the governor-general's reputation as a reformer began to disintegrate. Once the liberal and exile community in Eastern Siberia showed itself to be unwilling to carry out his instructions blindly, his relations with it quickly ruptured and collapsed. In a letter to Zavalishin, L′vov uneasily confessed how very mistaken he and his comrades had been to trust Murav′ev and imagine that the governor-general would actually live up to his claims and serve as an agent of social reform.[109] Indeed, even a close assistant such as Veniukov, who in reverential tones had likened Murav′ev's "enlightened" administration in Eastern Siberia to the reigns of Catherine the Great and Louis XIV and remained remarkably devoted to the governor-general and his cause, felt constrained to modify his adulation after witnessing how Murav′ev violently berated a raftsman on the Amur for some minor infraction. He acted "so very much like a general in the style of Arakcheev, that something was torn away in my heart, and from that point on I became colder to the person in whom I had to that point seen almost nothing but good qualities . . . Such lack of cultivation, such insolence!"[110] Herzen, whose information about Eastern Siberia was rich and diverse enough to afford him a remarkably detailed and nuanced view of the situation there, steered between negative and positive evaluations of Murav′ev. He noted the governor-general's progressive accomplishments, and made every attempt to avoid pronouncing final judgment. Even he, however, was ultimately inclined to comment that the

[107] Zavalishin, *Zapiski*, II, p. 399 (quote); Kubalov, "Pervenets," pp. 75–77; Semevskii, "M. V. Butashevich-Petrashevskii," no. 3, p. 44.

[108] Petrashevskii, "Pros′ba"; Leikina, *Petrashevtsy*, p. 80. Even as a carefully controlled newspaper, *Amur* proved to be short-lived, and ceased publication in 1862. Koz′min, "M. B. Zagoskin," p. 201–202.

[109] L′vov, "Pis′mo," *passim*; Semevskii, "M. V. Butashevich-Petrashevskii," no. 3, p. 33; Leikina, *Petrashevtsy*, p. 60. L′vov's fellow *petrashevets* Speshnev, however, suffered no such pangs of conscience. Having proved much more willing to provide the governor-general with what he wanted, their relations continued unbroken, and the two remained in contact even after they had both left Eastern Siberia. Barsukov, *Graf . . . Amurskii*, I, p. 647.

[110] Veniukov, *Iz vospominanii*, I, pp. 223 (quote), 273. A. A. Arakcheev (1769–1834) was a general notorious for his despotic manner and ruthless enforcement of a harsh disciplinary regime.

governor-general apparently wanted to "transform Siberia à la Pierre le Grand."[111] In a word, through his actions Murav′ev came to seem to many not so much an opponent to Nicholas I's system as one of its loyal generals, and his fall from grace was so complete that by 1863 Kropotkin could report: "The fate of Murav′ev in Siberia is remarkable. These days you rarely meet someone who speaks of him with sympathy. After the universal enthrallment, a reaction has set in."[112]

This reaction against Murav′ev in Siberia was repeated in the imperial capital. By the early 1860s, that segment of opinion in upper governmental circles which during and after the Crimean War stood solidly behind Murav′ev had largely turned against him and his endeavors. Age-old doubts about the Amur's navigability, which had seemingly been put to rest by Nevel′skoi's findings a decade earlier, had not only re-emerged but were now apparently confirmed, and the image of a beckoning highway to the Pacific was supplanted by that of a "swamp no more than three feet deep." Murav′ev's administrative practices and his all-too-obvious ambitions excited strong condemnation, and the entire Amur affair increasingly came to be seen as nothing more than "Murav′ev's fancy (zateia)."[113] Disillusionment reached into the very highest levels of the government. "It is a shame," wrote Alexander II to Konstantin Nikolaevich, "that despite all of his good qualities, which no one appreciates more than I do, he constantly tries to achieve the power to make himself independent from the central authorities. I can in no way allow this."[114] It was the tsar's uncle, however, who as Minister of the Navy presented the most striking example of disenchantment, for at one time he had been Murav′ev's sole ally in St. Petersburg and the only source of protection for the maverick governor-general against the wrath of Nicholas I's hostile ministers.

We have already noted the differences between the two in terms of their evaluation of Russia's strategic imperatives in the Far East and the implications for the Amur annexation. On a what might be called a moral level, Konstantin Nikolaevich's conviction that what Russia needed most in the post-Crimean era was not loud "victories and conquests" nor "audacious acts which might raise our name high for the moment" but rather "modest and ordinary labor" was bound to incline him yet more negatively against such an obviously ambitious and self-promoting individual.[115] In the same manner that he worked to

[111] "Imperatorskii kabinet," p. 721 (quote); [Gertsen], "Neschastnye soldaty," p. 528; I-r, "Graf Murav′ev-Amurskii," pp. 910–911; [Gertsen], "Gonenie," p. 967. As elsewhere in Russia, *Kolokol* was followed with great interest by the authorities in Eastern Siberia, and Herzen's critical treatment of Murav′ev apparently caused the governor-general considerable distress. Barsukov, *Graf . . . Amurskii*, I, pp. 628–629; Kropotkin, *Dnevnik*, p. 40.

[112] Kropotkin, *Dnevnik*, p. 92. [113] Veniukov, *Iz vospominanii*, I, p. 237.

[114] Letter of 16 November 1858, in *Perepiska Imperatora Aleksandra II*, p. 73.

[115] Letter from Konstantin Nikolaevich to A. I. Bariatinskii, 24 June 1857, in "Fel′dmarshal Kniaz′ Bariatinskii," p. 131.

expose the dictatorial corruption in the affairs of the Russian–American Company, Konstantin Nikolaevich focused his attention on Murav′ev as well. Much of Zavalishin's and other criticism of affairs in the Far East was published in the journal of his Naval Ministry, and there can be little question that the idea for Maksimov's fact-finding mission to the Far East and resulting exposé of affairs on the Amur originated with him, despite the author's own insistence that he simply wanted to unravel the "complicated knot" of conflicting reports from the region.[116] Out of all this scrutiny, a governmental commission was established to review the state of affairs in the Far East, and might have taken some action against Murav′ev's successor Korsakov, had the Polish uprising of 1863 not worked to suppress for the time being the last remaining liberal sentiments in the Russian government.[117] As for Murav′ev himself, despite the nominal honors he was accorded upon leaving Eastern Siberia in 1861, his reception in St. Petersburg was chilly enough to induce him to move his residence to Paris. Repeated entreaties over the following decades to be taken back into governmental service were consistently rejected, and he died and was buried in Paris in 1881.[118]

Murav′ev's fall from grace epitomizes a peculiar paradox which was broadly characteristic for the Amur epoch as a whole. The governor-general, who originally gained favor and support as a member of the oppositionist–reformist wave sweeping Russia, and who was seen as a dynamic proponent of Russia's national integrity against the disgrace of Official Nationality, ultimately came under attack for all of the traditional qualities of tsarist officialdom that were so universally despised. These included *proizvol* or arbitrariness in the exercise of his duties, unrestrained conceit, and an express inclination to let concern about his own career and glory override all other considerations. He fell into disgrace, in other words, in the name of the very principles which he himself had stood for, and indeed in the eyes of the same liberals and reformers who had earlier viewed him as their champion. This irony was compounded, moreover, by the perception on the part of some contemporaries that precisely his obsession with the Amur river – which he always envisioned as a means for realizing in practice the hallowed principles of social reform and national rejuvenation – had in fact been the true source of his undoing as a liberal advocate. Some public indication of this might be read between the lines of an item in the *Irkutskie Gubernskie Vedomosti* in 1859, in which the plans to found an independent newpaper the following year were described. Conceding that "the

[116] Maksimov, *Na Vostoke*, p. 116 (quote). Upon dispatching Maksimov, Konstantin Nikolaevich is said to have remarked: "Ach, that Murave'ev! He loves to make everyone else look like a fool: let's let him see what it feels like to be in this role!" Veniukov, *Iz vospominanii*, I, p. 243 (quote); [Veniukov], "Primechanie," p. 2n; Romanov, "Amur – zlokachestvennaia iazva," pp. 389–390. On Maksimov, who went on to become a noted literary figure in Russia, see Clay, "Ethos and Empire," pp. 78–80, 116–127.

[117] Zavalishin, *Zapiski*, II, pp. 413–414.

[118] Butsinskii, *Graf . . . Amurskii*, pp. 64–65; Ravenstein, *Russians*, p. 416.

Amur river and everything about it is all the rage at the moment," the essay nonetheless took delicate exception to Murav′ev's intention to name the projected new publication after this river (which after all even in Irkutsk was considered remote) and offered alternatives that were less exotic and dramatic but apparently more geniunely civic-minded, such as *Sibirskii* or *Irkutskii Vestnik* (The Siberian or Irkutsk Herald).[119] A correspondent to *Kolokol*, who could write rather more openly, made this point quite explicit. "I consider the major problem in Murav′ev's governing of Eastern Siberia to be the Amur. He has concentrated all of his thoughts and feelings on this brain-child of his, and this has frequently led him to commit blunders and injustices."[120] Many years later, a former adjutant concurred, noting that the governor-general's liberalism lasted only "up to the point when he became distracted by political concerns, and by the Amur in particular."[121]

"In Asia we too are Europeans"

The waning of the Amur epoch was largely a result of the problems and disappointments which we have examined up to this point. There were however factors not immediately related to events in the Far East which were of considerable significance in directing the attention of the country away from the Amur, thus contributing to its fall from the national spotlight. The most important of these external factors was the Russian advance into Central Asia. Russia had been pressing outward on its southeastern border since the sixteenth century, slowly but steadily conquering and absorbing the steppe peoples and their territories. The culmination of this process – the annexation of what came to be called Russian Turkestan – was getting under way at the same time that Murav′ev was arguing for the annexation of the Amur, and it involved many of the individuals who were also active in one way or another

119 Cited in Kubalov, "Pervenets," p. 58.

120 "Po dele Irkutskoi dueli," p. 616.

121 Miliutin, "General-gubernatorstvo," pp. 619–620. The controversy about Murav′ev continued for decades. In 1891, I. P. Barsukov published a two-volume documentary biography of Murav′ev, which was lavishly obsequious in its tone (*Graf . . . Amurskii*). This work was then subjected to a long and highly critical review by P. N. Butsinskii, a historian of Siberia at Kharkov University, which was published as a separate volume in mid-1890s by the Academy of Sciences in St. Petersburg (*Graf . . . Amurskii*). The following year, a rebuttal to Butsinskii appeared, dedicated to Murav′ev's memory and written by I. V. Efimov, who had served under him in Irkutsk (*Gr. N. N. Murav′ev-Amurskii pered sudom Prof. Butsinskogo*). With the exception of historical works published in the 1920s, when criticism of the heroes of imperial Russia was tolerated and even encouraged (e.g. Steklov, *Mikhail Aleksandrovich Bakunin*, I, pp. 504ff: Murav′ev was "an Asiatic administrator, at heart a despot and a tyrant"), Soviet historiography has tended to portray Murav′ev in a strongly positive light. See especially Shtein, *N. N. Murav′ev-Amurskii*, p. 41; Kabanov, *Amurskii vopros*, pp. 89–91. Two American dissertations devoted to Murav′ev both echo Barsukov with distressing fidelity, and are uncritically positive in their evaluations. Stanton, "Foundations"; Sullivan, "Count N. N. Muraviev-Amurskii." The fate of the Murav′ev legend in the Far East itself took a rather dramatic turn in the early 1990s, when Murav′ev's remains were removed from Montmartre cemetary and reinterred at Vladivostok. See Stephan, *Russian Far East*, p. 50.

in the Far East. In 1851, the Treaty of Kul′dzha between Russia and China was negotiated by Yegor Kovalevskii, who was at the same time an important supporter of the annexation of the Amur. The treaty opened up Western China to Russian commerce by establishing the town of Kul′dzha as an entrepôt for tariff-free trade between the two countries, much on the model of the Treaty of Kiakhta concluded a century and a quarter earlier.[122] Nikolai Ignat′ev, who negotiated the Treaty of Peking in 1860, had been sent two years earlier on a mission to establish diplomatic relations with the emirates of Khiva and Bukhara. Ignat′ev's diplomatic efforts in Central Asia were unsuccessful, and from military outposts in the Kazakh steppe the Russians began their advance in the late 1850s. The city of Tashkent was occupied in 1865, Samarkand and Bukhara in 1868, and Khiva capitulated in 1873. For the rest of the century the Russians extended and consolidated their control over this territory.[123]

A number of historical accounts have suggested that different factions, one promoting Russian activity in the Far East and the other in Central Asia,[124] were effectively competing in the upper echelons of the foreign policy establishment already in the 1860s. For Russian public opinion, however, there was a clear and logical connection between these two theaters of geographical expansion. Indeed, the motivations for the advance into Central Asia paralleled in many respects those that had stimulated the acquisition of the Amur. These included the lure of supposedly rich agricultural land – the St. Petersburg newspaper *Birzhevye Vedomosti* (Stock Exchange Report) had no hesitation in transferring the distinction of being "the most blessed region in the world in terms of climate, soils, and mineral resources" from the Amur to Central Asia[125] – the prospect of trade providing Russian commerce with access into other parts of Asia, and finally, as in the Far East, acute concern over the penetration of British influence into northern India and fear of its continued spread. Also, very much as in the Amur region, energetic and ambitious military commanders acting on their own initiative played a major role in the annexation of Central Asian territories.[126] One further motivation which was unique to Central Asia was the desire to pacify once and for all the nomads of the Kazakh steppe, whose periodic raids had menaced Russian frontier settlements for centuries.[127]

Along with all of this, the conviction of a messianic mission, which had been so instrumental in determining Russian views of their activities in the Far East, was no less important in perceptions of Central Asia. In a memorandum

122 Veniukov, "Postupatel′noe dvizhenie," pp.119–120; Kabanov, *Amurskii vopros*, p. 110; Clubb, *China*, p. 77.

123 Gillard, *Struggle*, pp. 105ff; Strong, "Ignatiev Mission," p. 236; Pierce, *Russian Central Asia*, pp. 17–45; Sarkisyanz, "Russian Conquest," *passim*.

124 E.g. Ritchie, "Asiatic Department," pp. 170–171.

125 Cited in Dahlmann, "Zwischen Europa," p. 66.

126 Pierce, *Russian Central Asia*, pp. 17–18.

127 Zavalishin, "Po povodu zaniatiia Tashkenta," October, p. 5.

circulated among Russian diplomatic missions in Europe in December of 1864, the foreign minister Gorchakov addressed Russian intentions in Central Asia. He referred to Russia's legitimate need to pacify its borders, but also to its duties as a civilizer of the "half-savage nomad populations, possessing no fixed social organization" who inhabit these territories. "Very frequently of late years," he noted, "the civilization of these countries, which are her neighbors on the continent of Asia, has been assigned [by Europe] to Russia as her special mission."[128] Indeed, such a mission was identified by some as the principal and really only important motivation behind the move into Central Asia. The orientalist V. V. Grigor′ev, whose strident views on Russian messianism we have examined in chapter 2,[129] spoke out quite clearly on this point. He dismissed arguments that the Russian advance was necessitated either by the economic considerations of expanding Russian trade or by the imperative of pacifying Russia's borders. Yet already in the late 1850s he expressed the conviction that Russia would nevertheless move into and annex its adjacent Central Asian territories, "impelled (v silu) by that historical law according to which peoples at a higher stage of development subordinate to themselves their neighbors who are poorly developed spiritually and materially." Russia, he affirmed, will continue to advance in Central Asia until it meets its cultural equal, by which he obviously meant England.[130]

A particularly powerful and revealing expression of how Russia's civilizing role in Central Asia was understood came from the novelist Fedor Dostoevskii. His comments date from the 1880s and thus fall outside of the time frame of the present study, but they are directly relevant to our subject nonetheless, for they demonstrate the remarkable degree to which the interest in bringing enlightenment to Central Asia was a continuation and indeed to a large extent a repetition of the messianic vision we have seen articulated in the Far East. Dostoevskii followed the Russian advance into Central Asia in the 1860s and 1870s with great interest, and the fall of the Turkmen fortress at Geok Tepe to Russian forces in 1881 gave him an opportunity to reflect on its larger significance. Repeating the old theme of the pointlessness of further activity in Europe, Dostoevskii depicted Asia as the untapped field on which Russia could redeem itself. "Don't you see," he interrogated the readers of one of his regular columns in a St. Petersburg newspaper,

> that with a turn to Asia, with a new attitude toward it, something like what happened in Europe with the discovery of America can happen to us. For in fact Asia is for us

[128] [Gorchakov], "Circular dispatch," pp. 72–73; Sarkisyanz, "Russian Imperialism," p. 49; Kazemzadeh, "Russia," pp. 494–495; Pierce, *Russian Central Asia*, p. 20.

[129] See above, pp. 54–55.

[130] Quoted in Veselovskii, *Vasilii Vasil′evich Grigor′ev*, pp. 184 (quote), 237–239. Also see Sarkisyanz, "Russian Imperialism," for an insightful discussion of this particular view and its significance.

the same thing as undiscovered America was at that time. Our striving for Asia will uplift and resuscitate our spirit and strengths . . .

In Europe we were hangers-on and slaves, but in Asia we are masters. In Europe we were Tatars, but in Asia we too are Europeans. The mission, our civilizing mission in Asia will give us spirit and draw us out there, if only we would get on with it! Build just two railroads, for a start – one into Siberia and the other into Central Asia – and you will see the results immediately.

Dostoevskii wrote his essay as an hypothetical dialog between himself and an educated and thoroughly Europeanized Russian audience, which had no sympathy for their country's activities in the East and considered the move into Central Asia to be a complete waste. His audience objected at this point that by throwing itself into Asia, Russia would risk undermining its own advances in knowledge and science. He retorted:

what kind of science do we have at the moment? – we are unschooled smatterers and dilettantes. But in Asia we will become serious, necessity itself will compel us and insure that as soon as a spirit of enterprise takes shape we will at once become masters even in science, instead of the hangers-on (prikhovostni), which is all that we are presently. But the main point (and about this there is no doubt) is that from the first steps we will understand and assimilate our civilizing mission in Asia. It will raise our spirit, it will give us dignity and self-awareness – and we have very little, if any, of these qualities today.

But why, his critics persisted, should we lower ourselves by turning to Asia? For what reason must we humble ourselves in this way in front of Europe? Dostoevskii's reply was one that should already be thoroughly familiar to us. "But we are not at all lowering ourselves! . . . It's not that we will lower ourselves, but rather we will raise our level at once, that's what's going to happen! Europe is sly and clever, it is guessing what is going on and, believe, me, *will begin to respect us immediately*!"[131]

Dostoevskii's reference to the "discovery of America" is notable, for it is a evocation of the very metaphor first articulated in regard to Russian activities in the Far East. He envisioned Asia as a vast untapped field abounding in resources which would allow an enterprising and diligent nation to construct a new civilization. Dostoevskii had no specific idea of exactly what these resources were, or how they might be developed, but this was entirely beside the point. Indeed, it was precisely because his Asia was unknown, and consequently offered an unrestricted opportunity for him to manipulate his dreams and his hopes, that he found the prospect so irresistibly alluring. He attempted to tantalize his readers' imaginations with it as well: "Now do you know that

[131] Dostoevskii, "Geok Tepe," pp. 36–38 (emphasis added). On Dostoevskii and Asia, see Bohatec, *Der Imperialismusgedanke*, pp. 203–206; Sarkisyanz, "Russian Attitudes," p. 248; Kohn, *Pan-Slavism*, p. 172; Kohn, "Dostoevsky," p. 515; Riasanovsky, "Asia through Russian Eyes," pp. 17–18.

there are countries there that are less known to us than the interior of Africa? And do we know what sort of riches are to be found in the depths of these boundless countries?" As had been the case on the Amur, Dostoevskii acknowledged the preeminence of the American experience as a prototype, and in fact he could begin to articulate more exactly what he had in mind for Russia only by imagining what the Westerners themselves would do with this opportunity.

Oh, if only Englishmen or Americans lived in Russia instead of us! . . . Now they would discover our America . . . Oh, they would have opened up everything, the metal ores and minerals, the countless deposits of coal – they would have discovered everything and figured out how to exploit it. They would have applied science, and compelled the land to give them a fifty-fold return – this same land that we [Russians] still consider to be barren steppe, empty as the palm of your hand.

And finally, this building of a new America would have the all-important effect of enhancing and renewing a morally crippled Russia.

The aspiration for Asia . . . would serve moreover as an outlet for numerous troubled minds, all of whom are yearning for something . . . Build a drainage canal, and the mold and stench will disappear. And once they have set about their new tasks, they will no longer be bored, they will all be regenerated (pererodiatsia). [Even] a talentless individual, with a wounded and reeking conceit, would find his opportunity (iskhod) there. For it is often the case that someone lacking talent in one place is resurrected elsewhere practically as a genius.[132]

Dostoevskii's sense of Russia's mission in Asia, therefore, comes from precisely the same sources, is driven by the same dynamics, and has the same goals that we have seen in Russian messianism in general throughout this study, most recently and most clearly in the arguments of Petrashevskii and Semenov regarding the Amur. It sprang from the same sort of willful turn away from Europe, which was itself motivated both by a sense of inferiority and as well as the disagreeable feeling of being an unwanted presence among the Western fraternity. The resulting redirection of attention to Asia was intended to offer Russia an alternative field for creative work – an undiscovered America, in Dostoevskii's telling metaphor – upon which it could exercise its own strengths and independently develop its national qualities. Yet the scale of cultural values remained entirely European – and, most importantly, the ultimate goal of Russia's activities was to raise the country's qualities to a level that would be satisfactory because it would be truly European. Once again, the loudly proclaimed turn to the East was in fact nothing of the sort. The compelling attraction of Central Asia had nothing to do with Asia at all, and was rather indicated in the conviction, more succinctly formulated by Dostoevskii than anyone, that "in Asia we too are Europeans." Asia would

[132] Dostoevskii, "Geok Tepe," p. 37.

be the scene of Russia's regeneration and resurrection.[133] The only real difference between the novelist and the other individuals we have considered in this study was a mere detail of geography. His Russian "America," his arena for the transformation of the quality of Russian civilization, was in Central Asia, while for them it had been in the Far East. In regard to the ideological quality of the messianic conviction, this shift of geographical focus made no difference whatsoever. All the vision required was a remote and unknown region offering apparent opportunities for development and a native society apparently in need of salvation, and these Russia could locate on any number of its borderlands.[134]

In regard to virtually all of the other images that have comprised the subject of this study, however, the geographical shift was decisive. For most Russians, Central Asia simply appeared to be of substantially greater and more immediate importance to the country as a whole than the Far East, and there were good reasons for this evaluation. In contrast to the Amur region, the material advantages Turkestan offered were palpable: on the one hand a valuable raw material (cotton) for Russia's burgeoning textile industry, and on the other a teeming population of potential customers for Russian products.[135] Whether or not these these really were the determining factors in Russia's domination of the region up to the revolution, as Soviet historians vociferously claimed – they certainly were dominating factors thereafter – is irrelevant.[136] The point is that considerably less speculation and fantasy was necessary to appreciate them than had been the case with the Far East. Beyond this, Turkestan was not really a foreign region for the Russians in quite the same way that the Amur and Ussuri valleys had been. As just indicated, throughout its history Russia had had steady and intimate contact with the peoples of Central Asia. The most intense phase of this was the century-and-a-half of Mongol domination; indeed, in view of this experience it may be said that modern Russia was born out of the contact with the nomadic tribes of the steppe. The Russian state then took territorial shape as it expanded east and southeast, out of the Moscow lands to the Volga and beyond, defeating and absorbing the remnants of the Golden Horde. In this way, expansion against the Central Asian tribes was woven into the very fabric of Russia's historical experience, and combined in itself the contrasting qualities of a national crusade on the

[133] Unfailingly recalcitrant, Dmitrii Zavalishin riased a lonely voice in opposition to this vision of Russia's civilizing mission in Central Asia. "We should have . . . no pretensions to civilize Asia, when the main element of civilization – science – is still so weakly developed among us. In the same way, we have no right to tease (obol'shchat') ourselves with the dream of spreading the true belief" when there are still so many pagans and Moslems in Russia's internal provinces. Zavalishin, "Po povodu zaniatiia Tashkenta," October, p. 6.

[134] Russian views of the Caucasus in the nineteenth century offer yet another example of just how "transportable" the messianic vision was. See especially Susan Layton's outstanding new study *Literature and Empire*; Rieber, *Politics*, pp. 70–72.

[135] Whitman, "Turkestan Cotton," *passim*; Dahlmann, "Zwischen Europa und Asien," pp. 59–61.

[136] Kazemzadeh, "Russia," pp. 493–494.

one hand and an apocalyptic struggle for existence itself on the other.[137] The advance into Russian Turkestan after 1860 was seen – rightly or wrongly – as a grand culmination of this historical process of expansion, and shared all the attendant associations.

The geographical shift of interest from the Far East to Central Asia can be seen in the careers of a number of individuals who had been involved in the affairs in the Far East. No sooner had the American Collins arrived home from his trip to the Amur than he presented a project to the American government for a second journey to Russia. This time, however, Collins wanted to go as head of an American "Commercial Commission to the Caspian" to survey the prospects for trade in Russian Turkestan. His plan was to descend the Volga to the Caspian – a "Russian lake," as he described it in Washington – and continue on to the Aral Sea, Bukhara, Tibet, Tashkent, and Teheran.[138] The geographer Veniukov, without question one of the most ardent and devoted of all those involved with the Amur, left Eastern Siberia after his Ussuri expedition in 1859 and was sent as a reconnaissance expert to Central Asia, to assist in the Russian advance against the Khanate of Kokand.[139] Finally, the last major explorer of this period in the Far East, Nikolai Przheval'skii, likewise proceeded directly to Central Asia upon leaving the Amur region. Unlike Veniukov, however, Przheval'skii had come to despise his assignment on the Amur. While still there, he wrote about the river as "a great slop pit, into which is poured everything base and revolting from all of Russia,"[140] and in describing the town of Nikolaevsk he even resurrected the graphic imagery of *The Inferno* which Herzen had used some 20 years earlier in describing Siberia as whole.

> In the same way that there was the inscription on the gates of Dante's hell: "All who enter here lose hope," every officer and official transferred out here can enter this in his diary, for the return from this place is as difficult as the exit from hell . . . In any event the moral death of anyone stationed here is inevitable, even if he was at the outset the very best sort of person . . . Don't think that this is an exaggeration or something I have thought up. My comments should be read by every person who passionately wants to come out here, as I once did.[141]

We can read in these words the death knell of the Amur epoch. Przheval'skii burned with impatience to get to Turkestan, and was able to begin the first of his four renowned Central Asian expeditions in 1870.

There was a final important example of the paradox by which the move into

[137] Kazemzadeh, "Russia," p. 489.

[138] Vevier, "Introduction," p. 23. The plan stirred no interest, and Collins retained his official appointment as "American Agent to the Amoor River" until 1869. Griffin, *Clippers*, p. 345.

[139] Veniukov, "Vospominaniia," pp. 304, 297. Veniukov could never accept the shift of attention to Central Asia, and he argued up to this death for the vital importance of the Amur annexation to Russia.

[140] Quoted in Dubrovin, *N. M. Przheval'skii*, p. 81.

[141] Quoted in Dubrovin, *N. M. Przheval'skii*, pp. 81–82; Rayfield, *Dream*, pp. 39–40.

Central Asia eclipsed the significance of the annexation of the Amur region in the name of the very considerations which had been associated originally with the Far East. It has been seen how in the late 1850s a contrast of sorts developed between what we have called maritime and continental perspectives regarding Russia's new acquisitions, the latter viewing the Amur region as a sort of gateway not so much east to the ocean as south overland into Mongolia and Northern China. The prospect of Russia's occupation of Turkestan offered further stimulation for this vision of exclusively continental dominion, and to many it seemed to represent the most natural and efficacious means to this end. Indeed, although we have noted important intimations of this perspective in the annexation of the Amur region, it was Central Asia which afforded the first opportunity to articulate clearly and fully the vision that was with few exceptions to dominate Russian strategic thinking down to 1917 and beyond: Russia as a continental land-based power dividing – peaceably or otherwise – its dominion of Asia with Britain as master of the seas. Dmitrii Zavalishin's brother Ippolit gave early expression to this sense of geopolitical destiny in 1862. "Sooner or later, one way or another," he wrote, "Russia and England will come together on the shores of the Indus and Ganges to solve Hamlet's eternal question: 'to be or not to be?'"

> Russia and England, one on dry land, the other on the sea, are equally imbued with the same spirit of expansion, because this spirit flows from their history and represents for them an historical inevitability . . . They will inevitably come together on the shores of the Indus and Ganges – and they will give a new look to the earth and will begin a new history of the world![142]

While isolated proponents of activity in the Far East continued into the 1860s and beyond to insist that by virtue of their size and population, the countries of Pacific East Asia present "enormous advantages over the khanates of Central Asia,"[143] their voices were effectively drowned out by those like Zavalishin who argued precisely the opposite.

"So here's where Russia begins!"

The frustration of the grand plans for the Amur region and the attendant redirection of the nation's attention to more propitious arenas of activity along Russia's Asian frontier inevitably served to raise a perceptual dilemma which we have already enountered in this study. To what extent and in what manner could the newly acquired territories in the Far East be properly considered to constitute a part of Russia? After all, prior to the period examined in this study, the Russians had had no protracted historical presence there and knew practically nothing about them. The monsoon climate and southern vegeta-

[142] Zavalishin, *Opisanie*, p. 2.
[143] Kpt, "Gde budet," p. 395. Also see Veniukov, "Vospominaniia," p. 297.

tion of much of this territory, along with its native cultures, were entirely foreign to anything in their experience up to that point. In short, as one critic wryly remarked, it was as if the Amur region had "dropped from heaven" into the Russians' lap, who were thereupon instructed to adopt it as their own land, and the results of this challenge were decidedly mixed. Well into the 1860s, peasant colonists bound for the Far East from European Russia could often do no better than to answer "to China" when asked exactly where they were headed, and once settled in the Amur region they were very aware that they had left Russia far behind. This was not necessarily a distressing prospect, as Przeval′skii learned from one settler he questioned on the Ussurii. "But, God willing," the peasant told him, "if we can get going and make a little headway, then we'll have lots of everything, and we'll make Russia right out here (tak i my zdes′ Rossiiu zdelaem)."[144] Yet however suitable such a buoyant attitude might have been for the purposes of colonizing a new country, it simply underscored at the same time the essential apartness of the region from the Russian *rodina* or homeland.

To be sure, there was a certain preparedness and indeed an urge to recognize native Russia in the empty expanses of the Amur region. A variety of factors could be taken as evidence, of which probably the most prosaic was indicated in an incident recounted by Peter Kropotkin. Kropotkin told of the reactions of a group of oarsmen from the Transbaikal region who, having begun a descent of the upper Amur, encountered a woman selling fresh pickles. The season for such a staple item had apparently not yet arrived in their home region. "Have you had them for a long time, then?" they asked.

> "For a long time, here on the Amur, for a long time, *batiushka*." The oarsmen were delighted, for finally they had come back to Russia: pickles were available at the right time of the year. Later, when they discovered a hazelnut tree, they were even more delighted, even though the nuts were damp: "Vot ona Rosseia [sic]-to gde nachalas'!" (So here's where Russia begins!).[145]

We may note that these reactions were not without their own ambiguity, for they could just as easily be turned around: if, after all, the Russianness of the Amur rested on its supply of marinated cucumbers and nuts, then realistically speaking the region could not have been very Russian at all. Rather more significant were attempts to recognize Russia not in foodstuffs but in the presence and operation in the Far East of vital national qualities. As we have seen, the mood of reform and reconstruction that dominated in post-Nicholaen Russia put a premium on certain sorts of national activity, and as long as the

[144] Potanin, Review of R. Maak, *Puteshestvie na Amur*, p. 82; Przheval′skii, "Ussuriiskii krai," p. 261.

[145] Kropotkin, "Iz Vostochnoi Sibiri" (no. 43: December 1863), p. 7. We are reminded that Goncharov, returning from a sojourn in various Asian countries, had similarly taken advantage of his return to "Russia" by requesting pickles during his visit to the Amur region. See above, p. 147n.

Amur appeared to be an arena where such activities could be meaningfully conducted, for a great many people its Russianness was not really an issue. After his experience in Yakutsk, Ivan Goncharov expressed this notion perhaps more emphatically than anyone. The prospect of an Eastern Siberian transformed through the progressive activities of the Russians was an inspiring one, and it alone was sufficient to imbue the region with a quality that for him was altogether native. "Regardless of the fact that the town is populated with Yakuts," he wrote, "I was pleased nonetheless when I drove through a jumble of one-story wooden homes that were greying with age: despite it all, this is Rus′, although Siberian Rus′!"[146] This attitude was characteristic for the Amur as well. What made these regions Russian was not their food, their geography, nor even – as Goncharov clearly indicates – the national origin of their inhabitants. It was rather the circumstance that they were being *osvoeny*, that is occupied and brought into the realm of civilization and "historical movement" (to use the phrase of the day) by the Russians.[147]

Approaching the problem in a rather different way, Peter Semenov tried to make something of the same point. His argument rested not so much on the Russianness of the regions in question – although he would have concurred wholeheartedly with Goncharov – but rather on the more fundamental naturalness and appropriateness of the advance to the Pacific in terms of Russia's larger geohistorical destiny.

> The occupation and colonization of the Amur brilliantly brings to an end the remarkable movement of the Slavic tribe, which began in the sixteenth century and which strived – against all the obstacles it encountered and with singular determination – in a direction diametrically opposed to all [other] national migrations: namely from the west to the east, from the shores of the Volga to the coasts of the Pacific Ocean. The history of the exploration and settlement of all Siberia, from the sixteenth century to the present day, but particularly the history of the occupation of the Amur in the seventeenth century, clearly demonstrates that the entire Slavic migration from the west to the east was a natural phenomenon, which flowed gradually out of the life of the Russian nation.

There was an obvious teleology in this vision, a sort of inverse Manifest Destiny which we have seen elsewhere in his writings. On the basis of it, he came to the firm conclusion that the Amur region represented an "immediate and inalienable continuation (prodolzhenie) of Russia."[148]

Attitudes proved to be volatile, however. In view of the sharply disappointing results of the attempts to colonize the Russian Far East, the vision of it as a site for progressive Russian activity was undermined, and with this the desire to nativize these new lands was perceptually undermined as well. Moreover, the comparative light cast by the move into Central Asia gave some opportu-

[146] Goncharov, *Fregat "Pallada"* (1986), p. 522.
[147] Veniukov, "Vospominaniia," p. 270.
[148] Semenov, "Amur," pp. 187–189.

nity for critical reflection about just what the annexation of the Amur region meant for Russia at the most fundamental level. Semenov's insistence on the legitimacy of the Far East as a "natural" outlet for Russian imperial expansion was quickly countered, and indeed no less a partisan of the Amur annexation than Dmitrii Zavalishin was among the first to express his doubts. Russia had historically needed to expand, the Decembrist wrote in 1865, because the presence of hostile tribes on its borders had always threatened its existence. Expansion of this sort – most recently exemplified in Turkestan – was therefore proper and "natural." "Only in its far-eastern regions, in its actions regarding China and America, did Russia dilute its natural motives with other factors – political concerns and desire for personal advantage – but for this reason there were complications there."[149] Zavalishin thus took issue, implicitly but unmistakably, with Semenov's suggestion that the Amur annexation "flowed" naturally and organically out of Russia's primordial movement to the east, and he characterized it instead as an abrupt and quirky thrust. And precisely because it did not conform to but rather violated Russia's genuinely organic pattern of "step-by-step moving forward," it was *in principle* an unnatural movement for Russia, and for this reason harmful. The Pan-Slav Nikolai Danilevskii, hardly an opponent of Russian expansion, put this same point even more emphatically. He insisted that legitimate territrial growth could only take the form of a spatially uninterrupted flowing-out from the center to the edges, and in all events had to remain cohesive and contiguous. "The spread of Russian settlement by jumps across the ocean or across great distances," he wrote in his manifesto *Russia and Europe* in 1869, "does not work, even if it is supported by the government. Our American colonies did not work out, and somehow the Amur isn't working out either."[150]

Zavalishin and Danilevskii may have been sceptical about the suitability of the Far East as an arena for Russian expansion, but neither would have thought to question the essentially legitimacy of the project of imperial expansion itself. The legitimacy of this project was questioned at the time, however, and in such a way as to further underscore the tenuousness of the connection between the Amur region on one hand and European Russia on the other. In chapter 5 we noted an ambivalence in Alexander Herzen's views of Russia's advance on the Pacific.[151] Captivated by precisely the same prospect of *osvoenie* and civilization-building that was so alluring for Goncharov, no one had been more excited than the editor of *Kolokol* by Russia's activities in the Far East. He passionately heralded the advance to the Pacific as an incontestable demonstration of Russia's continued vitality after the defeat of the Crimean War, and expressed satisfaction that Russia had finally reached its "natural boundary" on the Pacific.[152] Yet at the same time, the entire historical process

[149] Zavalishin, "Po povodu zaniatiia Tashkenta," October, p. 5.
[150] Danilevskii, *Rossiia*, p. 532. [151] See chapter 5.
[152] Gertsen, "Pis′mo k Dzhuzeppe Matstsini," p. 350.

of Russian territorial expansion, of which the Amur annexation was the most recent expression, was for him imbued with a thoroughly negative significance, for it had always involved the incorporation of foreign peoples and regions under the despotic aegis of Russian imperial autocracy. As far as he was concerned, Russia's existence as an empire was something essentially foreign to the country: an "all-devouring Germanic–Tatar *Moskovshchina*," in the rather fustian expression of one of his correspondents, made up of a misshapen and perverse blend of European absolutism and Oriental tyranny artifically grafted onto Russia by Peter the Great and maintained exclusively through the brute force of arms.[153]

These sentiments, muted momentarily in his exhilaration over the treaties of Aigun and Peking, were brought to the surface a few years later by the revolt of the Poles against the Russians in the early 1860s. In the light of developments on Russia's western border, his enthusiasm for the country's expansion in the Far East dissolved rapidly into an impassioned attack on the empire as a whole. In stirring terms, he proclaimed his solidarity with the insurgent Poles in 1863:

> We stand with the Poles because we are for Russia . . . We stand with them because we are firmly convinced that the absurdity (nelepost′) of an empire stretching from Sweden to the Pacific, from the White Sea to China, cannot bring any good to the peoples which Petersburg leads on a chain. The universal empires of the Chingizes and Tamerlanes belong to the most primitive and savage periods of development, to those times when the entire glory of the state was made up of military strength and vast territories. They could exist only with inescapable slavery at the bottom and unlimited tyranny at the top. Whether or not there was a need in the past for our imperial formation makes no difference today . . . *Yes, we are against the empire, because we are for its people!*[154]

Elsewhere, Herzen noted that it was not the genuine Russian nation (narod) at all but rather the Petrine system and its infamous Table of Ranks that had "settled entire countries, colonized Siberia, oozed (prisochit′sia) to the Pacific, to Persia, and to Sweden."[155] He did not mention the recent occupation of the Amur in particular, but it would seem that in the dark light of Murav′ev's infamies, he had come to view it as well as an example of that imperial "oozing" which he now found so thoroughly disagreeable and illegitimate. Where once *Kolokol* had written with warm enthusiasm about Russia's activities on the Amur, it was now inclined to depict them as a misuse and waste of the nation's resources.[156]

Herzen's argument was that the Russian empire should be dismantled and that all captive regions held in it against their will should be given their

[153] The expression was N. I. Kostomarov's. [Kostomarov], "Ukraina," p. 503.
[154] I-r, "Proklamatsiia 'Zemli i voli'," p. 1318 (emphasis in original).
[155] I-r, "Rossiia i Pol'sha," no. 67, pp. 556 (quote), no. 66, p. 543.
[156] "Finansovoe Polozhenie," p. 1816.

freedom. This was a profoundly unpopular position in Russia, and was to cost him much of his following among the Russian educated public.[157] Nevertheless, he expressed it repeatedly. He was preoccupied with Russia's western borderlands – Ukraine, Lithuania, Belorussia, and of course Poland[158] – but at the same time made it clear that he did not consider only the "advanced" European periphery as suitable for independent statehood. In this spirit, he explicitly included Siberia in his vision of post-Nicholaen liberation. The fact that this latter region differed from the others in that it lacked a dominant non-Russian nationality did not make it less deserving of its freedom.

> We support the right not only of each nationality, having separated from others and possessing natural borders, to independence, but also of each geographical region (polozhenie). If Siberia were to separate tomorrow from Russia, we would be the first to welcome its new life. The [territorial] cohesiveness of the state does not at all coincide with the well-being of the people.[159]

Siberia, by virtue of its unique geographical configuration, should have the right to statehood along with Poland, the Caucasus, the Baltic territories, and Finland. By sacrificing its imperial immensity, the framework would be created for Russia itself to develop on a more genuinely national basis.

What are we to make of Herzen's statements concerning the emancipation of Siberia? He was not speaking for a larger segment of Russian public opinion – as we have suggested that in earlier cases he was – and in any event his position was hardly consistent with his own earlier sentiments. Yet it is perhaps not so important to rationalize his ambivalence as simply to appreciate that it was there and to understand what it meant. In those moments when he was searching for an indication of Russia's enduring national vigor, then Murav'ev's dynamism and his success in advancing Russia's position on the Pacific made the Amur region seem quintessentally Russian, indeed the most important part of the entire country. Yet this enthusiasm proved to be extremely short-lived, and within a few short years the full separation of the region – and with it the loss of that link to the "Mediterranean of the future" and bridge to the United States that he had described in such ardent tones – was not only conceivable but actually the only legitimate alternative. And while Herzen's support of the radical dissolution of the empire was not shared by most of his compatriots, his underlying ambivalence about the Russian Far East was. For European Russians, the Amur never really ceased to occupy a position that was at best peripheral. When a variety of circumstances came together in a fortuitous way, as was the case in the late 1840s and 1850s, the region could move quickly to the center of national attention, but as these

[157] I-r, "Rossiia i Pol'sha," no. 66, p. 541.

[158] I-r, "Proklamatsiia 'Zemli i voli'," p. 1318; I-r, "Rossiia i Pol'sha," no. 66, pp. 539, 541; Izdatel' *Kolokola*, "Russkim offitseram v Pol'she," p. 1214.

[159] Izdatel' *Kolokola*, "Russkim offitseram v Pol'she," p. 1214.

circumstances evolved and changed it turned out that there was really nothing out there to hold anyone's interest or concern. In the final analysis, it seems clear that the very marginality and lack of knowledge which had made it possible to invest such extravagant hopes and expectations in the Amur region in the first place was precisely what allowed the region subsequently to fall back into complete and utter obscurity.

Conclusion

In framing the intentions of this study, the point was made at the outset that it would differ from narrative histories of Russia's incorporation of the Amur region, among other things by virtue of the fact that it would not attempt to explain the "why" of the annexation. To be sure, the story I have told offers insights – some of them telling – into this important question, but no argument as such is made, for the main point of the work involves something rather different. More than anything, *Imperial Visions* is an excavation of a geographical vision, which seeks to reconstruct and analyze the images through which Russians thought about and signified the Far East in the period under consideration. The rationale for this seemingly rather rarefied exercise, which the present work shares with a larger literature on envisioning, derives from the supposition that the images of a remote and little-known region can provide insight into the mind and culture of the individuals, groups, and societies that entertained them. Geographical visions, that is to say, can be taken as cultural artefacts, and as such they unintendedly betray the predilections and prejudices, the fears and hopes of their authors. Examining how a society perceives, ponders, and signifies a foreign place, – in other words, is a fruitful way of examining how the society – or parts of it – perceives, ponders, and signifies itself. A useful metaphor, already employed in this work, likens the object-region to a canvas, upon which an image or vision is projected to produce something which in certain ways is effectively a self-portrait.

The essential framework for this study was the evolving national mood or *Zeitgeist* in Russia during the second and the third quarters of the nineteenth century. This mood was shaped by two distinct but interrelated factors: on the one hand the development of a vigorous nationalist movement, and on the other an ever-more powerful and more universal current of opposition to the regime of Nicholas I. Russian thinking about the Far East was located squarely in this nexus of nationalist and oppositionist sentiment, and was animated by some of its most cherished principles as well as its most profound tensions and ambivalences. The prospect of a Russian advance on the Amur was inspiring significant nationalist–oppositionist elements in the 1840s, well before the actual occupation of the region was undertaken in the middle of the following decade. The acquisition of the Amur was seen as an important means of overcoming the hyper-conservative stagnation of Nicholas I's reign, a feeling much intensified by the adamant refusal of the tsar's ministers even to consider such a move. Moreover, the Amur was perceived not merely as a means for moving beyond a festering status quo, but additionally as an arena where the illustrious national future or futures evoked in the doctrine of Russian nationalism could be constructed. The Far East appeared to offer the Russians an opportunity to turn away from Europe and demonstrate – for the world, but more importantly for themselves – that their vital national energies and capacity for independent accomplishment had been neither dissipated through three decades of decline under Nicholas I nor irreversibly intimidated and deterred by the ignominious defeat of the Crimean War.

As we have seen, on the Amur the Russians glimpsed two different ways of bringing to life the ambitious principles and desires that animated their new nationalist preoccupations. While these alternatives were closely joined in a common project, they relied upon representations of the region that diverged in their emphases and indeed were not entirely compatible. On the one hand, the Amur was envisioned as a vast, virgin, and essentially empty territory, where the Russians could bring to life a new society by fostering the development of modern civilization, agriculture, and commerce. The immediate, if not entirely exclusive prototype for what the Russians had in mind was the remarkable example of the colonization of an entire continent by the young United States, and the notion that they could reenact this experience and create an "America" of their very own on the banks of the Amur was – for a while, at least – positively intoxicating. This alluring prospect was both stimulated and supported by the designation of the Amur as an Asian or Siberian Mississippi. At the same time, however, the Amur appeared to offer the first opportunity to satisfy a rather different imperative in Russia's newly defined sense of national identity – namely, that of bringing civilization to the savage peoples of Asia. Logically, this latter notion was conditioned upon the region being not empty but rather well populated with indigenes waiting to receive the enlightenment and salvation that the Russians were now seeking to bestow. In this casting, the Amur was not to be an "America" at all but instead securely retained its Asian identity. Moreover, it was not the Amur alone that rendered Russia's far-eastern advance significant to the achievement of this mission of salvation, for with this river region Russia obtained access to yet larger and more densely settled regions of dark Asia which similarly stood in need of Russia's beneficient attentions.

In this manner, the Russians sought to work through their sense of national identity and to articulate a vision of their national future by relocating both in their imaginations out to the entirely novel geographical theater of Siberia's remote eastern fringe. As they did so, however, we have seen how a variety of tensions and even contradictions streamed to the surface. The notion of a mission in Asia, for example – the logic of which could not at first have seemed more straightforward and self-evident – proved in the event to be tortuously complex. Were the Russians really seeking to turn away from the West, as many claimed to be doing, and concentrate their activities in those parts of Asia which Nature and history had prepared especially for them? Or by turning to the East were they in fact merely relocating the original contest with Europe out to a new and untested venue, which was appealing only because they believed it would be more favorable to themselves? By the same token, the entire proposition of the uniqueness and superiority of Russian civilization to that of the West, so fundamental to the nationalist doctrine, was significantly undermined by the widespread assumption that what Russia would bring to Asia was in fact nothing other than "Europeanism" in its best and most geniune variant. In a rather different connection, we have seen how Russian thinking about the Amur region brought to light a yet more fundamental ambivalence in the nationalist vision – namely, a marked indeterminacy regarding precisely how Russia itself was to be defined and delimited geographically. When Russian nationalism sought confirmation of Russia's greatness and glory, there was no obstacle whatsoever to its reveling in the geographical immensity and the ethnographic diversity of the empire. The line between nation and empire was blurred, and because the empire was by its very nature something that could and did expand, the physical parameters of the nation necessarily possessed a certain elastic quality as well. With the acquisition of the Amur region and the belief this spawned that a renewed Russia could be constructed in its uninhabited expanses, the perceptual boundaries of the national homeland were flexible enough to relax for a brief moment even to include these new and in every respect profoundly foreign territories. Yet the Russianness of the Amur region depended vitally upon the optimistic vision of a bright future on the Pacific, and thus this quality dissipated as soon as the vision itself began to darken.

All of these images and visions were in some way generic, in the sense that they did not really depend fundamentally upon qualities specific to the Amur region. There was an important dimension to Russian thinking about the Far East, however, which did recognize these special qualities and tried to assign them some significance in a broader national framework. The fact that the region had been and indeed essentially remained a *terra incognita* for the period of this study had the positive effect of enabling the sort of imaginative envisioning just noted, but at the same time was the source of a certain confusion when the Russians tried to assign a meaning to the various features of the region. The notion that with the Amur region Russia received a vital artery connecting it to the burgeoning commercial civilization on the Pacific

Ocean, for example, was an important element of the prospect of national rejuvenation which Russia's new territorial acquisition was supposed to bring. The problem, as we have seen, was to identify exactly how this connection was supposed to be effected. The initial belief that the river itself could provide the link was quickly surpassed and discarded, and ultimately the Amur came to be seen instead as something of an obstacle to Russia's secure esconcement on the Pacific. Related to this uncertainty was another dilemma – namely, to identify the precise geographical site in Russia's new acquisition where its real value lay. Again, the Amur river itself was the first alternative, and again it was virtually discounted in the space of the few years that this study encompasses. The focus of Russian attention migrated almost immediately south of the Amur river, either to the coasts of the Tatar straits and the waters beyond or alternatively to the continental interior of northern Manchuria. This latter bifurcation, in turn, gave rise to attempts to articulate broader perspectives on Russia's future development as a maritime or, alternatively, a land-based, continental power.

Yet despite the undeniable extravagance and even fantasy of much of this thinking about the Amur region, it would be a serious mistake to assume that the Russian imagination ever enjoyed anything resembling an absolute freedom to recreate the Amur region in its own image. The nationalists desired this freedom, it is true, and at the outset their line of speculation suggested that they even assumed they had it. In the final analysis, however, the material reality of the far-eastern territories invaded and to a significant extent exploded all of their "imaginative geographies." This circumstance does not invalidate the metaphor of the Amur region as a blank canvas, but it does make necessary a certain qualification – namely, that it was a canvas with its own contours and special qualities, all of which affected the sorts of things that could be drawn upon it. Indeed, in places this canvas would not even absorb the colors properly, such that the carefully adorned images quickly beaded into pathetic and meaningless pools of paint. This point is especially important to note in view of the inclination in the literature on envisioning to grant the "gaze" of the observer a sort of hegemonic license in regard to the object-region, a license which suggests a kind of absolute power and control. This relationship of power and domination has been elaborated in the work of Edward Said and many others, and it is implicit more generally in the inclination to use verbs like "creating" or more commonly "inventing" to describe the process of envisioning.[1] The material of the present study, however, points to a rather different conclusion. The Russians, to repeat, may have sought to invent the Amur in their quest for a national utopia, but they were manifestly unable to do so. Even in their "ardent fantasy," in their most impassioned and

[1] The present author acknowledges with regret his own earlier use of this language: "Inventing Siberia." For a brief but astute critique of this "inventionist approach" see Wicker, "Introduction," pp. 3–4.

creative outbursts, they could not entirely elude the realities of the lands in question. What we have witnessed was not at all the "invention" of the Amur region but rather the playing-out of an elemental tension – between the yearning for a brilliant national future and a boundless faith on the one hand and the grim material circumstances of the area in which this future was to be realized on the other. For much of our study these two elements were widely separated by the region's remoteness and the lack of knowledge about it, and it was in the broad intermediary space between them that all of the images we have examined took shape. Ultimately, however, they could not be kept entirely apart.

If we follow the fate of the Amur region beyond the period of this study, we see that there was to be no revival of anything resembling the euphoria which attended its annexation. Indeed, down to the revolution and beyond, the region was never to emerge entirely from under the gloomy pall which settled over it by the mid-1860s. This is not to suggest that Russian activity in the Far East ceased, for it did not, but it took place in other regions. The trend from the early nineteenth century of Russian movement south along the Pacific coast did not, as we have seen, come to an end with the annexation of the Amur, but reached here only a sort of middle ground and continued unabated. Vladivostok, at the southern limit of Russia's territorial acquisition, was founded in 1860 and immediately replaced Nikolaevsk as Russia's principal military and economic center on the Pacific coast. The Russian advance did not stop with this, however, and some three decades later the Russians continued to press south, as they joined the Western powers and Japan in the incipient contest for domination of the Korean peninsula. In 1897 the Russian obtained a 25-year lease on the Liaotung peninsula and the port of Port Arthur (Lüshun). This move, however, brought the determined resistance of the Japanese, who challenged the Russian position militarily and easily succeeded in eliminating the Russian Pacific fleet in 1904–1905. This led to the Russian retreat back to its earlier forepost at Vladivostok, which remains to the present Russia's main Pacific base.

The Amur itself was peripheral and largely irrelevant to these developments. Attempts at agricultural colonization were persistently disappointing, and the settlements in the region continued to have an overwhelmingly military character, which meant that their survival remained dependent as before on the constant infusion of resources from elsewhere. In 1881, an aging but still vituperative Zavalishin claimed that the region still could not feed itself, and the point was confirmed 10 years later by the more reliable voice of Nikolai Iadrinstev.[2] The overwhelming marginality of the region from the standpoint of the Russian center became clearly apparent in the 1890s, when the decision was taken made to connect European Russia to the Pacific with a Trans-Siberian railway. The Amur was judged to be so unimportant, however,

[2] Zavalishin, "Amurskoe delo," p. 94; Iadrintsev, *Sibir' kak koloniia*, pp. 66, 386.

and the natural conditions in the region so forbidding, that it was avoided altogether, and the easternmost leg of the railroad was laid from the Transbaikal region directly across northern Manchuria, through the city of Harbin (founded at this time) and on to Vladivostok. The decision of the government to route a major segment of such an economically, politically, and strategically vital rail line across what was effectively foreign territory, rather than along Russia's own Amur and Ussuri rivers, speaks more eloquently than anything to the absolute lack of significance of the region in its eyes.

It was only after the defeat by the Japanese in 1905 that national attention was redirected back to the Amur region, for it was now becoming clear that if Russia was going to maintain a significant presence on the Pacific at all, it could only be through fortifying its position there. This eventuality, however, merely raised a dilemma of an entirely different order, namely the appropriateness of a Russian presence of any sort in the Far East. This question surfaced during debates in the State Duma in 1907–1908 regarding the construction of a railway line from the Transbaikal region along the Amur and Ussuri to the Pacific, which was the first necessity if the region was indeed to be fortified. The whole of Russia's far-eastern activities, it was argued by some, represented nothing more than an imperialist venture, which diverted resources and energies from other, more vital parts of the country and in this way worked against Russia's general interests. "Not so very long ago," pronounced Baron P.L.Korf, "we began to be told that once we have emerged on the Pacific we should rule it, that the Pacific Ocean should become practically a Russian lake." (If it was a coincidence that the verb Korf used – *gospodstvovat'* – was the same one Murav'ev had employed in urging his view of Russian domination in the Far East on Nicholas I, it was an extremely telling one.[3]) "I don't understand what this notion is founded on," he continued, and begged to disagree. The loss of the Amur would represent nothing more than "the amputation of a finger" for Russia, while similar losses on the European frontier "would in general be more painful."[4] Korf did not speak for the majority, which concurred with the protestations of Peter Semenov-Tian-Shanskii and other delegates that the Amur region was an "inalienable (neot"emlemyi) part of Russia" and that the duty of the government was to "undertake all measures to defend it."[5] The resolution to construct the railroad was thus approved. It was highly significant nevertheless that objections such as those offered by Korf could be expressed at this forum, and that the chairperson of the Council of Ministers supported the new railroad by emphasizing that the Amur region – apparently in contrast to Turkestan – was Russia's "only colonial possession."[6] There still seemed to be something artifical or at least excessively

[3] See chapter 4. [4] Session 32 (30 May 1908), *Gosudarstvennyi Sovet*, pp. 1445–1448.
[5] Session 32 (30 May 1908), *Gosudarstvennyi Sovet*, p. 1377
[6] Session 33 (31 May 1908), *Gosudarstvennyi Sovet*, pp. 1528–1529. For a fuller discussion of the question of the Amur railroad, including a more detailed examination of the Duma debates, see Marks, "Burden of Siberia."

tenuous about the bonds connecting the Amur region to the rest of Russia: something unnatural and indeed potentially harmful.

The persistence of such pronounced skepticism, however, was offset by the persistence of some of the optimistic images and visions that had been associated with the Amur region from the outset. Indeed, at least some of these images have shown a remarkable resilience in the Russian imagination, and have resurfaced time and again. One example is the belief that the center of the world's economic and cultural activity was shifting to the Pacific, an expectation to which Alexander Herzen had given expression in the early 1850s. Almost 80 years later a revolutionary of rather different profile, Leon Trotsky, evoked the same image in a speech on the eve of his fall from power, using the military metaphor which he developed during his days as the head of the Red Army:

> Siberia is the workers' state's link (vykhod) to the Pacific ocean. And the Pacific ocean and its coasts are more and more becoming the arena of modern history. Today Siberia is the remote rear line of the Soviet Union. But the history of the next 10 to 20 years may give the order: about face! Then the front will be on the Pacific and the rear in the West, behind the Urals.

For 1,000 years, he went on, Europe was at the center of human history, but the First World War brought the end of this predominance. Again echoing Herzen, Trotsky maintained that economic and political domination of the world scene had shifted to the New World on the North American continent. "The fact is incontrovertible," he insisted.

> Europe has moved into second place, and the Atlantic ocean is losing its significance to the Pacific . . . There's no getting away from it! The dominance of the United States over Europe, of the Pacific over the Atlantic, will grow ever greater.

Trotsky did not welcome the ascendance of America, as had Herzen, and his excitement was stirred instead by the prospect of insurrection in China. Rather than the Pacific being a "long bridge" between the United States and Russia, he now proclaimed Siberia to be "a bridge between Moscow and Canton."[7] Nevertheless, his notion of the imminent ascendancy of the Pacific and of the fundamental significance of the development for Russia are identical to what we have examined in this study. And the fact that it was always imminent, always a vision of the future, allowed it to be recast again and again.

Related to this expectant view of the ascendancy of the Pacific were the beliefs that the Amur was the vital artery connecting it to Russia, and that by securing this artery the natural wealth of the Far East could support the growth of a new civilization. We have witnessed how painfully deceptive this prospect proved to be in the aftermath of the annexation, but it has lived on

[7] Trotskii, "O Sibiri," pp. 14–15.

and gone through various reincarnations. Indeed, the most recent of these has taken place in our own time, and can serve as a fitting final note in our study. The 1970s saw the undertaking in the USSR of one of the most ambitious construction projects of the twentieth century, namely the BAM or Baikal–Amur Mainline railway. This railway, linking the town of Ust-Kut north of Lake Baikal to the Pacific ocean coast at Sovetskaia Gavan′, was to run roughly parallel to the Trans-Siberian, some 125–185 miles to the north. The climatic, topographic, and geological obstacles which this project had to overcome were legendary, and the Soviet government sought to generate enthusiasm for it in terms of the very same sort of future-oriented utopian vision of progressive social development and natural resource wealth which we have followed throughout this study. As the then-General Secretary of the Communist Party Leonid Brezhnev put it in 1971, the BAM railroad "will bring enormous wealth to the motherland. It will create the biggest industrial towns we have ever seen. Cities and towns will grow [along it] like flowers."[8] Three years later, he elaborated upon this cheering prospect in a speech at a political conference:

> The construction of this railroad, which will cut across the Siberian massif with its inexhaustible natural wealth, opens a way for the creation of a new and major industrial region: along it will grow smaller settlements and cities, industrial enterprises and mines. Of course, new lands will come under the plow and be brought into agricultural production.

As he pronounced this final point, the General Secretary was interrupted by excited applause.[9] His audience was no doubt unaware of the fact that precisely the same prospect of agricultural abundance and "a golden chain of flourishing colonies" had been depicted for an equally or perhaps even more enthusiastic audience of compatriots some some 120 years earlier.[10] There was only one small difference: Brezhnev was pinning his hopes on a railroad, while the imagination of the earlier period had focused upon a river. In terms of the overall quality of the expectations, however, this detail had no significance whatsoever.

The *déjà vu* emanating from the late Soviet period was not, unfortunately, limited to the enthusiasm of the initial expectations. In precisely that same way that the visions of the mid-nineteenth century proved to be so empty and misleading, so the great promise of the BAM railway would give way eventually to utter failure. The "completion" of the line was announced in 1984, but much of the construction work had yet to be carried out, and the volume of traffic it has handled since has never approached the original estimates and expectations. Moreover, since the collapse of the Soviet Union and the drying

[8] Quoted in Specter, "Siberian Railroad," p. A4.

[9] *Velikii podvig*, p. 61. It was not only the Soviets who were excited by such optimistic prognostications. See Kirby, "Siberia," pp. 150ff.

[10] Veniukov, "Vospominaniia," p. 270

up of investment in the region from the center, the railway zone has been losing population steadily and whatever settlements were established there are withering.[11] Thus the fate of the BAM project uncannily resembles that of the initial attempts to develop the Amur region. The morbid implications for the Amur region as a whole were summed up in a study by an American geographer a year before the collapse of the USSR, which presented a series of depressing but thoroughly familiar conclusions.[12] The region still retains its "parasitic" relation to the European center of the country in that it continues to be unable to support itself and consumes more than it contributes, at any rate in a material sense. The high hopes of the late Soviet period for foreign commerce in the region – in particular trade with natural-resource-hungry Japan – have all come to naught. Finally, with the author's conclusion that the region's enduring significance for the rest of the country is not economic but rather military and strategic, we have a contemporary confirmation of Gennadii Nevel'skoi's prognostication from 1855 that the Amur was "long fated to remain as an army camp." It would be tempting indeed to conclude that a cycle has come full circle, were it not for the lurking suspicion that the tension between fanciful yearning and somber reality has even now not been entirely resolved, and that somehow Russia's dream of a Siberian Mississippi may well make yet another appearance.

[11] Specter, "Siberian Railroad," p. A4.

[12] Dienes, "Economic and Strategic Position."

Bibliography

Acton, Edward. *Alexander Herzen and the Role of the Intellectual Revolutionary*. Cambridge: Cambridge University Press, 1979

Adamov, E. A. "Russia and the United States at the Time of the Civil War." *Journal of Modern History* II:1 (1930), pp. 586–611

Adrianov, A. V. "Tomskaia starina." *Gorod Tomsk*. Tomsk: Pechatnoe Delo, 1912, pp. 101–183

Afonas′ev, D. "Nikolaevsk na Amure." *Morskoi Sbornik* 75:12 (1864), pp. 91–147.

A. K. "Amurskii krai, ego proshedshee, nastoiashchee i budushchee." *Semeinye vechera* (*otdel dlia starshego* vozrasta) 2 (February 1864), pp. 131–144

Akhmatov, Ivan. *Atlas istoricheskii, khronologicheskii, i geograficheskii Rossiiskogo gosudarstva sostavlennyi na osnovanii istorii Karamzina*. 2 v. St. Petersburg, 1831

Aleskeev, A. I. "Amurskaia ekspeditsiia (1849–1855)." Unpubl. doctoral dissertation. Institute of History, Academy of Sciences, Moscow, 1969

"Gennadii Ivanovich Nevel′skoi." G. I. Nevel′skoi, *Podvigi russkikh morskikh ofitserov na krainem vostoke Rossii 1849–1855*. Khabarovsk: Khabarovskoe knizhnoe iz-vo, 1969, pp. 6–26

Delo vsei zhizni. Khabarovsk: Khabarovskoe knizhnoe iz-vo, 1972

Amurskaia ekspeditsiia 1849–1855. Moscow: Mysl′, 1974

Russkie geograficheskie issledovaniia na Dal′nem vostoke i v Severnoi Amerike (XIX -nachalo XX v). Moscow: Nauka, 1976

Gennadii Ivanovich Nevel′skoi, 1813–1876. Moscow: Nauka, 1984

Alekseev, M. P. "Odin iz russkikh korrespondentov Nik. Vitsena (k istorii poiskov morskogo puti v Kitai i Indiiu v XVII veke)." *Sergeiu Fedorovichu Ol′denburgu k 50-letiiu nauchno-obshchestvennoi deiatel′nosti 1882–1932*. Leningrad: AN SSSR, 1934, pp. 51–60

"Pushkin i Kitai." G. F. Kungurov, ed., *A. S. Pushkin i* Sibir′, pp. 108–145. Irkutsk: Vostsiboblgiz, 1937

Allen, Robert V. *Russia Looks at America: The View to 1917*. Washington, DC: Library of Congress, 1988

Alter, Peter. *Nationalism*. Tr. S. McKinnon-Evans. London: Edward Arnold, 1989

Amerikantsy v Rossii i russkie v Amerike. Prazdnestva i rechi Amerikantsev i Russkikh v N′iuiorke, Peterburge, Moskve, Nizh. Novgorode, Kostrome, i podrobnoe opisanie . . . St. Petersburg, 1866

"Amurskaia Kompaniia." *Entsiklopedicheskii slovar′ sostavlennyi russkimi uchenymi i literatorami*, IV, pp. 179–180. St. Petersburg: I. I. Glazunov, 1861–1863
"Amurskii krai." *Birzhevye Vedomosti* 3 (3 January 1870), pp. 1–2
Anderson, Benedict. *Imagined Communities. Reflections on the Origins and Spread of Nationalism*. 2nd ed. London: Verso, 1991
Andree, Richard. *Das Amur-Gebiet und seine Bedeutung. Reisen in Theilen der Mongolei, den angrenzenden Gegenden Ostsibiriens, am Amur und seinen Nebenflüssen*. Leipzig: Spamer, 1867
Anisimov, S. S. *Puteshestviia P. A. Kropotkina*. Moscow–Leningrad: AN SSSR, 1943
Annenkov, P. V. *The Extraordinary Decade. Literary Memoirs*. Tr. I. R. Titunik. Ann Arbor: University of Michigan Press, 1968
Anschel, Eugene, ed. *The American Image of Russia*. New York: Frederick Ungar, 1974
Anuchin, D. N. *Iaponiia i Iapontsy*. Moscow: Tip. I. N. Kushnerev, 1907
Aref′ev, V. "M. V. Butashevich-Petrashevskii v Sibiri." *Russkaia Starina*, 109:1 (January 1902), pp. 177–186
Arendt, Hannah. *The Origins of Totalitarianism*. 2nd ed. New York: Harcourt Brace & Co, 1973
Arsen′ev, Konstantin. *Statisticheskii ocherk Rossii*. St. Petersburg: Imp. Akademii Nauk, 1848
Ascher, Abraham. "National Solidarity and Imperial Power: The Sources and Early Development of Social Imperialist Thought in Germany, 1871–1914." Unpubl. Ph.D. dissertation. Columbia University, New York, 1957
Atkinson, Thomas Witlam. *Travels in the Regions of the Upper and Lower Amoor and the Russian Acquisitions on the Confines of India and China . . .* London: Hurst & Blackett, 1860
Atlas rossiiskoi, sostoiashchei iz deviatnadtsati spetsial′nykh kart predstavliaiushchikh Vserossiiskuiu imperiiu . . . St. Petersburg: Imp. Akademiia Nauk, 1745
Avril′, F. "Svedeniia o Sibiri i puti v Kitai." *Russkii Vestnik* 10 (1842), pp. 69–104
Azadovskii, M. K. "Stranichki kraevedcheskoi deiatel′nosti dekabristov v Sibiri." M. K. Azadovskii, ed., *Sibir′ i Dekabristy*. Irkutsk: Irk. Gub. Ispoln. Komiteta, 1925, pp. 77–112
"Neosushchestvlennyi zamysel pobega dekabristov iz Chity (neopublikovannaia glava zapisok Zavalishina)." *Dekabristy i ikh vremia*. Moscow: Izd. Politkatorzhan, 1928, I, pp. 216–228
Vospominaniia Bestuzhevykh. Moscow–Leningrad: Iz-vo AN SSSR, 1931
"Putevye pis′ma Dekabrista M. A. Bestuzheva." *Zabaikal′e. Literaturno-khudozhestvennyi al′manakh* 5 (1952), pp. 206–243
Babey, Anna Mary. *Americans in Russia, 1776–1917. A Study of the American Travellers in Russia from the American Revolution to the Russian Revolution*. New York: Comet Press, 1938
Baikalov, A. V. "The Conquest and Colonization of Siberia." *Slavic and East European Review* 10:30 (1932), pp. 557–571
Bailey, Thomas. *Americans in Russia: Russian–American Relations from Early Times to Our Day*. Ithaca: Cornell University Press, 1950
Bakai, N. N. *K voprosu ob izuchenii istorii Sibiri*. Krasnoiarsk: Tip. Kudriavtseva, 1890
Sibir′ i dekabrist G. S. Baten′kov. Tomsk: Krasnoe Znamia, 1927.

Bakhrushin, S. V. *Kazaki na Amure*. Leningrad: Iz-vo Brokgauz-Efron, 1927(?)

Bakunin, M. A. "Russkim, pol'skim i vsem slavianskim druziam." *Kolokol*, nos. 122–123 (15 February 1862), pp. 1021–1028

[Bakunin, M. A.]. *Pis'ma M. A. Bakunina k A. I. Gertsenu i N. P. Ogarevu*. St. Petersburg: Izd. V. Vrublevskogo, 1906.

[Balasoglo, A. P.]. "Vostochnaia Sibir'. Zapiska o kommandirovke na ostrov Sakhalin kapitan-leitenanta Podushkina." *Chteniia v imp. obshchestve istorii i drevnostei Rossiiskikh pri Moskovskom Universitete* kn. ii (April–June 1875): ch. v, pp. 101–188

"Ispoved'." P. E. Shcheglov, ed., *Petrashevtsy. Sbornik Materialov*. 3 v. Moscow–Leningrad: Gos. Iz-vo, 1927; II, pp. 212–261

"Proekt uchrezhdeniia knizhnogo sklada s bibliotekoi i tipografiei. V. Desnitskii, ed., *Delo Petrashevtsev* 3 v. Moscow: AN SSSR, 1937–1951; II, pp. 16–47

"Vozvrashchenie." V. E. Evgravof, ed., *Filosofskie i obshchestvenno-politicheskie proizvedeniia Petrashevtsev*. [Moscow]: Gosudarstvennoe Iz-vo Politicheskoi Literatury, 1953, pp. 517–520

Bantysh-Kamenskii, N. N. *Diplomaticheskoe sobranie del mezhdu Russkim i Kitaiskim gosudarstvami c 1619 po 1792-i god*. Kazan': Tip. Imp. Un-ta, 1882

Barratt, G. *The Rebel on the Bridge. A Life of the Decembrist Andrey Rosen 1800–1884*. London: Paul Elek, 1975

Barsukov, I. P., ed. *Graf Nikolai Nikolaevich Murav'ev-Amurskii po ego pis'mam, ofitsial'nym dokumentam, rasskazam sovremennikov i pechatnym istochnikam*. 2 v. Moscow: Sinodal'naia Tipografiia, 1891

Barsukov, N. P., ed. *Zhizn' i trudy M. N. Pogodina*. 22 v. St. Petersburg: Tip. M. M. Stasiulevicha, 1888–1906

Bartol'd, V. V. *Istoriia izucheniia Vostoka v Evrope i Rossii. Socheniniia*. 9 v. Moscow: Nauka, 1963–1977, IX, pp. 197–484

Basargin, N. V. *Zapiski*. Petrograd: Ogni, 1917

Bassin, Mark. "The Russian Geographical Society, the 'Amur Epoch', and the Great Siberian Expedition 1855–1863." *Annals of the Association of American Geographers* 73:2 (1983), pp. 240–256

"Expansion and Colonialism on the Eastern Frontier: Views of Siberia and the Far East in Pre-Petrine Russia." *Journal of Historical Geography* 14:1 (1988), pp. 3–21

"Inventing Siberia: Visions of the Russian East in the early Nineteenth Century." *American Historical Review* 96:3 (June 1991), pp. 763–794

"Russia between Europe and Asia: The Ideological Construction of Geographical Space." *Slavic Review* 50:1 (Spring 1991), pp. 1–17

"Turner, Solov'ev, and the 'Frontier Hypothesis': The Nationalist Signification of Open Spaces." *Journal of Modern History* 65 (September 1993), pp. 473–511

"Russia and Asia." Nicholas Rzhevsky, ed., *Cambridge Companion to Russian Culture*. Cambridge: Cambridge University Press, forthcoming

[Baten'kov, G. S.]. "Obshchii vzgliad na Sibir'." *Syn Otechestva* (1822–1823) ch. 81: no. XLI, pp. 3–13, no. XLIV, pp. 147–159; ch. 83: no. II, pp. 53–63; 84: no. X, pp. 107–128, no. XI, pp. 145–156

"Otnoshenie k Speranskomu i Arakcheevu." M. V. Dovnar-Zapol'skii, ed., *Memuary Dekabristov*. Kiev: S. I. Ivanov, 1906

Baumann, L. A. *Kratkoe nachertanie geografii dlia nachinaiushchikh obuchat'sia sei nauke* . . . Tr. V. Ivanov. Moscow: Universitetskaia Tipografiia u N. Novika, 1788

Becker, Seymour. "Contributions to a Nationalist Ideology: Histories of Russia in the First Half of the Nineteenth Century." *Russian History/Histoire russe* 13 (1986), pp. 331–353.

"The Muslim East in Nineteenth-Century Russian Popular Historiography." *Central Asian Survey* 5: 3/4 (1986), pp. 25–47

[Belinskii, V. G.]. Review of Iakinf, *Kitai v grazhdanskom i nravstvennom otnoshenii. Sovremennik* ch. 7: otdel iii (1848), pp. 44–49

"Literaturnye mechtaniia" [1834]. *Sobranie sochinenii v trekh tomakh.* 3 v. Moscow: OGIZ, 1948, I, pp. 7–90. Iz-vo AN SSSR; I, p. 10

"Vzgliad na russkuiu literaturu 1846 goda." *Polnoe Sobranie Sochinenii.* 13 v. Moscow: Iz-vo AN SSSR, 1956, pp. 7–56

[Belogolvyi, A. V.]. Speech in honor of N. N. Murav'ev. *Irkutskie Gubernskie Vedomosti* no. 35 (28 August 1858): otdel ii, pp. 3–4.*Polnoe Sobranie Sochinenii.* Moscow: AN SSSR, 1956; X: pp. 7–56

Belomor, A. "Tsu-Shimskii epizod." *Russkii Vestnik* CCXLVIII (April 1897), pp. 234–250, and CCXLIX (May 1897), pp. 59–86

Ber, K. M. "Ob etnograficheskikh issledovaniiakh voobshche i v Rossii v osobennosti." *Zapiski Russkogo Geograficheskogo Obshchestva* I (1846), pp. 93–115

Berdiaev, N. *The Origins of Russian Communism.* Tr. R. M. French. Ann Arbor: University of Michigan Press, 1960

The Russian Idea. Boston: Beacon Press, 1962

Berezovskii, A. A. *Tamozhennoe oblozhenie i porto-franko v Priamurskom krae.* Vladivostok: Tip. "Torgovo-Promyshlennogo Vestnika Dal'nego Vostoka," 1907

Berg, L. S. *Natural Regions of the USSR.* Tr. O. Titelbaum. New York: Macmillan, 1950

Berkh, V. N. "Otkrytie kozakom Moskvitinym Vostochnogo okeana, i puteshestvie Pis'miannogo Golovy Vasiliia Poiarkova." *Syn Otechestva* XXXV (1823), pp. 49–59

Pervoe morskoe puteshestvie rossiian . . . St. Petersburg: Imperatorskaia Akademiia Nauk, 1823

Berlin, Isaiah. "Russia and 1848." *Russian Thinkers.* New York: Viking Press, 1978, pp. 1–21

"The Birth of the Russian Intelligentsia." *Russian Thinkers.* New York: Viking Press, 1978, pp. 114–135

Besprozvannykh, E. L. *Priamur'e v sisteme russko-kitaiskikh otnoshenii: XVII-seredina XIX v.* Khabarovsk: Khabarovskoe Knizhnoe Iz-vo, 1986

Bialer, Seweryn, ed. *The Domestic Context of Soviet Foreign Policy.* Boulder: Westview, 1981

[Billinge, Mark D.]. "Geosophy." R. J. Johnston *et al.*, eds., *The Dictionary of Human Geography.* Oxford: Blackwell, 1981, p. 138

Billington, J. H. *The Icon and the Axe. An Interpretive History of Russian Culture.* London: Weidenfeld & Nicolson, 1966

Billington, R. A. *Land of Savagery, Land of Promise: The European Image of the American Frontier in the Nineteenth Century.* New York: Norton, 1981

Bishop, Peter. *The Myth of Shangri-La. Tibet, Travel Writing, and the Western Creation of Sacred Landscape*. Berkeley: University of California Press, 1989

Boden, Dieter. *Das Amerikabild im russischen Schrifttum bis zum Ende des 19. Jahrhunderts*. Universität Hamburg: Abhandlungen aus dem Gebiet der Auslandkunde, Bd. 71, Reihe B, Bd. 41. Hamburg: Cram, De Gruyter, & Co., 1968

Bogdanova, A. A. "Pushkin i Sibir′." *Uchennyi zapiski Novosibirskogo Pedagogicheskogo Instituta* (seriia istoricheskaia-filologicheskaia). Vyp. 7 (1948), pp. 3–28

Bogolepov, M. "Tamozhennye bastiony v Priamur′e." *Sibirskie Voprosy* II:6 (1906), pp. 3–16.

Bogucharskii, V. [V. Ia. Iakovlev]. "Dekabrist-literator A. O. Kornilovich." In his *Iz proshlogo russkogo obshchestva*. St. Petersburg: "Geral′d," 1904, pp. 100–118

Bohatec, J. *Der Imperialismusgedanke und die Lebensauffassung Dostoewskijs. Ein Beitrag zur Kenntnis des russichen Menschen*. Graz–Köln: Hermann Böhlaus, 1951

Bolkhovitinov, N. N. *Russko-Amerikanskie otnosheniia 1815–1832*. Moscow: Nauka, 1975

"Vydvizhenie i proval proektov P. Dobbella (1812–1821 gg.)." *Amerikanskii Ezhegodnik* (1976), pp. 264–282

"How it was decided to sell Alaska." *International Affairs* 8 (1988), pp. 116–126

"The Sale of Alaska in the Context of Russian–American Relations in the nineteenth century." Hugh Ragsdale, ed., *Imperial Russian Foreign Policy*. Cambridge: Cambridge University Press, 1993, pp. 193–215

Brandt, Ferdinand. *Botschaft für alle, welche auswandern wollen. Mittheilungen über Ansiedlung in Russland durch Kauf und Pachtung von Ländereien, Gütern, Gärtnereien . . .* St Petersburg: Im Verlage des Verfassers, 1865

"Brd" [Mikhail Petrashevskii]. "Neskol′ko myslei o Sibiri." *Irkutskie Gubernskie Vedomosti*, no. 9 (11 June 1857): otdel ii, pp. 1–7

Brockway, L. H. *Science and Colonial Expansion. The Role of the British Royal Botanic Gardens*. New York: Academic Press, 1979

Brooks, Jeffrey. *When Russia learned to Read. Literacy and Popular Literature, 1861–1917*. Princeton: Princeton University Press, 1985

Broughton, W. R. *A Voyage of Discovery to the North Pacific Ocean . . . in the years 1795, 1797, and 1798*. London: T. Cadell & W. Davies, 1804 [Reprint "Bibliotheca Australiana" (no. 13), Amsterdam, 1967]

Brower, Daniel R. and Edward J. Lazzerini, eds. *Russia's Orient. Imperial Borderlands and Peoples, 1700–1917*. Bloomington: Indiana University Press, 1997

Bulychov, I. [D.]. *Puteshestvie po Vostochnoi Sibiri*. St. Petersburg: Imp. Akademiia Nauk, 1856

Bunakov, E. V. "Iz istorii russko-kitaiskikh otnoshenii v pervoi polovine XIX v." *Sovetskoe Vostokovedenie* 2 (1956), pp. 96–104

Burachek, E. "Amurskoe sudokhodstvo." *Sovremennaia Letopis′* no. 21 (June 1864), pp. 1–4

"Gde dolzhen byt′ russkii port na Vostochnom okeane?" *Morskoi Sbornik* 80:9 (1865), pp. 3–14

Bury, J. P. T. "Nationalities and Nationalism." *New Cambridge Modern History*. 14 v. Cambridge: Cambridge University Press, 1957–1979; X, pp. 213–245

[Büsching, A. F.]. *D. Antona Frederika Bishinga iz sokrashchennoi ego Geografii tri glavy o geografii voobshche . . .* Moscow: Imperatorskii Moskovskii Universitet, 1766

Buszynski, Leszek. *Russian Foreign Policy after the Cold War*. Westport, Connecticut: Praeger, 1996

Butsinskii, P. N. *Graf N. N. Murav'ev-Amurskii. Sochinenie I. Barsukova*. St. Petersburg: Imp. Akademiia Nauk, 1895

Bykonia, G. F. "Vzgliady G. S. Baten'kova na russkoe zaselenie Sibiri v XVII–XVIII vv." *Dekabristy i* Sibir'. Novosibirsk: Nauka, 1977, pp. 65–73

Cameron, M. E., *et al. China, Japan, and the Powers: A History of the Modern Far East*. 2nd ed. New York: Ronald Press Co., 1960

Carr, E. H. *Michael Bakunin*. London: Macmillan, 1937

Carter, Paul. *The Road to Botany Bay: An Exploration of Landscape and History*. Chicago: University of Chicago Press, 1987

Charushin, A. "Voprosy kolonizatsii na Dal'nem Vostoke." *Vestnik finansov, promyshlennosti, i torgovli* 8 (19 February 1906), pp. 293–299

Chernosvitov, R. A. "O dorogovizne khleba v Eniseiske," *Irkutskie Gubernskie Vedomosti* 2 (8 January 1859), otdel ii, pp. 7–9

"Korrespondentsiia. Krasnoiarsk," *Irkutskie Gubernskie Vedomosti* 51 (17 December 1859), otdel ii, pp. 3–5

"Neskol'ko slov o tom, kto prilozhil k delu nachala svobody v torgovle: utopisty, ili liudi prakticheskie." *Irkutskie Gubernskie Vedomosti* 13 (26 March 1860), chast' neoff., pp. 1–8

Chernov, S. N. "K istorii fonda Sledstvennoi Kommissii i Verkhovnogo Ugolovnogo Suda po delu dekabristov," *Arkhivnoe Delo* III–IV (1925), pp. 108–113

[Chichagov, P. V.]. "Zapiski Admirala Chichagova . . ." *Arkhiv Admirala P. V. Chicagova. Vypusk pervyi*. St. Petersburg: S. Dobrodeev, 1885, pp. 39–125

Chinard, Gilbert. *L'Amérique et le rêve exotique dans la littérature française au XVIIe et XVIIIe siècle*. Paris: Hachette, 1913

Chulkov, G. "Rafail Chernosvitov," *Katorga i Ssylka* 3 (1930), pp. 80–96

Clay, C. B. "Ethos and Empire: The Ethnographic Expedition of the Imperial Russian Naval Ministry, 1855–1862." Unpubl. Ph.D. thesis. University of Oregon, Eugene, 1989

Clubb, O. E. *China and Russia: The "Great Game."* New York: Columbia University Press, 1971

Cole, A. B., ed. *Yankee Surveyors in the Shogun's Seas*. Princeton: Princeton University Press, 1947

Collins, Perry McDonough. *Siberian Journey down the Amur to the Pacific 1856–1857*. Madison: University of Wisconsin Press, 1962

Conzen, Michael P. *et al. A Scholar's Guide to Geographical Writing on the American and Canadian Past*. (University of Chicago Geographical Research Paper, no 235). Chicago: University of Chicago Press, 1993

Costin, W. C. *Great Britain and China 1833–1860*. Oxford: Clarendon Press, 1937

Curtin, Philip D. *The Image of Africa: British Ideas and Action 1780–1850*. Madison University of Wisconsin Press, 1964

de Custine, Marquis A. *La Russie en 1839*. 8 v. Brussels: Wouters, 1843

Dahlmann, Dittmar. "Zwischen Europa und Asien: Russischer Imperialismus im 19

Jahrhundert." Wolfgang Reinhard, ed. *Imperialistische Kontinuität und nationale Ungeduld im 19. Jahrhundert*, pp. 50–67. Frankfurt: Fischer, 1991

Danilevskii, N. Ia. *Rossiia i Evropa. Vzgliad na kul'turnye i politicheskie otnosheniia Slavianskogo mira k Germano-Romanskomu.* 5th ed. St. Petersburg: Brat'ia Panteleevye, 1895

"Das neue Armeekorps der Baikal-Kosaken und das See-Departement der Russen am Stillen Ocean." *Petermanns Mittheilungen* (1856), p. 387

"Das Projekt eines Telegraphen-Gürtels um die ganze Erde." *Petermanns Mittheilungen* (1855), pp. 91–92

Das veränderte Russland, in welchem die jetzige Verfassung der Kriegestand zu Land- und Wasser vorgestellt werden. Erster Theil. Frankfurt: Nicolaus Förster, 1721

Dement'ev, A. G. *Ocherki po istorii russkoi zhurnalistiki 1840–1850gg.* Moscow–Leningrad: Gosudarstvennoe izdatel'stvo khudozhestvennoi literatury, 1951

Dement'ev, A. G., *et al. Russkaia periodicheskaia pechat' (1702–1894).* Moscow: Gosudarstvennoe Iz-vo Politicheskoi Literatury, 1959

Dennett, T. *Americans in Eastern Asia.* New York: Barnes & Noble, 1941

Des veränderten Russlands zweiter Theil. Hannover: Nicol[aus] Försters & Sohns, 1739

Desnitskii, V. A. ed., *Delo Petrashevtsev.* 3 v. Moscow: AN SSSR, 1937–1951

"Die neuste Expeditionen im Amur-Land und auf der Insel Sachalin." *Petermanns Mittheilungen* (1861), pp. 314–319

"Die neuste russischen Erwerbungen im Chinesischen Reiche." *Petermanns Mittheilungen* (1858), pp. 175–186

Dienes, Leslie. "Economic and Strategic Position of the Soviet Far East: Development and Prospect." A. Rodgers, ed., *The Soviet Far East.* London: Routledge, 1990, pp. 269–301

Diment, Galya. "Exiled *from* Siberia: The Construction of Siberian Experience by Early Nineteenth-Century Irkutsk Writers." *Idem* and Yuri Slezkine, eds., *Between Heaven and Hell: The Myth of Siberia in Russian Culture*, pp. 47–66. New York: St. Martin's Press, 1993

Diment, Galya and Yuri Slezkine, eds., *Between Heaven and Hell: The Myth of Siberia in Russian Culture.* New York: St. Martin's Press, 1993

von Ditmar, Karl. *Reisen und Aufenthalt in Kamtschatka in den Jahren 1851–1855. Beiträge zur Kenntnis des Russischen Reiches* 3. Folge; VII–VIII (1890–1900)

Dneprov, E. D. "*Morskoi Sbornik* v obshchestvennom dvizhenii perioda pervoi revoliutsionnoi situatsii v Rossii." *Revoliutsionnaia Situatsiia v Rossii v 1859–61 gg.* 4 v. Moscow: Nauka, 1965; IV, pp. 229–258

[Dobell]. "O Pertopavlovskom porte (iz zapisok g. Dobelia). *Syn Otchestva* ch. XXVI: no. XLVII, pp.53–56

[Dobroliubov, N. A.]. "Puteshestvie na Amur." *Polnoe Sobranie Sochinenii.* 6 v. Moscow: "Khudozhestvennaia Literatura," 1934–1939; IV, pp. 402–418

Doklad komiteta ob ustroistve russkikh amerikanskikh kolony. 2 v. St. Petersburg: Tip. Departamenta Vneshnei Torgovli, 1863–1864

[Dolgorukov, P. V.]. "Pis'mo iz Peterburga." *Kolokol*, no. 237 (15 March 1867), pp. 1933–1936

"Donesenie Kollinsa ob Amure." *Irkutskie Gubernskie Vedomosti* 28 (10 July 1958), pp. 7–8

Dopolneniia k aktam istoricheskim. 12 v. St. Petersburg: Arkheograficheskaia kommissiia, 1846–1872

Dostoevskii, F. M. "Geok Tepe. Chto dlia nas Aziia." *Polnoe Sobranie Sochinenii*. 30 v. Leningrad: Nauka, 1972–1990; XXVII, pp. 32–40

Dovnar-Zapol'skii, M. V. *Idealy dekabristov*. Moscow: I. D. Sytin, 1906

D. R. [Dmitrii Romanov]. "Pis'ma s reki Amura." *Russkii Vestnik* no. 16:2 (August 1858), pp. 364–370, and no. 17:1 (September 1858), pp. 15–28

Dubrovin, N. F., ed. *N. M. Przheval'skii. Biograficheskii ocherk*. St. Petersburg: Voennaia Tipografiia, 1890

Dulles, F. R. *America in the Pacific. A Century of Expansion*. Boston: Houghton Mifflin, 1932

Dulov, A. V. "K. Marks i F. Engel's o Sibiri 50–60-kh gg. IXI v." *Zapiski Irkutskogo oblastnogo kraevedcheskogo muzeia* (1965), pp. 8–21

"Sibirskaia publitsistika M. V. Petrashevskogo," *Trudy Irkutskogo Gosudarstvennogo Instituta im. A. A. Zhdanova* (Seriia istoricheskaia), 59:2 (1970), pp. 3–15

"Neizvestnye stat'i M. A. Bakunina v gazete 'Amur'." L. M. Goriushkin, ed., *Ssylka i katorga v* Sibiri. Novosibirsk: Iz-vo Nauka, 1975, pp. 161–171

Echeverria, D. *Mirage in the West: A History of the French Image of American Society to 1815*. Princeton: Princeton University Press, 1957

Efimov, I. V. *Gr. N. N. Murav'ev-Amurskii pered sudom Prof. Butsinskogo*. St. Petersburg: P. P. Soikin, 1896

Egorov, B. F. *Petrashevtsy*. Leningrad: Nauka, 1988

Eima, Ksavie. "Issledovaniia ob Amure i torgovle Vost. Sibiri." *Zhurnal Ministerstva Torgovli* 8:11 (1859), pp. 43–78

"Ekspeditsii russkikh v priamurskii krai," *Russkii Khudozhestvennyi Listok* 4 (1 February 1859), pp. 15–16; 5 (10 February 1859), pp. 17–18; 9 (20 March 1859), pp. 25–26

"Ekspeditsiia g. Ottona Eshe na Amur." *Vestnik Imperatorskogo Russkogo Geograficheskogo Obshchestva* 24:11 (1858), pp. 39–44

"Elena Pavlovna." *Entsiklopedicheskii Slovar' (Granat)*. 46 v. Moscow: "Granat," nd, XX, pp. 27–28

Eley, G. "Defining Social Imperialism: Use and Abuse of an Idea," *Social History* 3 (October 1976), pp. 265–290

Engels, F. Letter to Marx, 23 May 1851. In *Marx-Engels Werke*. 42 v. Berlin: Institut für Marxismus-Leninismus beim ZK der SED, 1957–1985, XXVII, pp. 265–268

Entsiklopedicheskii Slovar' (Granat). 50 v. Moscow: Glavnaia kontora tv-va Br. Granat, 1909–1947

Epov, A. "Dekabristy v Zabaikal'e." A. V. Kharchevnikov, ed., *Dekabristy v Zabaikal'e. Neizdannye materialy*. Chita: Pechatnoe Delo, 1925

Esin, B. I. *Russkaia dorevoliutsionnaia gazeta 1702–1917. Kratkii Ocherk*. Moscow: Iz-vo Moskovskogo Universiteta, 1971

von Etzel, A. and H. Wagner. *Reisen in den Steppen und hochgibirgen Sibiriens und der angrenzenden Länder Central Asiens*. Lepizig: O. Spamer, 1864

Evans, John. *The Russo-Chinese Crisis: N. P. Ignatiev's Mission to Peking, 1859–1860*. Newtonville, Massachusetts: Oriental Research Partners, 1987

Evans, John L. *The Petrasevskij Circle 1845–1849*. The Hague: Mouton, 1974

Evgen′ev-Maksimov, V. *"Sovremennik" v 40–50 gg. Ot Belinskogo do Chernyshevskogo*. Leningrad: Iz-vo Pisatelei v Leningrade, 1934

Fairbank, J. K. *Trade and Diplomacy on the China Coast. The Opening of the Treaty Ports 1842–1854*. Cambridge, Massachusetts: Harvard University Press, 1953

Fedotov, G. P. "The Russian." *Russian Review* 13:1 (January 1954), pp. 3–17

"Fel′dmarshal Kniaz′ Bariatinskii." *Russkii Arkhiv* XXVII:1 (1889), pp. 107–141

"Finansovoe Polozhenie." *Kolokol* 222 (15 June 1866), pp. 1815–1818

Fisher, Raymond H. *The Russian Fur Trade, 1550–1700*. Berkeley: University of California Press, 1943

Bering's Voyages: Whither and Why. Seattle: University of Washington Press, 1977

Flynn, James T. "Shishkov, Aleksandr Semenovich." *Modern Encyclopedia of Russian and Soviet History*. 59 v. Gulf Breeze, Florida: Academic International Press, 1976–1996; XXXV, pp. 1–3

Forsyth, James. *A History of the Peoples of Siberia. Russia's North Asian Colony, 1851–1990*. Cambridge: Cambridge University Press, 1992

Foust, Clifford M. "Russian Expansion to the East through the Eighteenth Century." *Journal of Economic History*, XXI:4 (December 1961), pp. 467–482

Muscovite and Mandarin. Russia's Trade with China and its Setting. Chapel Hill: University of North Carolina Press, 1969

"Fr.Aug. Lühdorfs Schilderung der Wichtigkeit des russischen Besitzes vom Amur-Strom und seine Reise von dessen Mündung bis Moskau." *Petermanns Mittheilungen* (1858), pp. 334–336

Friesen, P. M. *The Mennonite Brotherhood in Russia, 1789–1910*. Tr. J. B. Toews. Fresno, California: Board of Christian Literature, 1978

[Frolov, A. F.]. "Vospominaniia i zametki Aleksandra Filippovicha Frolova." *Russkaia Starina* 34:5 (May 1882), pp. 465–482 and 34:6 (June 1882), pp. 701–714

Frye, Richard S. "Oriental Studies in Russia." W. S. Vucinich, ed., *Russia and Asia. Essays on the Influence of Russia on the Asian* Peoples. Palo Alto: Hoover Institution Press, 1972, pp. 30–52

Fuller, W. C. *Strategy and Power in Russia 1600–1914*. New York: Free Press, 1992

Gagemeister, Iu. A. *O rasprostranenii Rossiiskogo gosudarstva s edinoderzhaviia Petra I-ogo do smerti Aleksandra I-ogo*. St. Petersburg: N. Grech, 1835

[Gakman, Iogann]. *Kratkoe zemleopisanie Rossiiskogo gosudarstva, izdannoe dlia narodnykh uchilishch Rossiiskoi imperii . . .* St. Petersburg: no pub., 1787

[Gakman, Iogann and Fedor Iankovich]. *Vseobshchee zemleopisanie, izdannoe dlia narodnykh uchilishch Rossiiskoi* Imperii . . . 2nd ed. 2 v. St. Petersburg: [Vil′kovskii], 1795–1798

Gedensh[trom], M. [M]. "Ob Amure." *Moskovskie vedomosti* no. 70 (1837), p. 518

Gedenshtrom, M. M. *Otryvki o Sibiri*. St. Petersburg: Meditsinskii Departament Min. Vn. Del, 1830

Geograficheskо-Statisticheskii Slovar′ Rossiiskoi Imperii. 5 v. St. Petersburg: V. Bezobrazov, 1863–1885

Gerbi, Antonello. *The Dispute of the New World. The History of a Polemic, 1750–1900*. Tr. J. Moyle. Pittsburgh: University of Pittsburgh Press, 1973

Gerschenkron, Alexander. "The Problem of Economic Development in Russian Intellectual History of the Nineteenth Century." E. J. Simmons, ed., *Continuity and Change in Russian and Soviet Thought*. New York: Russell & Russell, 1955, pp. 11–39

"Early Phases of Industrialization in Russia. Afterthoughts and Counterthoughts." W. W. Rostow, ed., *The Economics of Take-Off into Sustained Growth*. London: Macmillan, 1963, pp. 151–169

"Russia: Agrarian Policies and Industrialization 1861–1917." In his *Continuity in History and Other* Essays. Cambridge, Massachusetts: Harvard University Press, 1968, pp. 140–248

Gersevanov, N. M. [?]. "Zamechaniia o torgovykh otnosheniiakh Sibiri k Rossii." *Otchestvennye Zapiski* XIV (1841), otdel iv, pp. 23–34

Gerstfeldt, G. "Über die Zukunft des Amur-Landes." *Petermanns Mittheilungen* (1860), pp. 93–106

[Gertsen, A. I.]. "Neschastnye soldaty ostavlennye na golodnuiu smert'." *Kolokol* no. 63 (15 February 1860), pp. 527–528

"Gonenie na krymskikh tatar." *Kolokol* no. 117 (22 December 1861), pp. 966–967

"Russkim ofitseram v Pol'she." *Kolokol* no. 147 (15 October 1862), pp. 1213–1215

"Amerika i Rossiia." *Kolokol* no. 228 (1 October 1866), pp. 1861–1862

Gersten, A. I. "Pis'mo iz provintsii" [1836]. *Sobranie Sochinenii v 30 tomakh*. 30 v. Moscow: AN SSSR, 1954–1965; I, pp. 131–133

Letter to N. Kh. Ketcher (20 August 1838). *Sobranie Sochenenii v 30 tomakh*. 30 v. Moscow: AN SSSR, 1954–1965; XXI, pp.386–387

"Vil'iam Pen" [1839]. *Sobranie Sochinenii v 30 tomakh*. 30 v. Moscow: AN SSSR, 1954–1965; I, pp. 196–250

Byloe i dumy. *Sobranie Sochinenii v 30 tomakh*. 30 v. Moscow: AN SSSR, 1954–1965; VIII–XI

"Kreshchenaia sobstvennost'" [1853]. *Sobranie Sochinenii v 30 tomakh*. 30 v. Moscow: AN SSSR, 1954–1965; XII, pp. 94–117

"La Russie et le vieux monde" [1854]. *Sobranie Sochinenii v 30 tomakh*. 30 v. Moscow: AN SSSR, 1954–1965; XII, pp. 134–200

"Pis'mo k Dzhuzeppe Matstsini o sovremennom polozhenii Rossii" [1857]. *Sobranie Sochinenii v 30 tomakh*. 30 v. Moscow: AN SSSR, 1954–1965; XII, pp. 348–356

Letter to N. I. Sazonov and N. Kh. Ketcher (18 July 1835). *Sobranie Sochinenii v 30 tomakh*. 30 v. Moscow: AN SSSR, 1954–1965; XXI, pp. 43–47

Geyer, Dietrich. "Russland als Problem der vergleichenden Imperialismusforschung." R. v. Thadden, *et al.*, eds., *Das Vergangene und die Geschichte. Festschrift für Reinhard Wittram zum 70. Geburtstag*. Göttingen: Vandenhoeck & Ruprecht, 1973, pp. 337–368

Der russische Imperialismus. Studien über den Zusammenhang von innerer und auswärtiger Politik 1860–1914. Göttingen: Vandenhoeck & Ruprecht, 1977

"Modern Imperialism? The Tsarist and Soviet Examples." W. J. Mommsen and J. Osterhammel, eds., *Imperialism and After*. London: Allen & Unwin, 1986, pp. 49–62

Gibner. Ia. *Zemnovodnogo kruga kratkoe opisanie iz starie i novye geografii*. Tr. P. B. Inokhodtsev. Moscow: no pub., 1719

Gibson, James R. "Russia on the Pacific: the Role of the Amur." *Canadian Geographer* 12:1 (1968), pp. 15–27

"Sables to Sea Otters. Russia enters the Pacific." *Alaska Review* III:2 (Fall–Winter 1968–1969), pp. 203–217

Feeding the Russian Fur Trade: Provisionment of the Okhotsk Seaboard and the Kamchatka Peninsula, 1639–1856. Madison: University of Wisconsin Press, 1969

"Russian occupancy of the Far East 1639–1750." *Canadian Slavonic Papers* 12 (1970), pp. 60–77

"The Significance of Siberia to Tsarist Russia," *Canadian Slavonic Papers* 14 (1972), pp. 442–453

Imperial Russia in Frontier America: The Changing Geography of Supply of Russian America 1784–1867. New York: Oxford University Press, 1976

"European Dependence upon American Natives: The Case of Russian America." *Ethnohistory* 25:4 (Fall 1978), pp. 359–385

"Russian Expansion in Siberia and America." *Geographical Review* 70:2 (April 1980), pp. 127–136

"The Sale of Russian America to the United States." *Acta Slavica Iaponica* I (1983), pp. 15–37

Gil′ferding, A. F. "Mnenie zapadnykh slavian ob Amure i ego kolonizatsii." *Amur* no. 26 (28 June 1860), pp. 372–375

"Pis′mo k G. Rigeru v Pragu o Russko-Pol′skikh delakh." *Sobranie Sochinenii A. Gil′ferdinga*. 4 v. St. Petersburg: V. Golovin, 1868–1874; II, pp. 277–288

Gillard, David. *The Struggle for Asia 1828–1915. A Study in British and Russian Imperialism*. London: Methuen, 1977

Gleason, Abbott. "Republic of Humbug: The Russian Nativist Critique of the United States, 1830–1930." *American Quarterly* 44:1 (March 1992), pp. 1–23

Gnevucheva, V. F. *Materialy dlia istorii ekspeditsii Akademii Nauk v XVIII–XIX vekakh*. Trudy Arkhiva AN SSSR, Vyp. 4. Moscow–Leningrad: AN SSSR, 1940

Gogol′, N. V. *Mertvye dushi. Poema* [1842]. Moscow: Khudozhestvennaia Literatura, 1979

Golder, Frank. "Russian–American Relations during the Crimean War." *American Historical Review* XXXI:3 (1926), pp. 462–476

Gollwitzer, H. *Europe in the Age of Imperialism 1880–1914*. Tr. D. Adam and S. Baron. New York: Harcourt, Brace & World, 1969

Goncharov, I. A. "Po Vostochnoi Sibiri. V Iakutske i v Irkutske." *Russkoe Obozrenie* II:1 (January 1891), pp. 5–29

Fregat "Pallada." Sobranie Sochinenii. 8 v. Moscow: Gos. Iz-vo Khudozhestvennoi Literatury, 1952–1955; II–III

Fregat "Pallada." Leningrad: Nauka, 1986

Gooding, J. "Speransky and Baten′kov." *Slavic and East European Review* 66:3 (1988), pp. 400–425

[Gorchakov, A. M.]. "Circular dispatch addressed by Prince Gortchakov to Russian Representatives abroad." Dated 21 November 1864. *Parliamentary Papers*, Great Britain. *Accounts and Papers* LXXV:37 (1873), C-704, no. 2, appendix, pp. 70–75

"Gorchakov, Aleksandr Mikhailovich." *Sovetskaia Istoricheskaia Entsiklopediia*. 17 v. Moscow: "Sovetskaia Entsiklopediia," 1961–1976; IV, pp. 599–600

Gosudarstvennyi Sovet. *Stenograficheskie Otchety. 1907–1908 gg*. Sessiia tret′ia. St. Petersburg: Gosudarstvennaia Tipografiia, 1907–1908

Grebnitskii, N. A. "Svedeniia o torgovle v Amurskom krae." *Izventiia Sibirskogo Otdela Russkogo Geograficheskogo Obshchestva* X: 1–2 (1879), pp. 59–62

Greenberger, Allen J. *The British Image of India: A Study in the Literature of Imperialism 1880–1960*. London: Oxford University Press, 1969

Greene, Jack P. *The Intellectual Construction of America. Exceptionalism and Identity from 1492 to 1800*. Chapel Hill: University of North Carolina Press, 1993

Greenfeld, Liah. *Nationalism: Five Roads to Modernity*. Cambridge, Massachusetts: Harvard University Press, 1992

Griffin, Eldon. *Clippers and Consuls. American Consular and Commercial Relations with Eastern Asia, 1845–1860*. Ann Arbor: Edwards Bros., 1938

Grigor′ev, V. V. *Ob otnoshenii Rossii k Vostoku*. Odessa: n.p., 1840

"O torgovykh snosheniiakh mezhdu tuzemtsami severo-vostochnogo berega Azii i severo-zapadnoi Ameriki." *Zhurnal Ministerstva Vnutrennikh Del* XXXV:7 (1851), pp. 102–112

Grimsted, P. K. *The Foreign Ministers of Alexander I*. Berkeley: University of California Press, 1969

de Grunwald, Constantin. *Tsar Nicholas I*. Tr. B. Patmore. London: Douglas Saunders, 1954

G. S. [Grigorii Spasskii]. "Istoricheskie i statisticheskie zapiski o mestakh, lezhashchikh pri reke Amure." *Sibirskii Vestnik* ch. I (1824):kn. 6, pp. 175–186; ch. II:kn. 7, pp. 187–200; kn. 8, pp. 201–218; kn. 9–10, pp. 219–238; kn. 11, pp. 239–256; kn. 12, pp. 257–264; ch.III:kn. 13–14, pp. 265–272

Gubanov, F. "Pis′mo Sibiriaka na 'Pis′mo s Amura'." *Russkii Mir* 37 (8 August 1859), pp. 746–748

Gurevich, A. V., ed. *Vostochnaia Sibir′ v rannei khudozhestvennoi proze*. Irkutsk: Irkutskoe Oblastnoe Iz-vo, 1938

"Gustav Radde's asiatische Reisen und Sammlungen." *Petermanns Mittheilungen* (1860), p. 275

Gvozdev, A. "Otvet na nemnogie, no ves′ma trevozhnye slukhi ob Amure." *Sanktpeterburgskie Vedomosti* 35 (16 February 1860), p. 159

Hauner, Milan. *What is Asia to Us? Russia's Asian Heartland Yesterday and Today*. London: Routledge, 1992.

Hayes, C. J. H. *The Historical Evolution of Modern Nationalism*. New York: Macmillan, 1931

A Generation of Materialism 1871–1900. New York: Harper & Row, 1963

Hecht, David. *Russian Radicals look to America 1825–1894*. Cambridge Massachusetts: Harvard University Press, 1947

Hill, S. S. *Travels in Siberia*. 2 v. London: Longman, 1854

Hoetzsch, Otto. *Russland in Asien. Geschichte einer Expansion*. Stuttgart: Deutsche Verlagsanstalt, 1960

Hoffman, Erik P. and Frederick J. Flern, Jr., eds. *The Conduct of Russian Foreign Policy*. London: Butterworth, 1971

Hokanson, Katya. "Literary Imperialism, *Narodnost′*, and Pushkin's Invention of the Caucasus." *Russian Review* 53:3 (July 1994), pp. 336–352

Holborn, Hajo. "Russia and the European Political System." I. J. Lederer, ed., *Russian*

Foreign Policy. Essays in Historical Perspective. New Haven: Yale University Press, 1962, pp. 377–416

Holl, Bruce. "Avvakum and the Genesis of Siberian Literature." Galya Diment and Yuri Slezkine, eds., *Between Heaven and Hell: The Myth of Siberia in Russian Culture*, pp. 33–46. New York: St. Martin's Press, 1993

Honour, Hugh. *Chinoiserie: The Vision of Cathay*. London: John Murray, 1961

The New Golden Land. European Images of America from the Discoveries to the Present Time. London: Allen Lane, 1975

Hudson, G. F. "The Far East." *The New Cambridge Modern History*. 14 v. Cambridge: Cambridge University Press, 1957–1979; X, pp. 685–713

Hugo, Victor. "Les Orientales" [1829]. *Odes et Ballades*. Paris: Garnier et Flammarion, 1968

Hunczak, Taras. "Pan-Slavism or Pan-Russianism." In *idem*, ed., *Russian Imperialism from Ivan the Great to the Revolution*. New Brunswick: Rutgers University Press, 1974, pp. 82–105

Hunt, M. H. *The Making of a Special Relationship. The United States and China to 1914*. New York: Columbia University Press, 1983

Huttenbach, H. R. "The Origins of Russian Imperialism." Taras Hunczak, ed., *Russian Imperialism from Ivan the Great to the Revolution*. New Brunswick: Rutgers University Press, 1974

I. A. Review of R. Maak, *Puteshestvie na Amur . . . Otechestvennye Zapiski* 127:11 (1859), otdel iii, pp. 9–22

Iadrintsev, N. M "Sibir′ pered sudom russkoi literatury" [1865]. *Literaturnoe nasledstvo Sibiri* 5 (1980), pp. 21–28

Sibir′ kak koloniia v geograficheskom, etnograficheskom, i istoricheskom otnoshenii. 2nd ed. St. Petersburg: Izd. I. M. Sibirakova, 1892

Iakovlev, L. *Rasskazy ob Amure i o zelenom klinu*. Moscow: I. D. Sytin, 1902

Iakovleva, P. T. *Pervyi russko-kitaiskii dogovor 1689 goda*. Moscow: AN SSSR, 1958

Iakushin, V. E. *Gosudarstvennaia vlast′ i proekty gosudarstvennoi reformy v Rossii*. St. Petersburg: Al′tshuler, 1906

Iamaguchi. "Materialy po istorii snoshenii Rossii s Iaponiei." *Zapiski Priamurskogo Otdela Imperatorskogo Obshchestva Vostokovedeniia* III (1915), pp. 1–8

Ignat'ev, A. V. "The Foreign Policy of Russia in the Far East at the Turn of the Nineteenth and Twentieth Centuries." Hugh Ragsdale, ed., *Imperial Russian Foreign* Policy. Cambridge: Cambridge University Press, 1993, pp. 247–267

Ignat′ev, N. P. *Otchetnaia zapiska podannaia v Aziatskii departament v Ianvare 1861 goda N. P. Igat′evym* . . . St. Petersburg: V. V. Komarov, 1895

Imberkh, A. K. "Pis′mo A. K. Imberkha D. I. Zavalishinu, 29 Maia 1859." *Sbornik starinnykh bumag, khraniashchikhsia v muzee P. I. Shchukina*. 10 v. Moscow: A. I. Mamontov, 1900–1902; IX, pp. 217–236

"Imperatorskii kabinet i Murav′ev-Amurskii." *Kolokol* no. 87–88 (15 December 1860), pp. 721–742

I. N. [Ivan Noskov]. "O sibirskoi zheleznoi doroge." *Irkutskie Gubernskie Vedomosti* 43 (23 October 1858), otdel ii, pp. 1–5

"Inostrannye Izvestiia." *Sanktpeterburgskie Vedomosti* 118 (3 June 1858), pp. 697–698

I-r [A. I. Gertsen]. "Rossiia i Pol′sha." *Kolokol* no. 66 (15 March 1860), pp. 539–544; no. 67 (1 April 1860), pp. 555–558

"Graf Murav′ev-Amurskii i ego poklonniki." *Kolokol* no. 109 (15 October 1861), pp. 910–912
"Proklamatsiia 'zemli i voli'." *Kolokol* no. 160 (1 April 1863), p. 1318
Iskander [A. I. Gertsen]. "Amerika i Sibir′." *Kolokol* no. 29 (1 December 1858), pp. 233–235
"Rossiia i Pol′sha." *Kolokol* nos. 32–33 (1 January 1859), pp. 257–260
"Rossiia i Pol′sha." *Kolokol* no. 66 (15 March 1860), pp. 539–544, no. 67 (1 April 1860), pp. 555–558
"Kontsy i nachala (pis′mo vtoroe)." *Kolokol* no. 156 (15 February 1863), pp. 1297–1299
"Istoricheskie akty o podvigakh Erofeia Khabarova, na Amure, v 1649–1651 gg." *Syn Otechestva* 1 (1840), pp. 85–126
"Istoricheskie izyskaniia o naselenii russkikh na Amure." *Biblioteka dlia Chteniia* 3 (1859), pp. 22–33
Istoriia reki Amura, sostavlennaia iz obnarodovannykh istochnikov, s planami r. Amura. St. Petersburg: v tip. E. Veimara, 1859
Iurgenson, P. B. *Nevedomymi tropami Sibiri.* Moscow: Mysl′, 1964.
Ivanov, I. E. "Shishkov, Aleksandr Semenovich." *Sovetskaia Istoricheskaia Entsiklopediia.* 17 v. Moscow: "Sovetskaia Entsiklopediia," 1961–1976; XVI, pp. 287–288
"Iz pis′ma nemetskogo kuptsa (Otto Eshe), byvshogo na Amure." *Zhurnal dlia aktsionerov* no. 98 (1858), p. 1030
"Iz Sibiri." *Kolokol* no. 131 (1 May 1862), pp. 1089–1092
Izdatel′ *Kolokola* [A. I. Gertsen]. "Russkim offitseram v Pol′she." *Kolokol* no. 147 (15 October 1862), pp. 1213–1215
James, J. A. *The First Scientific Exploration of Russian America and the Purchase of Alaska.* Evanston: Northwestern University Press, 1942
Jefferson, Thomas. *The Writings of Thomas Jefferson.* 20 v. Washington, DC: The Thomas Jefferson Memorial Organization, 1907
Jensen, Ronald J. *The Alaska Purchase and Russian–American Relations.* Seattle: University of Washington Press, 1975
J. G. S. [James G. Swan]. "Explorations of the Amoor River." *Hunt's Merchant Magazine* XXXIX (August 1858), pp. 176–182
Johnston, R. J. *Geography and Geographers. Anglo-American Human Geography since 1945.* 2nd ed. London: Arnold, 1983
Kabanov, P. I. *Amurskii vopros.* Blagoveshchensk: Amurskoe Knizhnoe Iz-vo, 1959
Kappeler, Andreas. "Die Anfänge eines russischen China-Bildes im 17. Jahrhundert." *Saeculum* 31 (1980), pp. 27–43
Kapterev, N. F. *Kharakter otnoshenii Rossii k pravoslavnomu Vostoku v XVI–XVII stoletiiakh.* 2nd ed. Sergeev Posad: M. S. Elov, 1914
Karamzin, N. M. *Istoriia Gosudarstva Rossiiskogo.* 12 v. St. Petersburg: Izdanie E. Evodokimova, 1892
Karmannoi atlas Rossiiskoi imperii. St. Petersburg: Imp. Academiia Nauk, 1773
Karpov, G. V. *Issledovatel′ zemli Sibirskoi P. A. Kropotkin.* Moscow: Gos. Iz-vo Geograficheskoi Literatury, 1961
Karpovich, Michael. "Pushkin as a Historian." S. H. Cross and E. J. Simmons, eds.,

Centennial Essays for Pushkin, pp. 181–200. Cambridge, Massachusetts: Harvard University Press, 1937

Kashik, O. I. "Torgovlia v Vostochnoi Sibiri v XVII-nachale XVIII vv." *Voprosy istorii Sibiri i Dal'nego Vostoka*. Novosibirsk: Iz-vo Sibirskogo Otdeleniia AN SSSR, 1961, pp. 187–198

Katkov, M. N. *Sobranie peredovykh statei Moskovskikh Vedomostei. 1863 god*. Moscow: V. V. Chicherin, 1897

Kaufman, A. A. "Pereselenie." *Entsiklopedicheskii* slovar' (*Brokgauz-Efron*). St. Petersburg: Tip-lit. I. A. Efrona, 1898, XXXIII, pp. 265–281

"Po Amuru i Priamur'iu." In his *Po novym mestam. Ocherki i putevye zametki 1901–1903*. St. Petersburg: Obshchestvennaia Pol'za, 1905, pp. 1–101

Kazemzadeh, Firuz. "Russia and the Middle East." I. J. Lederer, ed, *Russian Foreign Policy. Essays in Historical Perspective*. New Haven: Yale University Press, 1962, pp. 489–530

Keeble, Curtis, ed. *The Soviet State: The Domestic Roots of Soviet Foreign Policy*. Boulder: Westview, 1985

Kennedy, P. M. "German Colonial Expansion: Has the 'Manipulated Social Imperialism' been Ante-Dated?" *Past and Present* 54 (February 1972), pp. 134–141

Kerner, Robert. "Russian Expansion to America, Its Bibliographical Foundations." *Papers of the Bibliographical Society of America* 25 (1931), pp. 111–129

The Urge to the Sea. The Course of Russian History: The Role of Rivers, Portages, Ostrogs, Monastaries, and Furs. Berkeley: University of California Press, 1942

Kharnskii, K. *Kitai s drevneishikh vremen do nashikh dnei*. Khabarovsk–Vladivostok, 1927

Khlebnikov, K. "Zapiski o Kalifornii." *Syn Otechestva* 124: II (1829), pp. 208–227, 276–288, 336–347, 400–410; 125: III (1829), pp. 25–35

Khomiakov, A. S. *Ermak. Tragediia v piati deistviiakh. Polnoe sobranie sochineniia A. S. Khomiakova*. 4 v. Moscow: I. N. Kushnerev, 1900–1914, IV, pp. 305–418

Kiernan, V. G. *Lords of the Human Kind. European Attitudes towards the Outside World in the Imperial Age*. Harmondsworth: Penguin, 1972

Kipp, J. W. "The Grand Duke Konstantin Nikolaevich and the Epoch of the Great Reforms, 1855–1866." Unpubl. Ph.D. thesis. Pennsylania State University, College Park, Pennsylania, 1970

Kirby, Stuart. "Siberia: Heartland and Framework." In Curtis Keeble, ed., *The Soviet State: The Domestic Roots of Soviet Foreign* Policy. Boulder: Westview, 1985, pp. 140–160

Kirillov, I. *Tretii Rim. Ocherk istoricheskogo razvitiia idei russkogo messianizma*. Moscow: I. M. Matistov, 1914

Kirilov, P. "Zametki o Sibiri." *Syn Otechestva* 25 (1842), otdel viii, pp. 81–86

Kirt, K. "Zhizn' i deiatel'nost' Aleksandra Fedorovicha Middendorfa." K. Kirt and E. V. Kurman, *A. F. Middendorf (1815–1894); E. A. Middendorf (1851–1916)*. Tartu: Obshchestvo Estestvoispytatelei pri AN Estonskoi SSR, 1963, pp. 5–14

"Kisilev, Pav. Dmitr."*Bol'shaia Entsiklopediia*. 22 v. St. Petersburg: "Prosveshchenie," 1896; X, pp. 762–763

"Kisilev, Pavel Dmitrievich." *Sovetskaia Istoricheskaia Entsiklopediia*. 17 v. Moscow: "Sovetskaia Entsiklopediia," 1961–1976; 7, pp. 293–294

Kiukhel'beker, V. K. *Dnevnik V. K. Kiukhel'bekera*. Leningrad: Priboi, 1929.
Kleinbort, L. [M.]. *Russkii imperalizm v Azii*. St. Petersburg: Znanie, 1906
Klevetskii, M. I. *Rukovodstvo k geografii s upotrebleniem zemnogo shara i landkart* . . . 3 v. St. Petersburg: [Tip. Sukhoputn. kad. korpusa], 1773
K . . . lin., A. "Neskol'ko slov po povodu pisem s Amura g. Tol . . . zina." *Syn Otechestva* 45 (October 1859), pp. 1200–1201
Kohn, Hans. "Messianism." In his *Revolutions and Dictatorships: Essays in Contemporary History*. Cambridge, Massachusetts: Harvard University Press, 1941, pp. 11–37
The Twentieth Century. A Midway Account of the Western World. New York: Macmillan, 1949
Pan-Slavism: Its History and Ideology. Notre Dame: University of Notre Dame Press, 1953
"Dostoevsky and Danilevsky: Nationalist Messianism." E. J. Simmons, ed., *Continuity and Change in Russian and Soviet* Thought. Cambridge, Mssachusetts: Harvard University Press, 1955, pp. 500–517
The Idea of Nationalism. A Study of its Origins and Background. New York: Macmillan, 1956
"Russia: The Permanent Mission." In his *Political Ideologies of the Twentieth Century*. 3rd ed. New York: Harper & Row, 1966, pp. 89–118
Political Ideologies of the Twentieth Century. 3rd ed. New York: Harper & Row, 1966
Prelude to Nation-States. The French and German Experience, 1789–1815. Princeton, New Jersey: Van Nostrand, 1967
"Introduction." Taras Hunczak, ed.,*Russian Imperialism from Ivan the Great to the Revolution*. New Brunswick: Rutgers University Press, 1974, pp. 3–17
"Kolonisation im Amur-Land." *Petermanns Mittheilungen* (1864), p. 33
"Konstantin Nikolaevich." *Entsiklopedicheskii slovar' (Granat)*. 46 v. Moscow: "Granat," nd; XXV, pp. 67–69
Koot, J. T. "The Asiatic Department of the Russian Foreign Ministry and the Formation of Policy toward the Non-Western World, 1881–1894." Unpubl. Ph.D. thesis. Harvard University, Cambridge, Massachusetts, 1980
Kornilov, A. M. *Zamechaniia o Sibiri*. St. Petersburg: Tip. Karla Kraiia, 1828
Korsak, A. K. *Istoriko-statisticheskoe obozrenie torgovykh snoshenii Rossii s Kitaem*. Kazan': Ivan Dubrovin, 1857
Korsh, V. P. "Bassein reki Amura i ego znachenie." *Moskovskie Vedomosti* no. 138 (17 November 1851), pp. 1327–1328; no. 139 (20 November 1851), pp. 1339–1340
[Kostomarov, N. I.]. "Ukraina." *Kolokol* no. 61 (15 January 1860), pp. 499–503
Ocherk torgovli Moskovskogo gosudarstva v XVI-XVII vv. St. Petersburg: Izd. N. Tiblena, 1862. [Russian Reprint Series, vol. 3. The Hague, 1966]
Kovalevskii, E. P. *Sibir'. Dumy*. St. Petersburg: Tip. Pliushara, 1832
Puteshestvie v Kitai. 2 v. St. Petersburg: Korolev & Co, 1853
"Kovalevskii, Egor Petr." *Bol'shaia Entsiklopediia*. 22 v. St. Petersburg: "Prosveshchenie," 1896; XI, p. 120
Koyré, Alexandre. *La philosophie et la problème national en Russie au début du XIXe siècle*. Bibliothéque de l'Institut Français de Léningrad, X. Paris: H. Champion, 1929
Koz'min, N. N. "M. B. Zagoskin i ego znachenie v istorii razvitiia Sibirskoi obshchest-

vennosti." In his *Ocherki proshlogo i nastoiashchego Sibiri.* St. Petersburg: Pechatnyi Trud, 1910, pp. 161–229

Kpt "Gde budet nash glavnyi torgovyi i voennyi port na Velikom okeane?" *Birzhevaia Vedomost'* no. 127 (9 May 1863), p. 395; no. 131 (12 May 1863), pp. 403–404

Krasheninnikov, S. P. *Opisanie zemli Kamchatki.* Moscow/Leningrad: Iz-vo Glavsevmorputi, 1949

Kristof, Ladis K. D. "The Russian Image of Russia: An Applied Study in Geopolitical Methodology." C. Fisher, ed., *Essays in Political Geography.* London: Methuen & Co., 1968, pp. 345–387

The Historical and Political Role of a Nation in Space: A Study of Concepts, Perceptions, Images, and Identifications. Unpubl. Ph.D. thesis. University of Chicago, 1969

[Krizhanich, Iurii]. "Povestvovanie o Sibiri (perevod s Latinskoi rukopisi XVII stoletiia)." *Sibirskii Vestnik* 17–18 (1822), pp. 1–26, 77–98, 149–170, 221–244

Krizhanich, Iurii. "Historia de Sibiria . . ." plus Russian tr. A. A. Titov, ed., *Sibir'v XVII veke.* Moscow: Izd. G. Iudina, 1890, pp. 115–121

Politika. Tr. V. V. Zelenin and A. L. Gol'denberg. Moscow: Nauka, 1965

Kropotkin, P. A. "Iz Vostochnoi Sibiri (Chita, polovina iiunia 1863 g; Amur nizhe Blagoveshchenska, 22 iiunia)." *Sovremennaia Letopis'* no. 42 (December 1863), pp. 10–12

"Iz Vostochnoi Sibiri (Amur, nizhe Blagoveshchenska, 27 iiulia 1863 g.). *Sovremennaia Letopis'* no. 43 (December 1863), pp. 6–8

"Iz Vostochnoi Sibiri. (Selo Khabarovka, 3-ogo Avgusta 1863 g.)." *Sovremennaia Letopis'* no. 44 (December 1863), pp. 12–14

"Iz Vostochnoi Sibiri. Amur vyshe Khabarovki, 20-ogo avgusta." *Sovremennaia Letopis'* no. 45 (December 1863), pp. 4–7

"Iz Irkutska." *Sovremennaia Letopis'* no. 14 (April 1864), pp. 11–13

"Iz Irkutska (31-ogo marta 1864)." *Sovremennaia Letopis'* no. 19 (May 1864), pp. 9–12; no. 20 (June 1864), pp. 7–9

"Iz Vostochnoi Sibiri (Khabarovka, 8-go iiulia)." *Sovremennaia* Letopis' no. 34 (October 1864), pp. 6–8

"Iz Vostochnoi Sibiri (St. Mikhailo-Semenovskaia)." *Sovremennaia Letopis'* no. 42 (December 1864), pp. 14–15

"Poezdka iz Zabaikalia na Amur chrez Man'chzhuriiu." *Russkii Vestnik* 57:6 (1865), pp. 663–681

"Dve poezdki v Man'chzhuriiu v 1864 godu . . ." *Zapiski Sibirskogo Otdela Imperatorskogo Russkogo Geograficheskogo Obshchestva* VIII:1 (1865), pp. 1–120

"The Russians in Manchuria." *Forum* XXXI (May 1901), pp. 267–274

Dnevnik P. A. Kropotkina. Moscow: Gosudarstvennoe Iz-vo, 1923

Memoirs of a Revolutionist. New York: Dover, 1971

von Krusenstern, A. J. *Voyage Round the World in the Years 1803, 1804, 1805, and 1806* . . . Tr. R. B. Hoppner. London: T. Davidson, 1813 [Reprint "Bibliotecha Australiana," XXVIII-XXXIX, Amsterdam, 1968]

Kubalov, B. G. *Dekabristy v Vostochnoi Sibiri. Ocherki.* Irkutsk: Irkutskoe Gubernskoe Arkhivbiuro, 1925

A. I. Gertsen i obshchestvennost' Sibiri (1855–1862). Irkutsk: Irkutskoe Knizhnoe Iz-vo, 1958

"Pervenets chastnoi sibirskoi pechati gazeta 'Amur'' (1860–1862)." *Zapiski Irkutskogo Oblastnogo Kraevedcheskogo Muzeia* II (1961), pp. 55–87

Letters to M. K. Azadovskii. *Literaturnoe Nasledstvo Sibiri* I (1969). Novosibirsk: Zapadno-Sibirskoe Knizhnoe Iz-vo, pp. 271–281

Kucherov, Alexander. "Alexander Herzen's Parallel between the United States and Russia." J. S. Curtis, ed., *Essays in Russian and Soviet History in Honor of Geroid Tanquary Robinson*, pp. 34–47. Leiden: E. J. Brill, 1963

Kuleshov, V. I. *"Otechestvennye Zapiski" i literatura 40-kh godov XIX veka*. Moscow: Iz-vo MGU, 1958

Kun, Miklosh. "Gertsen, Ogarev, i Bakunin: novye materialy o deiatel'nosti russkoi revoliutsionnoi emigratsii v gody pervoi revoliutsionnoi situatsii v Rossii." Monica Partridge, ed., *Alexander Herzen and European Culture*, pp. 105–132. Nottingham: Astra Press, 1984

Kungurov, G. F. "V. G. Belinskii o tvorchestve pisatelei-sibiriakov." *Uchenye Zapiski Irkutskogo Gosudarstvenogo Pedagogicheskogo Instituta* 12 (1957), pp. 21–24

Kuniaev, S. Iu. "Velikii put'." A. Perlovskii, ed., *Sibirskie stroki*. Moscow: Molodaia Gvardiia, 1984, pp. 5–10

Kurts, B. G. *Russko-kitaiskie snosheniia v XVI, XVII, i XVIII stoletiiakh*. [Kiev]: Gosudarstvennoe iz-vo Ukrainy, 1929

Kushner, Howard I. *Conflict on the Northwest Coast. American–Russian Rivalry in the Pacific Northwest 1790–1867*. Westport, Connecticut: Greenwood Press, 1975

Kuznetsov, A. S. "Sibirskaia programma tsarizma 1852 g." *Ocherki istorii Sibiri* 2 (1971), pp. 3–27

[Lamanskii, V. I.]. "Mnenie anglichan i amerikantsev o russkom vladychestve na Amure." *Vestnik Imperatorskogo Russkogo Geograficheskogo Obshchestva* ch. 25 (1859), otdel v, pp. 53–56

Langer, W. L. *The Diplomacy of Imperialism 1890–1902*. 2nd ed. New York: Knopf, 1951

Lantzeff, George V. *Siberia in the Seventeenth Century: A Study of the Colonial Administration*. University of California Publications in History, XXX. Berkeley: University of California Press, 1943

Lantzeff, George V. and Richard A. Pierce. *Eastward to Empire*. Montreal: McGill–Queens University Press, 1973

Lapérouse, Jean François de Galaup. *Voyage de Lapérouse autour de monde pedant les années 1785, 1786, 1787, et 1788*. Paris: Les Libraries Associés, 1965

Lapteva, L. P. "Gil'ferding, Aleksandr Fedorovich." *Slavianovedenie v dorevoliutsionnoi Rossii. Bio-bibliograficheskii slovar'*. Moscow: Nauka, 1979, pp. 121–125

Laserson, Max M. *The American Impact on Russia, Diplomatic and Ideological, 1784–1917*. New York: Macmillan, 1950

Laverychev, V. Ia. "K voprosu ob osobennostiakh imperializma v Rossii." *Istoriia SSSR* 1 (January–February 1971), pp. 75–90

Layton, Susan. "The Creation of an Imaginative Caucasian Geography." *Slavic Review* 45 (1986), pp. 470–485

Russian Literature and Empire. Conquest of the Caucasus from Pushkin to Tolstoy. Cambridge: Cambridge University Press, 1994

Leikina, V. R. *Petrashevtsy*. Moscow: Obshchestvo polit. katorzhan i ssyl'no-poselentsev, 1924.

Lejeune, Dominique. *Les sociétés de géographie et l'expansion colonial au XIXe siècle*. Paris, Albin Michel, 1993

Lemberg, Eugen. *Nationalismus*. 2 v. Hamburg: Rowohlt, 1964

Lemke, M. "Krest'ianskie volneniia 1855 g." *Krasnaia Letopis'* 7 (1923), pp. 131–177.

Lenin, V. I. "'Krest'ianskaia reforma' i proletarski-krest'ianskaia revoliutsiia" [1911]. *Polnoe sobranie sochinenii*. 5th ed. 55 v. Moscow: Gosudarstvennoe Iz-vo Politicheskoi Literatury, 1961; 20, pp. 171–180

Lensen, G. A. *Russia's Japan Expedition of 1852 to 1855*. Gainesville: University of Florida Press, 1955

The Russian Push towards Japan. Russo-Japanese Relations, 1697–1875. Princeton: Princeton University Press, 1959

Leonov, N. I. *Aleksandr Fedorovich Middendorf*. Moscow: Nauka, 1967

Levin, Sh.M. "Krymskaia voina i russkoe obshchestvo." In his *Ocherki po istorii russkoi obshchestvenoi mysli. Vtoraia polovina XIX-nachalo XX veka*, pp. 293–406. Leningrad: Nauka, 1974

Ley, David and Marwyn Samuels, "Introduction: Contexts of Modern Humanism in Geography." David Ley and Marwyn Samuels, eds., *Humanistic Geography. Prospects and* Problems. Chicago: Maaroufa Press, 1978

Lichtheim, George. *Imperialism*. New York: Praeger, 1971, pp. 1–18

Lin, T. C. "The Amur Frontier Question between China and Russia, 1850–1860." *Pacific Historical Review* III: 1 (1934), pp. 1–27

Lincoln, Abraham. "Second Annual Message" [1862]. *Messages and Papers of the Presidents*. VII, pp. 3327–3343. New York: Bureau of National Literature, 1897

"Third Annual Message" [1863]. *Messages and Papers of the Presidents*. VII, pp. 3380–3392. New York: Bureau of National Literature, 1897

"Fourth Annual Message" [1864]. *Messages and Papers of the Presidents*. VII, pp. 3444–3449. New York: Bureau of National Literature, 1897

Lincoln, W. Bruce. "The Circle of the Grand Duchess Yelena Pavlovna, 1847–1861." *Slavic and East European Review* XLVIII: 112 (July 1970), pp. 373–387

Nicholas I. Bloomington: Indiana University Press, 1978

Petr Petrovich Semenov-Tian-Shanskii: The Life of a Russian Geographer. Russian Biography Series, no. 8. Newtonville, Massachusetts: Oriental Research Partners, 1980

In the Vanguard of Reform. Russia's Enlightened Bureaucrats 1825–1861. De Kalb: Northern Illinois University Press, 1982

[Liudorf, F. A.]. "Amurskii krai, ego nastoiashchee polozhenie i potrebnosti." *Zapiski dlia Chteniia* November–December (1868), pp. 208–227

Livingstone, David N. *The Geographical Tradition. Episodes in the History of a Contested Enterprise*. Oxford: Blackwell, 1992

Lobanov-Rostovskii, A. *Russia and Asia*. Ann Arbor: George Wahr, 1951

Lomonosov, M. V. "Kratkoe opisanie raznykh puteshestvii po severnym moriam i pokazanie vosmozhnogo prokhodu sibirskim okeanom v Vostochnuiu Indiiu." *Polnoe sobranie sochinenii*. 11 v. Moscow-Leningrad: AN SSSR, 1950–1983; VI, pp. 417–498

"Oda na den′ vosshestviia na vserosiiskii prestol Ee Velichestva Gosudaryni Imperatritsy Elisavety Petrovny 1747 goda." *Polnoe sobranie sochinenii.* 11 v. Moscow–Leningrad: AN SSSR, 1950–1983; VIII, pp. 196–207

Loubat, J. F. *Narrative of the Mission to Russia, in 1866 . . .* New York: D. Appleton & Co., 1879

Lowenthal, D. and M. Bowden, eds. *Geographies of the Mind. Essays in Historical Geosophy in Honor of John Kirkland Wright.* New York: Oxford University Press, 1976

Lunin, M. S. *Dekabrist M. S. Lunin. Sochineniia i pis′ma.* Peterburg [sic]: Trudy Pushkinskogo Doma pri Akademii Nauk, 1923

L′vov, F. "Pis′mo D. I. Zavalishinu, 4 Avgust [1860]." *Sbornik starinnykh bumag, khraniashchikhsia v muzee P. I. Shchukina.* 10 v. Moscow: A. I. Mamontov, 1900–1902; X, pp. 243–251

L′vov, G. "S Dal′nego Vostoka." *Russkie Vedomosti* no. 216 (18 September 1908), pp. 2–3; no. 220 (23 September 1908), pp. 2–3

Maak, Richard. *Puteshestvie na Amur, sovershennoe po rasporiazheniiu sibirskogo otdela imperatorskogo russkogo geograficheskogo obshchestva v 1855 godu R. Maakom.* St. Petersburg: K. Vul′f, 1859

Puteshestvie po doline reki Usuri. St. Petersburg: Bezobrazov, 1861

MacKenzie, David. *The Lion of Tashkent: The Career of General M. G. Chernaiev.* Atlanta: University of Georgia Press, 1974

Madsen, Deborah, ed. *Visions of America since 1492.* London: Leicester University Press, 1994

Maggs, Barbara W. *China in the Literature of Eighteenth Century Russia.* Unpubl. Ph.D. dissertation. University of Illinois, Champaign–Urbana, 1973

[Magnitskii, M. L.]. "Sud′ba Rossii." *Raduga* kn. VI–VIII (1833), pp. 391–414

Maier, Lothar. "Gerhard Friedrich Müller's Memorandum on Russian Relations with China and the Reconquest of the Amur." *Slavonic and East European Review* 59:2 (April 1981), pp. 219–240

Maikov, L. N., ed. *Rasskazy Nartova o Petre Velikom. Sbornik otdeleniia russkogo iazyka i slovesnosti imperatorskoi Akademii Nauk* LII, no. 8. St. Petersburg: Imp. Akademiia Nauk, 1891

Maksimov, S. V. *Na Vostoke. Poezdka na Amur (v 1860–1861 gg.), dorozhnye zametki i vospominaniia.* St. Petersburg: tip. "Obshchestvennaia Pol′za," 1864.

Na Vostoke. Poezdka na Amur. Dorozhnye zametki i vospominaniia. 2nd ed. St. Petersburg: S. V. Zvonarev, 1871.

Maksimovich, K. [I.]. "Kratkii otchet o puteshestvii v Amurskii Krai i Iaponiiu v 1859–1864 gg." *Zhurnal Ministerstva Gosudarstvennykh Imushchestv* 86 (July 1864), pp. 433–442.

Maksimovich, L. M. *Novyi i polnyi geograficheskii slovar′ Rossiiskogo gosudarstva . . .* 4 v. Moscow: Universitetskaia tipografiia, 1788–1789

Malia, Martin. *Alexander Herzen and the Birth of Russian Socialism.* New York: Grosset & Dunlap, 1965

Malozemoff, Andrew. *Russian Far Eastern Policy 1881–1904. With Special Emphasis on the Causes of the Russo-Japanese War.* Berkeley: University of California Press, 1958

Mamonov, [M. A.]. "Iz bumag grafa Mamonova." A. K. Borozdin, ed, *Iz pisem i pokazanii* dekabristov. St. Petersburg: M. V. Pirozhkov, 1906, pp. 143–157

Mancall, Mark. "Major-General Ignatiev's Mission to Peking 1859–1860." *Papers on China* [Cambridge, Massachusetts] 10 (1956), pp. 55–96

"The Kiakhta Trade." C. D. Cowan, ed., *The Economic Development of China and Japan. Studies in Economic History and Political Economy*. London: Allen & Unwin, 1964, pp. 19–48

Russia and China: Their Diplomatic Relations to 1728. Cambridge, Massachusetts: Harvard University Press, 1971

"Russia and China: the Structure of the Contact." W. S. Vucinich, ed., *Russia and Asia. Essays on the Influence of Russia on the Asian Peoples*, pp. 313–337. Palo Alto: Hoover Institution Press, 1972

March, Andrew L. *The Idea of China*. New York: Praeger, 1974

Marks, Steven G. "The Burden of Siberia: The Amur Railroad Question in Russia, 1866–1916). *Sibirica*, 1:1 (1993–1994), pp. 9–28

Road to Power. The Trans-Siberian Railroad and the Colonization of Asian Russia 1850–1917. Ithaca: Cornell University Press, 1991

Mart′ianov, P. "Pis′mo k Aleksandru II." *Kolokol* no. 132 (8 May 1862), pp. 1093–1097

Martos, A. I. *Pis′ma o Vostochnoi Sibiri*. Moscow: Universitetskaia Tipografiia, 1827

Marx, Karl. "Revue" [1850]. *Marx–Engels Werke*. 42 v. Berlin: Institut für Marxismus–Leninismus beim ZK der SED, 1957–1985; VII, pp. 213–225

"Die Hegira Bakunins." *Marx–Engels Werke*. 42 v. Berlin: Institut für Marxismus–Leninismus beim ZK der SED, 1957–1985; XVIII, pp. 442–444

Marx, Karl and Friedrich Engels. *Marx–Engels Werke*. 42 v. Berlin: Institut für Marxismus–Leninismus beim ZK der SED, 1957–1985

Materialy dlia istorii russkikh zaselenii po beregam Vostochnogo Okeana. St. Petersburg: Morskoe Ministerstvo, 1861

Matveev, Z. N. "Dekabrist Zavalishin ob Amurskom dele." *Izvestiia Primorskogo Gubernskogo Arkhivnogo Biuro* I:2 (1923), pp. 15–22

"Maximowicz's Reise auf dem unteren Ssungari." *Petermanns Mittheilungen* (1862), pp. 167–170

Mazour, A. G. "Dmitry Zavalishin: Dreamer of a Russian–American Empire." *Pacific Historical Review* 5:1 (1936), pp. 26–37

"Doctor Yegor Scheffer: Dreamer of a Russian Empire on the Pacific." *Pacific Historical Review* 6:1 (1937), pp. 15–20

McAleavy, H. *The Modern History of China*. London: Weidenfeld & Nicolson, 1967

McDonald, David M. "A Lever without a Fulcrum: Domestic Factors and Russian Foreign Policy, 1905–1914." Hugh Ragsdale, ed., *Imperial Russian Foreign Policy*. Cambridge: Cambridge University Press, 1993, pp. 268–311

McKay, D. V. "Colonialism in the French Geographical Movement 1871–1881." *Geographical Review* XXXIII (1943), pp. 214–232

McNally, Raymond T. *The Major Works of Peter Chaadaev. A Translation and Commentary*. Notre Dame: University of Notre Dame Press, 1969

McPherson, H. M., ed. "The Interest of William McKendree Gwin in the Purchase of Alaska, 1854–1861." *Pacific Historical Review* 3 (1934), pp. 28–38

"Projected Purchase of Alaska, 1856–1860." *Pacific Historical Review* 3 (1934), pp. 80–87

Melamed, E. I. *Russkie universitety Dzhordzha Kennana. Sud'ba pisatelia i ego knig.* Irkutsk: Vostochno-Sibirskoe Knizhnoe Iz-vo, 1988

"Men'shikov, Aleksandr Sergeevich." *Sovetskaia Istoricheskaia Entsiklopediia.* 17 v. Moscow: "Sovetskaia Entsiklopediia," 1961–1976; IX, pp. 355–356

Merk, Frederick. *History of the Westward Movement.* New York: Alfred A. Knopf, 1978

"Mestnoe obozrenie." *Amur* (20 March 1860), pp. 265–266

Middendorf, A. F. *Puteshestvie na sever i vostok Sibiri.* 2 v. St. Petersburg: Imp. Akademiia Nauk, 1860–1877

Miliukov, P. "Eurasianism and Europeanism in Russian History." *Festschrift Th. G. Masaryk zum 80. Geburtstag.* 2 v. Bonn: Friedrich Cohen, 1930; I, pp. 225–236

Miliutin, B. A. "General-gubernatorstvo N. N. Murav'eva v Sibiri." *Istoricheskii Vestnik* XXXIV (November 1888), pp. 317–364; (December 1888), pp. 595–635

Miller, D. H., ed. *The Alaska Treaty.* Kingston: Limestone Press, 1981

Miller, G. [Gerhard Müller]. "Istoriia o stranakh, pri reke Amure lezhashchikh, kogda onye sostoiali pod rossiiskim vladeniiam." Tr. I. Golubtsov. *Ezhemesiachnye sochineniia k pol'ze i uveseleniiu* VI:2 (July 1757), pp. 3–39; (August 1757), pp. 99–130; (September 1757), pp. 195–227; (October 1757), pp. 291–328

Miller, Hunter. "Russian Opinion on the Cession of Alaska." *American Historical Review* 48 (1943), pp. 521–531

Mirzoev, V. G. *Istoriografiia Sibiri (domarksistskii period).* Moscow: Mysl', 1970

Modestov, V. "Zagranichnye vospominaniia." *Istoricheskii Vestnik* XII (April 1883), pp. 103–124

Modzalevskii, L. B. "Pushkin i G. T. Khlebnikov." G. F. Kungurov, ed., *A. S. Pushkin i Sibir.* Moscow-Irkutsk: Vostsiboblgiz, 1937, pp. 146–154

von Mohrenschidt, Dmitri. *Toward a United States of Russia: Plans and Projects of Federal Reconstruction of Russia in the Nineteenth Century.* Rutherford: Fairleigh Dickinson University Press, 1981

Mommsen, Wolfgang J. *Das Zeitalter des Imperalismus.* Frankfurt: Fischer, 1969

"Der moderne Imperialismus als innergesellschaftliches Problem." In *idem*, ed., *Der moderne* Imperialismus. Stuttgart: W. Kohlhammer, 1971, pp. 14–30

Imperialismustheorien. Ein Überblick über die neueren Imperialismusinterpretationen. 3rd ed. Göttingen: Vandenhoeck & Ruprecht, 1987

"The Varieties of the Nation State in Modern History: Liberal, Imperialist, Fascist, and Contemporary Notions of Nation and Nationality." Michael Mann, ed., *The Rise and Decline of the Nation State.* Oxford: Blackwell, 1990, pp. 210–226

Monas, Sidney. *The Third Section. Police and Society in Russia under Nicholas I.* Cambridge, Massachusetts: Harvard University Press, 1961

"Foreward." Alexander Yanov, *The Origins of Autocracy. Ivan the Terrible in Russian History.* Tr. S. Dunn. Berkeley: University of California Press, 1981, pp. xiii–xiv

Mordvinov, Aleksandr. "Vozrazheniia na zamechaniia o torgovykh otnosheniiakh Sibiri k Rossii (Stat'iu g-na Gersevanova)." *Otechestvennye Zapiski* XXII (1842), otdel v, pp. 31–56

Mosse, W. E. *Alexander II and the Modernization of Russia*. New York: Collier Books, 1962

[Müller, Gerhard]. "Nachrichten von dem Amur-Flusse, aufgesetzt im Anfang des 1741sten Jahres." *Magazin für die neue Historie und Geographie, angelegt von D. Anton Friedrich Büsching* II (1769), pp. 483–518

"Rassuzhdenie o predpriiatii voiny s kitaitsami . . ." [1763]. Bantysh-Kamenskii, N. N., ed., *Diplomaticheskoe sobranie del mezhdu Russkim i Kitaiskim gosudarstvami c 1619 po 1792-i god.* Kazan′: Tip. Imp. Un-ta, 1882, pp. 378–393

Murav′ev, N. "Konstitutsiia." M. V. Dovnar-Zapol′skii, ed., *Memuary dekabristov*. Kiev: S. I. Ivanov, 1906, pp. 58–71

"Nachalo nashikh krugosvetnykh plavanii." *Zapiski gidrograficheskogo departamenta morskogo ministerstva* ch. VII (1849), pp. 501–507

Nadezhdin, N. I. "Ob etnograficheskom izuchenii narodnosti russkoi." *Zapiski Russkogo Geograficheskogo Obshchestva* II (1847), pp. 61–115

"Nash krainyi Vostok." *Vsemirnaia Illiustratsiia* VII (1872), no. 167, pp. 178–179; no. 168, pp. 190–191; no. 169, p. 206; no. 170, pp. 226–227; no. 171, p. 243

Nazimov, N. "Otvet na stat′iu. Po povodu statei ob Amure." *Morskoi Sbornik* 38:12 (December 1858), pp. 115–119

Nebol′sin, P. I. *Pokorenie Sibiri. Istoricheskoe issledovanie.* St. Petersburg: I. Glazunov, 1849

Nesselrode, Charles. *Lettres et papiers du chancelier Comte de Nesselrode 1760–1856*. 11 v. Paris: A. Lahure, 1904–1912

"Neueste Nachrichten über die Amur-Länder . . ." *Petermanns Mittheilungen* (1858), pp. 474–475

Nevel′skoi, G. I. *Podvigi Russkikh morskikh ofitserov na krainem vostoke Rossii 1849–55 gg. Pri-amurskii i pri-ussuriiskii krai.* St. Petersburg: tip. L. S. Suvorina, 1897

Niebuhr, Reinhold. *The Structure of Nations and Empires*. New York: Charles Scribners Sons, 1959 [rpt. 1977]

N. N. "Proekt ustava obshchestva parakhodstva i torgovli na reke Amure." *Irkutskie Gubernskie Vedomosti* no. 11 (25 July 1857), otdel ii, pp. 5–11

N. N. Kh.-O. "O parakhodstve na Amure." *Irkutskie Gubernskie Vedomosti* no. 14 (15 August 1857), otdel ii, pp. 2–4; no. 17 (5 September 1857), otdel ii, pp. 5–10

Noskov, Ivan. "Izvestiia iz Shankhaia." *Amur* 1 (1 January 1860), pp. 10–11

"Amurskii krai v kommercheskom, promyshlennom, i khoziaistvennom otnoshenii." *Biblioteka dlia Chteniia* 34:4 (1865), pp. 1–26; 34:5 (1865), pp. 1–36

Novyi entsiklopedicheskii slovar′ (Brokgauz-Efron). 29 v. St. Petersburg: Tip. Aktsionernogo Obshchestva "Brokgauz-Efrona," 1911–1916

"O Petropavlovskom porte (iz zapisok g. Dobelia)," *Syn Otechestva* XXVI:47 (1815), pp. 53–56

"O torgovle s Kitaem." *Vestnik Imperatorskogo Russkogo Geograficheskogo Obshchestva* kn. 11 (1858), otdel v, pp. 48–53

Obolenskii, E. P. "Vospominanie o K. F. Ryleeve." P. I. Bartenev, ed., *Deviatnadtsatyi Vek*. 2 v. Moscow: F. Iuganson, 1872; I, pp. 312–332

Ocherk dvadtsatipiatiletnei deiatel′nosti sibirskogo otdela Imperatorskogo Russkogo Geograficheskogo Obshchestva. Irkutsk: VSRGO, 1876

Ocherk istorii ministerstva inostrannykh del 1802–1902. St. Petersburg: R. Golike i A. Vil'borg, 1902

Odoevskii, A. I. "Kn. M. N. Volkonskoi." *Polnoe sobranie sochinenii*. Leningrad: Sovetskii Pisatel', 1958, p. 87

Odoevskii, V. F. *Russkie nochi*. Nachdruck der Moskauer Ausgabe von 1913. *Slavische Propyläen*; Bd. 24. Munich: Wilhelm Fink Verlag, 1967

Offord, Derek. *Portraits of early Russian Liberals. A Study in the Thought of T. N. Granovsky, P. V. Annenkov, A. V. Druzhinin, and K. D. Kavelin*. Cambridge: Cambridge University Press, 1985

Ogarev, N. "Pis'mo k avtoru 'Vozrazheniia na stat'iu *Kolokola*'." *Kolokol* no. 38 (15 March 1859), pp. 307–313

Ogarev, N. P. "Na novyi god." *Kolokol* no. 89 (1 January 1861), pp. 745–752

Ogarev, N. "Otvet na otvet Velikorussu." *Kolokol* 'no.108 (1 October 1861), pp. 901–905.

O'Gorman, Edmundo. *The Invention of America: An Inquiry into the Historical Nature of the New World and the Meaning of its History*. Bloomington: Indiana University Press, 1961

Okun', S. B. "Kolonial'naia politika tsarizma v Okhtsko-Kamchatskom krae v XVIII veke." *Bor'ba Klassov* 12 (December 1934), pp. 82–89

"Politika russkogo tsarizma na Tikhom okeane v XIX veke." *Bor'ba Klassov* 9 (1935), pp. 34–43

"Sibirskii komitet." *Arkhivnoe Delo* 1:38 (1936), pp. 92–103

"K voprosu o russko-amerikanskikh otnosheniiakh v nachale XIX veka." *Nauchnyi Biulleten' LGU* 8 (1946), pp. 19–22

The Russian–American Company. Tr. C. Ginsburg. Cambridge, Massachusetts: Harvard University Press, 1951

Opisanie Primorskogo pereselencheskogo raiona. Spravochnaia knizhka dlia khodokov i pereselentsev. St. Petersburg: "Sel'skii Vestnik," 1911

"Otchet Imperatorskogo Russkogo Geograficheskogo Obshchestva za 1851 god." *Vestnik Imperatorskogo Russkogo Geograficheskogo Obshchestva* ch. IV: kn. 2 (1852), pp. 15–84

"Otchet Imperatorskogo Russkogo Geograficheskogo Obshchestva za 1852 god." *Vestnik Imperatorskogo Russkogo Geograficheskogo Obshchestva* ch. VII (1853), pp. 11–94

"Otchet Imperatorskogo Russkogo Geograficheskogo Obshchestva za 1855 god." *Vestnik Imperatorskogo Russkogo Geograficheskogo Obshchestva* ch. XVI (1856), pp. 1–62

"Otto Esche's Expedition nach dem Amur." *Petermanns Mittheilungen* (1858), pp 161–162

Ouvaroff. "Projet d'une académie asiatique." *Etudes de philologie et de critique*. 2nd ed Paris: Typographie de Firmin Didot Frères," 1845, pp. 1–48

P. "Svedeniia o khode torgovli v Shankhae." *Amur* no. 8 (23 February 1860), pp 109–110; no. 9 (1 March 1860), pp. 120–122

Palmer, Aaron H. *Memoir, geographical, political, and commercial, on the present state productive resources, and capabilities for Commerce of Siberia, Manchuria, and the Asiatic Islands of the Northern Pacific; and on the importance of opening commer-*

cial intercourse with those countries, etc. 30th Congress (Senate), 1st session, Miscellaneous Document no. 80, 1848

Pal′mer, A. Kh. *Zapiska o Sibiri, Manchzhurii i ob ostrovakh severnoi chasti Tikhogo Okeana*. St. Petersburg: Tip. M. Stasiulevicha, 1906

"Panin, Viktor Nikitich." *Sovetskaia Istoricheskaia Entsiklopediia*. 17 v. Moscow: "Sovetskaia Entsiklopediia," 1961–1976; X, pp. 785–786

Papers relating to Foreign Affairs. Washington, DC: Government Printing Office, various years

Parry, Albert. "Yankee Whalers in Siberia." *Russian Review* V:2 (Spring 1946), pp. 36–49

Parry, Benita. *Delusions and Discoveries: Studies on India in the British Imagination 1880–1930*. Berkeley: University of California Press, 1972

Parshin, V. P. *Poezdka v Zabaikal′skii krai*. 2 v. Moscow: v tip. N. Stepanova, 1844

Pasetskii, V. M. *Geograficheskie issledovaniia dekabristov*. Moscow: Nauka, 1977

Pavlov, P. N. *Promyslovaia kolonizatsiia Sibiri v XVII veke*. Krasnoiarsk: Krasnoiarskoe Gos. Ped. Iz-vo, 1974

P. B[artenev]. "Ob ukraino-slavianskom obshchestve (iz bumag D. P. Golokhvastova)." *Russkii Arkhiv* XXX: ii (1892), pp. 334–359

Pereira, N. G. O. *Tsar-Liberator: Alexander II of Russia, 1818–1881*. Newtonville, Massachusetts: Oriental Research Partners, 1983

Perepiska Imperatora Aleksandra II s Velikim Kniazem Konstantinom Nikolaevichem . . . Moscow: Terra, 1994

"Perovskii, Lev Alekseevich." *Sovetskaia Istoricheskaia Entsiklopediia*. 17 v. Moscow: "Sovetskaia Entsiklopediia," 1961–1976; XI, p. 51

"Perovskii, Vasilii Alekseevich." *Sovetskaia Istoricheskaia Entsiklopediia*. 17 v. Moscow: "Sovetskaia Entsiklopediia," 1961–1976, XI; pp. 50–51

"Perry McD. Collins 'Berichte über seine Reise durch das Asiatische Russland, 1856 u. 1857.'" *Petermanns Mittheilungen* (1859), pp. 19–29

"Peschtschuroffs Aufnahme des Amur-Stromes im Jahre 1855." *Petermanns Mittheilungen* (1856), pp. 472–479

Pestel, P. I. *Russkaia Pravda. Nakaz Vremennomu Verkhovnomu Pravelniiu*. St. Petersburg: Kul′tura, 1906

[Petrashevskii, M. V.]. "Po povodu rechi A. V. Belogolovogo. Pisano ot Zabaikal′skikh kuptsov I. V. L. i A. K. i inykh iz torgovoi bratii, na Amure promyshliaiushchei." *Irkutskie Gubernskie Vedomosti* no. 39 (25 September 1858), otdel ii, pp. 5–10; no. 40 (2 October 1858), otdel ii, pp. 10–18

"Mestnoe obozrenie." *Amur* no. 3 (19 January 1860), pp. 33–37

Petrashevskii, M. V. "Pros′ba M. V. Petrashevskogo e. v. Gospodinu ministru vnutrennikh del." *Kolokol* no.92 (15 February 1861), pp. 770–775; no. 93 (1 March 1861), pp. 782–783

Petri, E. "Sibir′ kak koloniia." *Sibirskii Sbornik* II (1886), pp. 83–98

Petrovich, Michael B. *The Emergence of Russian Pan-Slavism 1856–1870*. New York: Columbia University Press, 1956

Pierce, Richard A. *Russian Central Asia 1867–1917. A Study in Colonial Rule*. Berkeley: University of California Press, 1960

Russia's Hawaiian Adventure 1815–1817. Kingston, Ontario: The Limestone Press, 1976

Pipes, R. E. "Domestic Politics and Foreign Affairs." I. J. Lederer, ed., *Russian Foreign Policy. Essays in Historical Perspective*. New Haven: Yale University Press, 1962, pp. 145–170

"Pis′ma o Manzhurii," *Syn Otechestva* kn. 6 (1849), otdel v, pp. 27–60

"Pis′mo N. N. Murav′eva-Amurskogo." *Russkii Vestnik* 195 (March 1888), pp. 413–418

P. L. "Chekhi i Amur." Den′ no. 2 (21 October 1861), pp. 12–13; no. 10 (16 December 1861), pp. 14–16

Plamenatz, John. "Two Types of Nationalism." E. Kamenka, ed., *Nationalism: The Nature and Evolution of An Idea*. Canberra: Australian University Press, 1973, pp. 22–37

"Po dele Irkutskoi dueli." *Kolokol* no. 73–75 (15 June 1860), pp. 614–617

Pogodin, M. P. "Rech′, proiznesennaia v Dume, na obede, v chest′ Amerikanskogo posol′stva" [1866]. *Rechi, proiznesennye M. P. Pogodinym v torzhestvennykh i prochikh sobraniiakh 1830–1872*. Moscow: Synodal′naia Tipografiia, 1872, pp. 262–264

"O russkoi politike na budushchee vremia" [1854]. *Istoriko-politicheskie pis′ma i zapiski vprodolzhenii Krymskoi voiny 1853–1856*. Moscow: V. M. Frish, 1874, pp. 231–244

"Pis′mo k Gosudariu Tsesarevichu, Velikomu Kniaziu, Aleksandru Nikolaevichu" [1838]. *Istoriko-politicheskie pis′ma i zapiski vprodolzhenii Krymskoi voiny*. Moscow: V. M. Frish, 1874

"Vtoroe pis′mo k Izdatel′iu gazety 'Le Nord'" [1856–1857]. *Stat′i politicheskie i pol′skii vopros*. Moscow: F. B. Miller, 1876, pp. 14–24

Pokrovskii, M. N. *Diplomatiia i voiny tsarskoi Rossii v XIX stoletii*. Moscow: "Krasnaia Nov′," 1923

Polevoi, B. P. "Opoznanie statei Petrashevtsa A. P Balasoglo o Sibiri, Dal′nem Vostoke i Tikhom Okeane." *Istoriia SSSR* 5:1 (1961), pp. 155–159

"Glavnaia zadacha pervoi Kamchatskoi ekspeditsii po zamyslu Petra I," *Voprosy geografii Kamchatki* 2 (1964), pp. 88–94

Polevoi, N. A. *Ermak Timofeich, ili Volga i Sibir′*. St. Petersburg: K. Krai, 1845

Ponomarev, V. N. "Russian Policy and the United States during the Crimean War." Hugh Ragsdale, ed., *Imperial Russian Foreign* Policy. Cambridge: Cambridge University Press, 1993, pp. 173–192

Popkin, Cathy. "Chekhov as Ethnographer: Epistemological Crisis on Sakhalin Island." *Slavic Review* 51 (Spring 1992), pp. 36–51

Popov, A. "Tsarskaia diplomatiia v epokhu taipinskogo vosstaniia." *Krasnyi Arkhiv* 21 (1927), pp. 182–199

Popov, N. A. "Vopros o pereseleniiakh v Rossiiu." *Sovremennaia Letopis′* no. 46 (1865), pp. 1–4; no. 48, pp. 3–5

"Popytka Iakutskikh kuptsov zavesti torgovliu s Amurom." *Amur* no. 76 (26 September 1861), pp. 708–710

Potanin, G. N. Review of R. Maak, *Puteshestvie na Amur*. *Russkoe Slovo* 1 (1860), otdel ii, pp. 82–99

"K kharakteristike Sibiri." *Kolokol* no. 72 (1 June 1860), pp. 604–606

"Zametki o Zapadnoi Sibiri." *Russkoe Slovo* (1861), pp. 189–214

Presniakov, A. E. *Emperor Nicholas I of Russia. The Apogee of Autocracy, 1825–1855*. Tr. J. C. Zacek. Gulf Breeze, Florida: Academic International Press, 1974

Proskuriakov, S[?]. *Podrobnyi uchebnyi atlas Rossiiskoi Imperii*. Np: Korpus topografov, 1841

Przheval'skii, N. M. "Ussuriiskii krai. Novaia territoriia Rossii." *Vestnik Evropy* kn. 5 (1870), pp. 236–267; kn. 6 (1870), pp. 543–583

"Avtobiografiia N. M. Przheval'skogo." *Russkaia Starina* 19:11 (November 1888), pp. 528–545

Pushchin, I. I. *Zapiski o Pushchine i pis'ma iz Sibiri*. Moscow: Vsesoiuz. Ob-vo Polit. Katorzhan i Ssyl'no-Poselentsev, 1925

Pushkarev, S. *The Emergence of Modern Russia 1801–1917*. Tr. R. H. McNeal and T. Yeddin. New York: Holt, Rinehart &Winston, 1963

Pushkin, A. S. Letter to P. Ia. Chaadaev (19 October 1836). *Sochineniia Pushkina*, ed. V. I. Saitov. 3 v. St. Petersburg: Imp. Akademii Nauk, 1906–1911; III, pp. 387–389

"Klevetnikam Rossii." *Polnoe sobranie sochinenii v 10 tomakh*. 10 v. 4th ed. Leningrad: Nauka, 1977–1979; III, pp. 209–210

"Zametki pri chtenii *Opisaniia zemli Kamchatki* S. P. Krasheninnikova." *Polnoe sobranie sochinenii v 10 tomakh*. 10 v. 4th ed. Leningrad: Nauka, 1977–1979; IX, pp. 321–349

Letter to N. I. Gnedich (23 February 1825). *Polnoe sobranie sochinenii v 10 tomakh*. 10 v. 4th ed. Leningrad: Nauka, 1977–1979; X, pp. 99–100

"*Puteshestvie v Kitai* E. Kovalevskogo." *Sovremennik* 12 (December 1853), otdel iv, pp. 39–48

Puteshestvie po Amuru na parakhode "Argun'." Moscow: N. Ernst, 1860.

Putilov, N. "Po povodu rasskaza 'Vot tebe Amur.'" *Amur* no. 32 (22 April 1861), pp. 251–253

Pypin, A. N. *Panslavizm v proshlom i nastoiashchem*. St. Petersburg: Kolos, 1913

Quested, R. K. I. *The Expansion of Russia in East Asia, 1857–1860*. Kuala Lumpur: University of Malaya Press, 1968

R. "Iz Irkutska." *Russkii Vestnik* XXI:2 (May 1859), pp. 145–147

Ra'anan, Uri and Kate Martin, eds. *Russia: A Return to Imperialism?* New York: St. Martin's Press, 1995

[Radde, Gustav]. "Pri-Amurskaia Oblast'." *Illiustratsiia*, 3:57 (1859), pp. 101–102

"Gustav Radde's Vorlesungen über Sibirien und das Amur Land." *Petermanns Mittheilungen* (1860), pp. 257–263, 386–394; (1861), pp. 261–268.

Radde, Gustav. Untitled article. *Irkutskie Gubernskie Vedomosti* 46 (12 November 1859), otdel ii, pp. 8–10

Bericht über Reisen im Süden von Ost-Sibirien . . ., aufgeführt in den Jahren 1855 bis incl. 1859. Beiträge zur Kenntnis des Russischen Reiches und der angrenzended Ländern Asien; Bd. 23 plus atlas. St. Petersburg: Kaiserliche Akademie der Wissenschaften, 1861

Reisen im Süden von Ost-Sibirien im den Jahren 1855–1859 incl. 2 v. St. Petersburg: Kaiserlich Akademie der Wissenschaften, 1862–1863

23,000 Mil' na iakhte "Tamara." Puteshestvie ikh imperatorskikh vysochestv velikikh Kniazei Aleksandra i Sergeia Mikhailovichei v 1890–1891 gg. 2 v. St. Petersburg: Eduard Goppe, 1892–1893

Raeff, Marc. *Siberia and the Reforms of 1822*. Seattle: University of Washington Press, 1956

Michael Speransky: Statesman of Imperial Russia. 2nd ed. The Hague: Martinus Nijhoff, 1969

Rammelmeyer, A. "V. F. Odoevskij und seine 'Russischen Nächte.'" V. F. Odoevskii, *Russkie nochi*. Nachdruck der Moskauer Ausgabe von 1913. *Slavische Propyläen*; Bd. 24. Munich: Wilhelm Fink Verlag, 1967, pp. v–xxvi

[Rastorguev, E.]. *Poseshchenie Sibiri v 1837 g ego imperatorskim Vysochestvom gosudarem naslednikom tsesarevichem*. St. Petersburg: Kh. Gintse, 1841

von Rauch, Georg. "J. Ph. Fallmeyer und der russische Reichsgedanke bei F. I. Tjutcev." In his *Studien über das Verhältnis Russlands zu Europa*. Darmstadt: Wissenschaftliche Buchgesellschaft, 1964, pp. 158–200

Ravenstein, E. G. *The Russians on the Amur . . .* London: Trübner & Co., 1861

Rawls, James J. "The California Mission as Symbol and Myth." *California History* 71:3 (Fall 1992), pp. 343–360, 449–451

Rayfield, Donald. *The Dream of Lhasa. The Life of Nicholay Przevalsky (1839–1888), Explorer of Central Asia*. London: Elek Books, 1976

Review of *Istoriia reki Amura. Morskoi Sbornik* 90:4 (April 1859), ch. V, pp. 50–51

Riasanovsky, Nicholas V. *Russia and the West in the Teachings of the Slavophiles. A Study of Romantic Ideology*. Cambridge, Massachusetts: Harvard University Press, 1952

"Russia and Asia: Two Views." *California Slavic Studies* I (1960), pp. 170–181

"The Emergence of Eurasianism." *California Slavic Studies* IV (1967), pp. 39–72

Nicholas I and Official Nationality in Russia, 1825–1855. Berkeley: University of California Press, 1969

"Asia through Russian Eyes." W. S. Vucinich, ed., *Russia and Asia. Essays on the Influence of Russia on the Asian Peoples*. Palo Alto: Hoover Institution Press, 1972, pp. 3–29

A Parting of Ways. Government and the Educated Public in Russia 1801–1855. Oxford: Clarendon Press, 1976

Riazanov, D. *Ocherki po istorii marksizma*. Moscow: Iz-vo "Moskovskii Rabochii," 1923

Rieber, Alfred J. *The Politics of Autocracy. Letters of Alexander II to Prince A. I. Bariatinskii, 1857–1864. Etudes sur l'histoire, l'économie, et la sociologie des pays slaves*, XII. Paris-Hague: Mouton, 1966

"The Historiography of Imperial Russian Foreign Policy: A Critical Survey." Hugh Ragsdale, ed., *Imperial Russian Foreign Policy*. Cambridge: Cambridge University Press, 1993, pp. 360–443

Ritchie, G. B. "The Asiatic Department during the Reign of Alexander II, 1855–1881." Unpubl. Ph.D. thesis. Columbia University, New York, 1970.

Ritter, Karl. *Zemlevedenie Azii*. Tr. P. P. Semenov *et al.* 6 v. St. Petersburg: no pub., 1856–1895

Roberts, H. L. "Russia and the West: A Comparison and Contrast." *Slavic Review* 23:1 (March 1964), pp. 1–30

Rogger, Hans. *National Consciousness in Eighteenth-Century Russia*. Cambridge, Massachusetts: Harvard University Press, 1960

"America in the Russian Mind – or Russian Discoveries of America." *Pacific Historical Review* XLVII:1 (February 1978), pp. 27–52

Romanov, D. [I]. "Proekt zheleznoi dorogi ot Amura k zalivu De-Kastri." *Irkutskie Gubernskie Vedomosti* no. 38 (18 September 1858), otdel ii, pp. 3–12; no. 40 (2 October 1858), otdel ii, pp. 1–10

"Po povodu statei ob Amure." *Sanktpeterburgskie Vedomosti* no. 265 (3 December 1858), pp. 1555–1557

"O sibirskoi zheleznoi doroge." *Irkutskie Gubernskie Vedomosti* no. 49 (4 December 1858), otdel ii, pp. 4–10; no. 50 (11 December), otdel ii, pp. 3–8

"Prisoedinenie Amura k Rossii (1636–1858)." *Russkoe Slovo* IV (1859), pp. 179–200; IV, pp. 329–388; VII, pp. 93–136; VIII, pp. 107–171

"Na stat'ig Zavalishina ob Amure." *Irkutskie gubernskie vedomosti* no.45 (5 November 1859), otdel ii, p. 3

"Proekt russko-amerikanskogo mezhdunarodnogo telegrafa." *Russkoe Slovo* IV (1859), otdel i, pp. 91–146

"S ust'ia Amura." *Russkoe Slovo* IV (1860), pp. 103–116

"Izvstiia s Amura." *Moskovskie Vedomosti* no. 69 (27 March 1860), pp. 535–536

"Amur." *Morskoi Sbornik* 47:6 (May 1860), pp. 173–187

"Amur – zlokachestvennaia iazva Rossii." *Amur* no. 27 (3 July 1860), pp. 387–395

"Poslanie k g. redaktoru *Morskogo Sbornika*." *Amur* no. 51 (20 December 1860), pp. 770–775

Prizabytyi vopros o telegrafa chrez Sibir', mezhdu starym i novym svetom. St. Petersburg: Imp. Akademiia Nauk, 1861

Poslednie sobytiia v Kitae i znachenie ikh dlia Rossii. Irkutsk: Irkutskaia Gubernskaia Tipografiia, 1861

Rozen, A. E. *Zapiski Dekabrista.* St. Petersburg: Tovarishchestvo "Obshchestvennaia Pol'za," 1917

Rozengeim, M. P. *Stikhotvoreniia.* St. Petersburg: Artilleriiskii Departament, 1858

Russkie ekspeditsii po izucheniiu severnoi chasti Tikhogo Okeana v pervoi polovine XVIII v. Moscow: AN SSSR, 1984

"Russkie na Amure." *Sovremennik* 71:10 (1858), otdel ii, pp. 266–273

Russkii biograficheskii slovar'. 25 v. St. Petersburg: Russkoe Istoricheskoe Obshchestvo, 1896–1918

Russko-indeiskie otnosheniia v XVII veke: sbornik dokumentov. Moscow: Iz-vo Vostochnoi Literatury, 1958

Russko-kitaiskie otnosheniia v 17 v. Materialy i dokumenty. 2 v. Moscow, 1969–1972

Ryleev, K. "Voinarovskii." *Polnoe sobranie stikhotvorenii.* Leningrad: Sovetskii Pisatel', 1971, pp. 185–225

S. Untitled communication, dated Irkutsk, 25 November 1858. *Irkutskie Gubernskie Vedomosti* no. 48 (27 November 1858), otdel ii, pp. 1–4

Sabir, C. *Le fleuve Amour. Histoire, Géographie, Ethnographie.* Paris: G. Kugelmann, 1861

Safronov, A. "O sibirskoi zheleznoi doroge." *Severnoe Pchelo* no. 135 (21 June 1858), pp. 593–594

Said, Edward W. *Orientalism.* New York: Pantheon Books, 1978

Sakulin, P. N. *Iz istorii russkogo idealizma. Kniaz' V. F. Odoevskii. Myslitel'- Pisatel'.* 2 v. Moscow: M. and S. Sabashnikovye, 1913

Sarkisyanz, Emanuel. "Russian Attitudes toward Asia." *Russian Review* 13:4 (October 1954), pp. 245–254

"Russian Conquest in Central Asia: Transformation and Acculturation." W. S.

Vucinich, ed., *Russia and Asia. Essays on the Influence of Russia on the Asian Peoples*. Palo Alto: Hoover Institution Press, 1972, pp. 248–288

"Russian Imperialism Reconsidered." In Taras Hunczak, ed., *Russian Imperialism from Ivan the Great to the Revolution*. New Brunswick: Rutgers University Press, 1974, pp. 45–81

Sarychev, G. A. *Puteshestvie flota kapitana Sarycheva po Severo-vostochnoi chasti Sibiri, Ledovitomu moriu i vostochnomu okeanu, v prodolzhenie os'mi let* . . . 2 v. St. Petersburg: Shnor, 1802

Satsuma, Gay. "'Scholarly Entrepreneur': Robert J. Kerner and Russian Eastward Expansion." Unpubl. ms.

Saunders, David B. "Historians and Concepts of Nationality in early Nineteenth-century Russia." *Slavonic and East European Review* 60:1 (January 1982), pp. 44–62

Savitskii, P. N. "Kontinent-Okean (Rossiia i mirovoi rynok)." *Rossiia. Osobii geograficheskii mir*. Prague: Evraziiskoe Knogoizdatel'stvo, 1927, pp. 3–24

"Iz proshlogo Russkoi geografii: periodizatsiia Russkikh geograficheskikh otkrytii." *Nauchnye trudy Narodnogo universiteta v Prage* IV (1931), pp. 272–298

Savvin, V. P. *Vzaimootnosheniia tsarskoi Rossii i SSSR s Kitaem*. Moscow–Leningrad: Gosudarstvennoe Izdatel'stvo, 1930

Schapiro, Leonard. *Rationalism and Nationalism in Russian Nineteenth-Century Political Thought*. New Haven: Yale University Press, 1967

Schimmelpennick van der Oye, David. *Ex Oriente Lux: Ideologies of Empire and Russia's Far East, 1895–1904*. Unpubl. Ph.D. dissertation. Yale University, New Haven, 1997

Schmid, Georg. "Goethe und Uwarow und ihr Briefwechsel." *Russische Revue. Vierteljahresschrift für die Kunde Russlands* [St. Petersburg]. XVII:1 (1888), pp. 131–182

Schmidt, F. "Reisen im Amur-Lande und auf der Insel Sachalin . . . Botanischer Theil." *Mémories de l'Academie Impériale des Sciences de St. Pétersbourg* VIIe Serie, XII:2 (1868)

Schwab, Raymond. *The Oriental Renaissance. Europe's Rediscovery of India and the East 1680–1880*. Tr. G. Patterson-Black and V. Reinking. New York: Columbia University Press, 1984

Seddon, J. H. *The Petrashevtsy. A Study of the Russian Revolutionaries of 1848*. Manchester: Manchester University Press, 1985

Selinov, V. I. "Proshloe Kamchatki v kruge istoricheskikh interesov Pushkina." F. G. Kungurov, ed, *A. S. Pushkin i Sibir*. Moscow–Irkutsk: Vostsiboblgiz, 1937', pp. 146–154

Semenov, P. P. "Opisanie Novoi Kalifornii, Novoi Mekhiki, i Oregona v fizicheskom, politicheskom, i etnograficheskom otnosheniiakh." *Vestnik Russkogo Geograficheskogo Obshchestva* II (1851), pp. 81–156; III, pp. 1–76, 189–266

"Obozrenie Amura v fiziko-geograficheskom otnoshenii." *Vestnik Imperatorskogo Russkogo Geograficheskogo Obshchestva* 15:6 (1855), pp. 227–254.

"Amur." *Entsiklopedicheskii slovar', sostavlennyi russkimi uchenymi i literatorami*. 5 v. St. Petersburg: I. I. Glazunov, 1861–1863; IV, pp. 180–191.

Review of A. Middendorf, *Puteshestvie na sever i vostok Sibiri . . . Zapiski Imperatorskogo Russkogo Geograficheskogo Obshchestva* 1 (1862), pp. 5–12

Semenov, P. P. ed. *Geograficheskо-Statisticheskii Slovar' Rossiiskoi Imperii*. 5 v. St. Petersburg: V. Bezobrazov, 1863–1885

"Rech' vitse-predsedatelia obshchestva, P. P. Semenova, po povodu trekhstoletiia Sibiri, chitannaia v zasedanii 8-ogo Dekabria." *Pravitel'stvennyi Vestnik* 278 (16 December 1882), pp. 2–3

"Znachenie Rossii v kolonizatsionnom dvizhenii evropeiskikh narodov." *Izvestiia Imperatorskogo Russkogo Obshchestva* XXVII:4 (1892), pp. 349–369

Semenov-Tian-Shanskii, P. P. *Istoriia poluvekovoi deiatel'nosti imperatorskogo russkogo geograficheskogo obshchestva 1845–1895*. 3 v. St. Petersburg: Bezobrazov, 1896

Memuary, vols. I, III–IV. Petrograd: Izdanie Sem'i, 1915–1917

Semevskii, V. I. *Iz istorii obshchestvennykh idei v Rossii v kontse 1840-kh godov*. Rostov-na-Donu: "Donskaia Rech'," 1905

"Baron Vladimir Ivanovich Shteingel'. Biograficheskii ocherk." *Obshchestvennye dvizheniia v Rossii v pervuiu polovinu XIX veka*. St. Petersburg: "Gerol'd," 1905, I, pp. 281–320

Politicheskaia i obshchestvennaia idei dekabristov. St. Petersburg: Tip. Pervoi St.Pet-oi Trudovoi Arteli, 1909

"M. V. Butashevich-Petrashevskii v Sibiri." *Golos Minuvshego* III (1915), no. 1, pp. 66–87; no. 2, pp. 18–57; no. 3, pp. 43–84

Semivskii, N. V. *Noveishie, liubopytnye i dostovernye povestvovaniia o Vostochnoi Sibiri, iz chego mnogoe donyne ne bylo vsem izvestno*. St. Petersburg: Voennaia Tip. Glavnogo Shtaba, 1817

Semmel, Bernard. *Imperialism and Social Reform. English Social–Imperial Thought, 1894–1914*. Cambridge, Massachusetts: Harvard University Press, 1960

The Liberal Ideal and the Demons of Empire. Theories of Imperialism from Adam Smith to Lenin. Baltimore: Johns Hopkins Press, 1993

"Seniavin, Lev Grigor'evich." *Russkii Biograficheskii Slovar'*. 27 v. St. Petersburg: Demakov, 1896–1913; IX, pp. 335–336

Serebrennikov, I. I. "The Siberian Autonomous Movement and its Future." *Pacific Historical Review* III:4 (1934), pp. 400–415

Seton-Watson, Hugh. "Russian Nationalism in Historical Perspective." Robert Conquest, ed., *The Last Empire. Nationality and the Soviet Future*. Palo Alto: Hoover Institution Press, 1986, pp. 4–29

Sgibnev, A. "Vidy russkikh na Amur i na torgovliu s iaponieiu v XVIII i pervoi polovine XIX st." *Amur* no. 35 (1860), pp. 517–520; no. 36, pp. 530–536; no. 38, pp. 563–567; no. 39, pp. 578–582; no. 40, pp. 593–598; no. 41, pp. 611–615; no. 42, pp. 626–630; no. 43, pp. 642–647; no. 45, pp. 672–677; no. 46, pp. 689–695; no. 47, pp. 705–709

Shafer, B. C. *Nationalism. Myth and Reality*. New York: Harcourt, Brace & World, 1955

Shatrova, G. P. *Dekabristy* i Sibir'. Tomsk: Iz-vo Tomskogo Universiteta, 1962.

Dekabrist D. I. Zavalishin. Problemy formirovaniia dvorianskoi revoliutsionnosti i evoliutsiia dekabrizma. Krasnoiarsk: Iz-vo Krasnoiarskogo Universiteta, 1984.

Shchapov, A. P. "Istoriko-geograficheskoe raspredelenie russkogo narodonaseleniia" [1864–1865]. *Sochineniia A. P. Shchapova*. 3 v. St. Petersburg: M. V. Pirozhkov, 1906–1908; II, pp. 182–364

"Novaia era. Na rubezhe dvukh tysiacheletii" [1863]. *Sobranie sochinenii.*

Dopolnitel′nyi tom k izdaniiu 1905–1908 gg. Irkutsk: Vostochnosibirskoe Oblastnoe Izdatel′stvo, 1937, pp. 3–19

Shchegolev, P. E. *Dekabristy*. Moscow-Leningrad: Gosudarstvennoe Izdatel′stvo. 3.v., 1926

Petrashevtsy. Sbornik materialov. 3 v. Moscow–Leningrad: Gosudarstvennoe Iz-vo, 1926–1928

Shchekatov, Afanasii. *Kartina Rossii*. 2 v. Moscow: Vol′naia tipografiia F. Liubiia, 1807

Shchukin, N. "Podvigi russkikh na Amure v XVII stoletii, opisannye na osnovanii podlinnykh bumag Iakutsk. i Nerchinsk. arkhivov." *Syn Otechestva* kn. 9 (1848), otdel i, pp. 1–52

Shelgunov, N. V. "Sibir′ po bol′shoi doroge." *Russkoe Slovo* V (1863), January, pp. 1–48; February, pp. 1–39; March, pp. 1–62

Shemelin, Fedor. *Zhurnal pervogo puteshestviia rossiian vokrug zemnogo shara*. 2 v. St. Petersburg: Meditsinskaia Tipografiia, 1816–1818

Shrenk, L. I. *Ob inorodtsakh Amurskogo kraia*. 3 v. St. Petersburg: Imp. Akademiia Nauk, 1883–1903

Shtein, M. G. *N. N. Murav′ev-Amurskii. Istoriko-Biograficheskii Ocherk*. Khabarovsk: Dal′nevostochnoe Gosudarstvennoe Iz-vo, 1946

Shteingel′, V. I. "Zapiski." *Obshchestvennye dvizheniia v Rossii v pervuiu polovinu XIX veka*. St. Petersburg, "Gerol′d," 1905, I, pp. 321–474

"Zapiski Barona V . . . a I . . . a Sh . . . a, predstavlennye Imperatoru Nikolaiu Pavlovichu." A. K. Borozdin, ed., *Iz pisem i pokazanii dekabristov*. St. Petersburg: M. V. Pirozhkov, 1906, pp. 55–72

"Pis′ma." *Letopisi Gosudarstvennogo Literaturnogo Muzeia* [Moscow] III (1938), pp. 367–399

Shtraikh, S. Ia. *Moriaki-dekabristy*. Moscow–Leningrad: Voenno-Morskoe Iz-vo, 1946

Shul′man, N. K. "Puteshestvie A. F. Middendorfa po Priamur′iu." *Uchenye Zapiski Blagoveshchenskskogo Pedagogicheskogo Instituta im. M. I. Kalinina* 12 (1968), pp. 83–85

Shumakher, P. V. *Stikhi i pesni*. Moscow: 1902

Shmurlo, E. "Vostok i zapad v russkoi istorii." *Uchenie Zapiski Imperatorskogo Iur′evskogo Universiteta* III:3 (1895), pp. 1–37

"Sibir′ – ta zhe Rus′." *Amur* no. 84 (23 October 1861), pp. 797–798

Sibiriak. "Zametki o priamurskom krae." *Russkii Khudozhestvennyi Listok* 15 (20 May 1859), pp. 45–46

Simon, Gerhard. "Russischer und sowjetischer Expansionismus in historischer Perspektive." Heinrich Vogel, ed., *Die sowjetische Intervention in Afghanistan. Entstehung und Hintergründe einer weltpolitischen Krise*. Baden-Baden: Nomos Verlagsgesellschaft, 1980, pp. 93–117

Skak, Mette. *From Empire to Anarchy: Postcommunist Foreign Policy and International Relations*. New York: St. Martin's Press, 1996

Skal′kovskii, K. A. *Suezskii kanal i ego znachenie dlia russkoi torgovli*. St. Petersburg: Obshchestvennaia Pol′za, 1870

Russkaia torgovlia v Tikhom Okeane. St. Petersburg: A. S. Suvorin, 1883

Skalon, V. N. *Russkie zemleprokhodtsy-issledovateli Sibiri XVII veka*. Moscow: Obshchestvo Ispytateli Prirody, 1952

Sladkovskii, M. I. *History of Economic Relations between Russia and China*. Tr. M. Roublev. Jerusalem: Israel Program for Scientific Translations, 1966

Istoriia torgovo-ekonomicheskikh otnoshenii narodov Rossii s Kitaem (do 1917 g). Moscow: Nauka, 1974

Slezkine, Yuri. *Arctic Mirrors. Russia and the Small Peoples of the North*. Ithaca: Cornell University Press, 1994

Slovtsov, P. A. *Istoricheskoe obozrenie Sibiri*. 2 v. St. Petersburg: K. Kraia, 1838–1844

Smith, Anthony D. *Nationalism in the Twentieth Century*. New York: New York University Press, 1979

Smith, Bernard. *European Vision and the South Pacific*. 2nd ed. New Haven: Yale University Press, 1986

Smith, Henry Nash. *Virgin Land: The American West as Symbol and Myth*. New York: Vintage Books, 1957

Smith, Woodruf D. "The Ideology of German Colonialism, 1840–1906." *Journal of Modern History* 46 (December 1974), pp. 641–662

The German Colonial Empire. Chapel Hill: University of North Carolina Press, 1978

Sochava, V. B. "Stranitsy iz proshlogo russkoi geografii (zhizn′ i deiatel′nost′ A. F. Middendorfa)." *Sibirskii Geograficheskii Sbornik* 2 (1966), pp. 215–236

"Soedinennye Shtaty v epokhu grazhdanskoi voiny i Rossiia." *Krasnyi Arkhiv* 38 (1930), pp. 148–154

[Soimonov, F. I.]. "Drevniaia poslovitsa: Sibir′ – zolotoe dno." *Sochineniia i Perevody, k pol′ze i uveseleniiu sluzhashchiia* ch. II (November, 1761), pp. 449–467

Sokolov, N. N. "P. A. Kropotkin kak geograf." *Trudy Instituta Istoriia Estestvoznaniia AN SSSR* IV (1952), pp. 408–442

Sokolov, V. N. *Dekabristy v Sibiri*. Novosibirsk: Novosibirskoe Oblastnoe Gosudarstvennoe Iz-vo, 1946

Sovetskaia istoricheskaia entslikopediia. 13 v. Moscow: "Sovetskaia Entsiklopediia," 1961–1971

[Spasskii, G. I.]. "Istoriia plavanii rossiian iz rek Sibirskikh v Ledovitoe More." *Sibirskii Vestnik* ch. 15 (1821), pp. 17–28, 79–90, 122–136; ch. 16 (1821), pp. 270–281; ch. 17 (1822), pp. 39–48, 117–128, 185–196; ch. 18 (1822), pp. 305–314, 379–398; ch. 19 (1822), pp. 167–180

"Skazanie Russkikh o reke Amure v XVII stoletie." *Vestnik imperatorskogo Russkogo geograficheskogo obshchestva* 7:2 (1853), pp. 14–41

Specter, Michael. "A Siberian Railroad, From 'Hero' to Disaster." *New York Times* 15 August, 1994, p. A4

[Speshnev, N. A.]. Untitled lead article. *Irkutskie Gubernskie Vedomosti* no. 11 (25 July, 1857), otdel ii, pp. 1–5, 12–13

Stankov, P. "Pochemu i kak Kamchatskaia morskaia baza byla perenesena s Kamchatki v Nikolaevsk-na-Amure (po arkhivnym materialam)." *Izvestiia Primorskogo Gubernskogo Arkhivnogo Biuro* I:2 (1923), pp. 22–31

Stanton, John W. "Foundations of Russian Foreign Policy in the Far East 1847–1875." Unpubl. Ph.D. thesis. University of California, Berkeley, 1932

Starr, S. Frederick. "Tsarist Government: The Imperial Dimension." J. R. Azrael, ed., *Soviet Nationality Policies and Practices*. New York: Praeger, 1978, pp. 3–38

Statistichekoe obozrenie Sibiri, sostavlennoe na osnovanii svedenii, pocherpnutykh iz aktov pravitel′stva i drugikh dostovernykh istochnikov. St. Petersburg: Shnor, 1810

Steadman, John M. *The Myth of Asia*. New York: Simon & Schuster, 1969

Steklov, Iu. *Mikhail Aleksandrovich Bakunin: ego zhizn′ i deiatel′nost′*. 4 v. Moscow: Iz-vo Kommunisticheskoi Akademii, 1926–1927

Stepanov, A. A. "Mikhail Ivanovich Veniukov." M. I. Veniukov, *Puteshestviia po Priamuriiu, Kitaiu, i Iaponii*. Khabarovsk: Khabarovskoe Knizhnoe Iz-vo, 1970, pp. 5–22

Stephan, John J. "The Crimean War in the Far East." *Modern Asian Studies* III:3 (1969), pp. 257–277

"Far Eastern Conspiracies? Russian Separatism on the Pacific," *Australian Slavonic and East European Studies* 4:1/2 (1990), pp. 135–152

The Russian Far East. A History. Palo Alto: Stanford University Press, 1994

Strakhovsky, L. *L'Empereur Nicholas I et l'esprit nationale russe*. Louvain: Librairie universitaire, 1928

Strong, J. W. "The Ignatiev Mission to Khiva and Bukhara in 1858." *Canadian Slavonic Papers* 17 (1975), pp. 236–259

Struve, B. V. *Vospominaniia o Sibiri 1848–1854 gg*. St. Petersburg: "Obshchestvennaia Pol′za," 1889

Sukhova, N. G. "Sibirskaia ekspeditsiia A. F. Middendorfa." *Vestnik Leningradskogo Universiteta*. Seriia geologicheskaia i geograficheskaia. 6:1 (1961), pp. 144–150

Karl Ritter i geograficheskaia nauka v Rossii. Leningrad: Nauka, 1990

Sullivan, J. L. "Count N. N. Muraviev-Amurskii." Unpubl. Ph.D. dissertation. Harvard University, Cambridge, 1955

Sumner, B. H. "Russia and Europe." *Oxford Slavonic Papers* 2 (1951), pp. 1–16

Svatikov, S. G. *Rossiia i Sibir′ (K istorii sibirskogo oblastnichestva v XIX v)*. Prague: Izd. Obshchestva Sibiriakov v ch.S. R., 1929

"Rossiia i Sibir′." *Vol′naia Sibir′* [Prague] VIII (1930), pp. 34–45

Sverbeev, N. "Opisanie plavaniia po reke Amuru ekspeditsii general-gubernatora Vost. Sibiri v 1854 g." *Zapiski Sibirskogo Otdeleniia Russkogo Geograficheskogo Obshchestva* 3 (1857), pp. 1–78

Sverdlov, N. V. "K istorii russko-amerikanskikh otnoshenii na Tikhom okeane i Dal′nem Vostoke v XIX-nachalo XX v." M. N. Tikhomirov, ed., *Sbornik statei po istorii Dal′nego Vostoka*. Moscow: AN SSSR, 1958, pp. 309–315

Svod instruktsii dlia Kamchatsko-Amerikanskoi ekspeditsii. St Petersburg, 1851

Sychevskii, E. P. "Russko-Kitaiskaia torgovliia na Amure v seredine XIX stoletiia." *Trudy Blagoveshchenskogo Gosudarstvennogo Pedagogicheskogo Instituta* 6 (1955), pp. 158–164

Taranovsky, Theodore. "Russian Foreign Policy: An Approach to Analysis." Unpubl. ms.

Tatishchev, S. S. *Imperator Aleksandr II. Ego zhizn′ i tsarstvovanie*. 2 v. St. Petersburg: A. S. Suvorin, 1903

Te-v, M. "Istoriia Reki Amura, sostavlennaia iz obnarodovannykh istochnikov." *Vestnik Promyshlennosti* IV:12 (1859), pp. 63–108

Thaden, Edward C. *Conservative Nationalism in Nineteenth Century Russia* Seattle: University of Washington Press, 1964

Tiutchev, F. I. "Rossiia i Germania" [1844]. *Polnoe Sobranie Sochinenii F. I. Tuitcheva* P. V. Bykov, ed. St. Petersburg: T-vo A. F. Marks, 1913

Stikhotvoreniia. Berlin: Slovo, 1921

Letter to M. P. Pogodin, 11 October 1855. *Literaturnoe Nasledstvo* 97 (1988), I, p. 422
"Pis'ma k raznym litsam." *Literaturnoe Nasledstvo*, 97 (1988), I, pp. 497–548
Tkhorzhevskii, S. S. *Iskatel'istiny.* Dokumental'naia povest'. Leningrad: Sovetskii Pisatel', 1974
Tol . . . zin, M. "Pis'ma s Amura (Pis'mo 1–2)." *Syn Otechestva* 14 (1859), pp. 375–380.
Treadgold, Donald W. *The Great Siberian Migration: Government and Peasant in Resettlement from Emancipation to the First World War*. Princeton: Princeton University Press, 1957
"Russia and the Far East." I. J. Lederer, ed., *Russian Foreign Policy. Essays in Historical Perspective*. New Haven: Yale University Press, 1962, pp. 531–574
Trotskii, L. D. "O Sibiri." *Severnaia Aziia* 3 (1927), pp. 5–18
Trubetskoi, S. P. "Predpolozhenie dlia nachertaniia Ustava Slaviano-Russkoi Imperii . . ." *Vostannie Dekabristov. Materialy*. Moscow–Leningrad: Gosudarstvennoe Iz-vo, 1925, I, pp. 108–132
Trudy sibirskoi ekspeditsii imperatorskogo Russkogo geograficheskogo obshchestva. Matematicheskii otdel. St. Petersburg: V. Bezabrazov, 1864
Trudy sibirskoi ekspeditsii imperatorskogo Russkogo geograficheskogo obshchestva. Fizicheskii otdel. 3 v. St. Petersburg: V. Bezabrazov, 1868–1881
Turner, F. J. "The Significance of the Mississippi Valley in American History." In his *The Frontier in American History*. New York: Henry Holt & Co., 1931, pp. 177–204
Ulam, Adam B. "Nationalism, Panslavism, Communism." I. J. Lederer, ed., *Russian Foreign Policy. Essays in Historical Perspective*. New Haven: Yale University Press, 1962, pp. 39–67
"Unichtozhenie masonskikh lozh' v Rossii, 1822." *Russkaia Starina* 18 (April 1877), pp. 641–664
[Unterberger, P. F.]. "Rech' pravitelia del Vost. Sibir. Otdel. Imperatorskogo Russkogo Geogr. Obshchestva po povodu . . . 300-letnego iubeleia prisoedineniia Sibiri k Russkoi derzhave." *Izvestiia Vostochno-Sibirskogo Otdela Imperatorskogo Russkogo Geograficheskogo Obshchestva* XXII:3 (1882), pp. 64–69
Urry, J. *None but Saints: The Transformation of Mennonite Life in Russia 1789–1889*. Winnipeg: Hyperion Press, 1989
Usol'tsev, A. F. "Svedeniia o reke Sungari ot ust'ia do g. Girina." *Zapiski Sibirskogo Otdela Imperatorskogo Russkogo Geograficheskogo Obshchestva* kn. VIII (1865), otdel i, pp. 173–227
Uvarov, S. *Desiatiletie ministerstva narodnogo prosviashcheniia 1833–1843*. St. Petersburg: Imp. Akademiia Nauk, 1864
Val'skaia, B. A. "Petrashevtsy v russkom geograficheskom obshchestve." *Ocherki istorii russkoi etnografii, fol'kloristiki i antropologii* VII (1977), pp. 54–65. *Trudy Instituta Etnografii im. N. N. Miklukho-Maklaia*, Novaia Seriia, 104
Vasil'ev, V. P. "Opisanie Man'chzhurii." *Zapiski Imperatorskogo Russkogo Geograficheskogo Obshchestva* kn. XII (1857), pp. 1–77
"Zapiska o Nindute." *Zapiski Imperatorskogo Russkogo Geograficheskogo Obshchestva* kn. XII (1857), pp. 79–109
"Opisanie bol'shikh rek, vpadaiushchikh v Amur." *Vestnik Imperatorskogo Russkogo Geograficheskogo Obshchestva* ch. 19 (1857), otdel ii, pp. 109–126

"Otkrytie Kitaia" [1884]. In his *Otkrytie Kitaia i dr. st. Akademika V. P. Vasil'eva*. St. Petersburg: Stolichnaia Tipografiia, 1900, pp. 1–33

"Kitaiskii progress" [1884]. In his *Otkrytie Kitaia*. St. Petersburg: Tipografiia Gorokhovaia, 1900, pp. 139–164

Velikii podvig partii i naroda. Materialy torzhestvennogo zasedaniia v Alma-Ate, posviashchennogo 20-letiiu osvoeniia tselinnykh i zalezhnykh zemel'. Moscow: Politizdat, 1974

Vend, V. *L'Amiral Nevelskoy et la conquête définitive du fleuve Amour*. Paris: Librarie de la *Nouvelle Revue*, 1894

Veniukov, M. I. "Obozrenie reki Ussuri i zemel' k vostoku ot nee do moria." *Vestnik Imperatorskogo Russkogo Geograficheskogo Obshchestva* ch. 25 (1859), otdel ii, pp. 185–242

"Kolonizatsiia Russkoi Azii." *Severnaia Pchela* no. 207 (19 September 1861), pp. 851–852

Starye i novye dogovory Rossii s Kitaem. St. Petersburg: Bezobrazov, 1861

[Veniukov, M. I.]. "Primechanie k budushchei istorii nashikh zavoevanii v Azii." *Kolokol. Pribavlenie k pervomu desiatiletiiu* (1 August 1867), pp. 1–10

Veniukov, M. I. "Ob uspekhakh estestvennoistoricheskogo izucheniia Aziatskoi Rossii v sviazi s geograficheskimi otkrytiiami v etoi strane za poslednie 25 let . . ." *Trudy Pervogo S"ezda Estestvoispytatel'ei v S.-Peterburge*. St. Petersburg: Imp. Akademiia Nauk, 1868, pp. 49–58

"Postupatel'noe dvizhenie Rossii v severnoi i vostochnoi Azii." *Rossiia i Vostok*. St. Petersburg: V. Bezobrazov, 1877, pp. 68–134

"Ocherk mezhdunarodnykh voprosov v Azii." *Rossiia i Vostok*. St. Petersburg: V. Bezobrazov, 1877, pp. 200–296

"Vospominaniia o zaselenii Amura v 1857–1858 gg." *Russkaia Starina* XXIV (1879): 1 (January), pp. 81–112; 2 (February), pp. 267–304

"Graf Nikolai Nikolaevich Murav'ev-Amurskii." *Russkaia Starina* XXXXIII:2 (February 1882), pp. 523–526

Iz vospominanii. 3 v. Amsterdam: no pub., 1895–1901

"Puteshestvie v Kitai i Iaponiiu." *Puteshestveiia po Priamuriiu, Kitaiu, i Iaponii*. Khabarovsk: Khabarovskoe Knizhnoe Iz-vo, 1970, pp. 139–221

Vernadskii, G. I. "O dvizhenii russkikh na vostok." *Nauchnyi Istoricheskii Zhurnal* I: 2 (1913), pp. 52–61

"Pushkin kak istorik." *Uchennye zapiski russkoi uchenoi kollegii v Prage* I: 2 (1924), pp. 61–79

Nachertanie russkoi istorii. ch. 1. Prague: Evraziiskoe Knigoizdatel'stvo, 1927

Vernadsky, George. "The Expansion of Russia." *Transactions of the Connecticut Academy of Arts and Sciences* 31 (1933), pp. 393–425

Veselovskii, N. I., ed. *Vasilii Vasil'evich Grigor'ev po ego pis'mam i trudam 1816–1881* St. Petersburg: tip. A. Transhelia, 1887

Vetrinskii, Ch. "Sorokovye gody XIX v." D. N. Ovsianiko-Kulikovskii, ed, *Istoriia Russkoi Literatury XIX v*. 5 v. Moscow: Mir, 1910; II, pp. 66–203

Vevier, C. "The Collins Overland Line and American Continentalism." *Pacific Historical Review* XXVIII (August 1959), pp. 237–253

"Introduction." P. M. Collins, *Siberian Journey down the Amur to the Pacific 1856–1857*. Madison: University of Wisconsin Press, 1962, pp. 3–39

Vil′brekht, A. and A. Maksimovich. *Obshchii uchebnyi atlas dlia upotrebleniia v gimnaziiakh Rossiiskoi imperii.* St. Petersburg: Departament Narodnogo Prosveshcheniia, 1823

Vilkov, O. N. *et al.*, eds. *Dekabristy i* Sibir′. Novosibirsk: Nauka, 1977

Voeikov, A. I. "Budet li Tikhii Okean glavnym torgovym putem zemnogo shara?" *Izvestiia Imperatorskogo Russkogo Geograficheskogo Obshchestva* XL (1904), pp. 481–556

Volkonskii, S. G. "Pis′ma." *Letopisi gosudarstvennogo literaturnogo muzeia* III (1938), pp. 89–153

Von Laue, Theodore. *Sergei Witte and the Industrialization of Russia.* New York: Columbia University Press, 1963

V. R. K. "Otryvki iz pisem morskogo ofitsera." *Morskoi Sbornik* XXIX:5 (May 1857), pp. 5–27

V ssylke i zakliuchenii. Vospominaniia dekabristov. Moscow: Tipografiia Russkogo Tovarishchestva, 1908

Vucinich, Alexander. *Science in Russian Culture. A History to 1860.* London: Peter Owen, 1965

Vvedenie v geografiiu, sluzhashchee ko iz″iasneniiu *vsekh landkart zemnogo shara s gosudarstvennymi gerbami, i opisnaie sfery . . .* 2nd ed. Moscow: Kompaniia Tipograficheskaia, 1790

Vyvyan, J. M. K. "Russia in Europe and Asia." *New Cambridge Modern* History. Cambridge: Cambridge University Press; X, 1960, pp. 357–388

"Vzgliad na Severo-Amerikanskie Soedinennye Shtaty." *Syn Otechestva* CLXXX (1836), pp. 172–192

Wakeman, F. *The Fall of Imperial China.* New York: Free Press, 1975

Walicki, Andrzej. *A History of Russian Thought from the Enlightenment to Marxism.* Palo Alto: Stanford University Press, 1979

Wallander, Celeste A., ed. *The Sources of Russian Foreign Policy after the Cold War.* Boulder: Westview, 1996

Watrous, Stephen. "The Regionalist Conception of Siberia, 1860 to 1920." Galya Diment and Yuri Slezkine, eds., *Between Heaven and Hell: The Myth of Siberia in Russian Culture.* New York: St. Martin's Press, 1993, pp. 113–132

Wehler, Hans-Ulrich. *Bismarck und der Imperialismus.* Cologne: Kiepenheuer & Witsch, 1969

"Bismarck's Imperialism." *Past and Present* 48 (August 1970), pp. 119–155

"Einleitung." In *idem*, ed., *Imperialismus.* 3rd ed. Königstein: Athenäum–Droste, 1979, pp. 11–36

"Sozialimperialismus." In *idem*, ed., *Imperialismus.* 3rd ed. Königstein: Athenäum–Droste, 1979, pp. 83–96.

Whitman, John. "Turkestan Cotton in Imperial Russia." *American Slavic and East European Review* XV (1956), pp. 190–205

Whittaker, C. H. "The Impact of the Oriental Renaissance in Russia: The Case of Sergej Uvarov." *Jahrbücher für Geschichte Osteuropas* (Neue Folge), 26 (1978), pp. 503–524

Wicker, Hans-Rudolf. "Introduction." *Idem*, ed., *Rethinking Nationalism and Ethnicity. The Struggle for Meaning and Order in Europe.* Oxford: Berg, 1997, pp.1–42

Williams, W. A. *American–Russian Relations 1781–1947*. New York: Rinehart, 1952
Winks, Robin W. and James R. Rush, eds. *Asia in Western Fiction*. Manchester: Manchester University Press, 1990
Wittram, Reinhard. "Das russische Imperium und sein Gestaltswandel." *Historische Zeitschrift* 187:3 (June 1959), pp. 568–593
Wolff, Larry. *Inventing Eastern Europe. The Map of Civilization on the Mind of the Enlightenment*. Stanford: Stanford University Press, 1994
Wright, John Kirtland. "Introduction." *Human Nature in Geography. Fourteen Papers, 1925–1965*. Harvard: Harvard University Press, 1966, pp. 1–10
"*Terrae Incognitae.* The Place of Imagination in Geography." *Human Nature in Geography. Fourteen Papers, 1925–1965*. Harvard: Harvard University Press, 1966, pp. 68–88
Zablotsskii-Desiatovskii, A. P. *Graf P. D. Kiselev i ego vremia*. 4 v. St. Petersburg: M. M. Stasiulevich, 1882
Zaborinskii, Akhilles I. "Graf Nikolai Nikolaevich Murav′ev-Amurskii v 1848–1856 gg." *Russkaia Starina* VI (1883), pp. 623–658
Zalevskii, M. N. *Imperator Nikolai Pavlovich i ego epokha (po vospominaniiam sovremennikov)*. [Frankfurt-am-Main: Posev], 1981
"'Zametki o Zapadnoi Sibiri′ Gr. Potanina . . ." *Amur* no. 25 (28 March 1861), pp. 196–199; no. 26 (1 April 1861), pp. 203–207
"Zametki ob Amure, o torgovle i bogatstvakh Vost. Sibiri." *Vestnik Promyshlennosti* 2:5 (November 1858), pp. 99–102
Zavalishin, D. I. "Po povodu statei ob Amure." *Morskoi Sbornik* XXXVIII (November, 1858), pp. 34–48.
"Amur. Stat′ia 1-aia. Kogo obmanyvaiut i kto okonchatel′no ostaetsia obmanutym." *Vestnik Promyshlennosti* IV:10 (1859), pp. 41–83
"Otvet g. Romanovu na ego vozrazheniia na stat′iu pod zaglaviem: po povodu statei ob Amure." *Morskoi Sbornik* XL:5 (May 1859), pp. 1–20
"Otvet g. Nazimovu po povodu stat′i ob Amure." *Morskoi Sbornik* XL:6 (1859), pp. 101–119
"Zamechaniia na izvlechenie iz otcheta g. O. Eshe i vypisku iz Triestskoi gazety." *Vestnik Promyshlennosti* IV:11 (1859), pp. 72–82 [also printed in *Morskoi Sbornik* XL:6 (1859), pp. 120–132]
"Po povodu zaniatiia Tashkenta." *Sovremennaia Letopis′* 35 (September 1865), pp. 5–6; 37 (October 1865), pp. 5–6
"Kaliforniia v 1824 g." *Ruskii Vestnik* 60 (1865), pp. 322–368
"O Sibirsko-Ural′skoi Zheleznoi Doroge." *Sovremennaia* Letopis′ 9 (8 March 1870), pp. 12–13
"Sukhorukov i Kornilovich." *Drevniaia i Novaia Rossiia* 6 (1878), pp. 170–172
"Amurskoe delo i vliianie ego na Vostochnuiu Sibir′ i gosudarstvo." *Russkaia Starina* XXXII:9 (September 1881), pp. 75–100
"Dokumenty, otnosiashchiesia k Sibiri, 1840–1864 gg." *Russkaia Starina* XXXII:10 (October 1881), pp. 387–418
Zapiski dekabrista D. I. Zavalishina. 2 v. St. Petersburg: "Sirius," 1906
"California in 1824" (ed. J. Gibson). *Southern California Quarterly* LV:4 (Winter 1973), pp. 369–412
Zavalishin, Ippolit. *Opisanie Zapadnoi Sibiri*. Moscow: Tip. Gracheva, 1862

Zenzinov, M. "Pis′mo k redaktoru iz Nerchinska." *Moskvitianin* 3:6 (1843), pp. 552–558
"Istoricheskie vospominaniia o reke Amure." *Moskvitianin* 4:7 (1843), pp. 106–118
Letter to M. P. Pogodin, 30 January 1843. *Literaturnoe Nasledstvo* 58 (1952), p. 160
"Pis′mo k redaktoru iz Nerchinska." *Moskvitiain* 6:11 (1844), pp. 234–243
"Iz Nerchinska." *Russkii Vestnik* 27:12 (June 1860) kn. 2, pp. 354–357
Untitled communication from Nerchinsk, dated 11 August 1860. *Amur* no. 39 (6 September 1860), pp. 529–530
Zhitkov, K. G. "Vitse-Admiral Ivan Fedorovich Likhachev." *Morskoi Sbornik* 11 (November 1912), pp. 1–47
Ziablovskii, E. F. *Kurs vseobshchei geografii po nyneshnemu Politicheskomu razdeleniiu* . . . 3 v. St. Petersburg: Imp. Akademii Nauk, 1818–1819
[Zubov, V. A.]. "Obshchee obozrenie torgovli s Azieiu . . ." *Syn Otechestva i Severnyi Arkhiv* ch. 6:XXXV (1829), pp. 95–105; XXXVI, pp. 164–176

Index

Cambridge Studies in Historical Geography

1 Period and place: research methods in historical geography. *Edited by* ALAN R. H. BAKER and M. BILLINGE
2 The historical geography of Scotland since 1707: geographical aspects of modernisation. DAVID TURNOCK
3 Historical understanding in geography: an idealist approach. LEONARD GUELKE
4 English industrial cities of the nineteenth century: a social geography. R. J. DENNIS*
5 Explorations in historical geography: interpretative essays. *Edited by* ALAN R. H. BAKER and DEREK GREGORY
6 The tithe surveys of England and Wales. R. J. P. KAIN and H. C. PRINCE
7 Human territoriality: its theory and history. ROBERT DAVID SACK
8 The West Indies: patterns of development, culture and environmental change since 1492. DAVID WATTS*
9 The iconography of landscape: essays in the symbolic representation, design and use of past environments. *Edited by* DENIS COSGROVE and STEPHEN DANIELS*
10 Urban historical geography: recent progress in Britain and Germany. *Edited by* DIETRICH DENECKE *and* GARETH SHAW
11 An historical geography of modern Australia: the restive fringe. J. M. POWELL*
12 The sugar-cane industry: an historical geography from its origins to 1914. J. H. GALLOWAY
13 Poverty, ethnicity and the American city, 1840–1925: changing conceptions of the slum and ghetto. DAVID WARD*
14 Peasants, politicians and producers: the organisation of agriculture in France since 1918. M. C. CLEARY
15 The underdraining of farmland in England during the nineteenth century. A. D. M. PHILLIPS
16 Migration in Colonial Spanish America. *Edited by* DAVID ROBINSON
17 Urbanising Britain: essays on class and community in the nineteenth century. *Edited by* GERRY KEARNS and CHARLES W. J. WITHERS
18 Ideology and landscape in historical perspective: essays on the meanings of some places in the past. *Edited by* ALAN R. H. BAKER and GIDEON BIGER
19 Power and pauperism: the workhouse system, 1834–1884. FELIX DRIVER
20 Trade and urban development in Poland: an economic geography of Cracow from its origins to 1795. F. W. CARTER
21 An historical geography of France. XAVIER DE PLANHOL
22 Peasantry to capitalism: Western Östergötland in the nineteenth century. GÖRAN HOPPE and JOHN LANGTON

23 Agricultural revolution in England: the transformation of the agrarian economy, 1500–1850. MARK OVERTON*

24 Marc Bloch, sociology and geography: encountering changing disciplines. SUSAN W. FRIEDMAN

25 Land and society in Edwardian Britain. BRIAN SHORT

26 Deciphering global epidemics: analytical approaches to the disease records of world cities, 1888–1912. ANDREW CLIFF, PETER HAGGETT and MATTHEW SMALLMAN-RAYNOR*

27 Society in time and space: a geographical perspective on change. ROBERT A. DODGSHON*

28 Fraternity among the French peasantry: sociability and voluntary associations in the Loire valley, 1815–1914. ALAN R. H. BAKER

29 Imperial Visions: nationalist imagination and geographical expansion in the Russian Far East, 1840–1865. MARK BASSIN

Titles marked with an asterisk * are available in paperback.

www.ingramcontent.com/pod-product-compliance
Ingram Content Group UK Ltd.
Pitfield, Milton Keynes, MK11 3LW, UK
UKHW040703180125
453697UK00010B/370